Praise

"A sympathetic, comprehensive, and unique study of the so-far unsuccessful but serious attempts by science-savvy theologians to reconcile science with Christianity."

—Victor J. Stenger, author of the *New York Times* bestseller *God: The Failed Hypothesis*

"This book is an exceptionally well-written, informed and witty smack down of Christian attempts to deny the fact of evolution or incorporate it into their faith. The authors show us in this masterful book, the likes of which I have never seen before, that the implications of evolution are devastating for the Bible and the doctrines based on it. Absolutely brilliant!"

—John W. Loftus, author of *Why I Became an Atheist* and *The Outsider Test for Faith*

"A rigorous (and humorous) examination of the questionable hypotheses offered by today's Christian apologists for jamming square pegs into round holes or trying to keep oil and water mixed when in fact doctrinal Christianity and evolution are at best on a blind date with one another. Marriage does not seem to be an option, not after reading this book! I especially loved the section entitled, 'Damage Control.'"

—Edward T. Babinski, editor of *Leaving the Fold: Testimonies of Former Fundamentalists*

"I love this book! It is an important addition to the field of Biblical studies, the most complete single-volume examination of evolution and the Bible that I've ever read. Anyone with any semblance of an open mind will see this book as the final nail in the coffin of theology that treats the Bible as a science text."

—Charles Shingledecker, author of *The Crazy Side of Orthodoxy*

"An intensely personal, quite extensive, highly entertaining and appropriately provocative perspective on why major biblical conceptions of creation and the discourses of their reception have become immensely problematic in the context of neo-Darwinian interpretations of evolution."

—Jaco Gericke, author of *The Hebrew Bible and Philosophy of Religion*

Evolving out of Eden

Christian Responses to Evolution

Robert M. Price
Edwin A. Suominen

tellectual.com

Tellectual Press
tellectual.com
Valley, WA

Copyright © 2013
by Robert M. Price and Edwin A. Suominen

All rights reserved.
Please do not make or keep unauthorized copies of this book.

ISBN 978-0-9851362-4-6

Tellectual Press™ *is an imprint of Tellectual LLC.*

Scripture quotations are from the King James Version unless otherwise indicated. Those cited with "NASB" are taken from the New American Standard Bible®, Copyright © 1960, 1962, 1963, 1968, 1971, 1972, 1973, 1975, 1977, 1995 by The Lockman Foundation (www.Lockman.org). Used by permission.

Cover image: Charles Darwin expels Adam and Eve from the Garden of Eden, holding a tablet inscribed not with the Law, but *On the Origin of Species* (highly abridged), which accuses the First Couple of never existing. Thanks to Carol Selby Price for this fine re-interpretation of *Adam and Eve Driven out of Eden* by Gustav Doré (1865).

Thanks to Edward Babinski, Samantha Bishop-Strand, Jerry Coyne, Jaco Gericke, Russell Kolts, John Loftus, Dave Mack (an excellent proofreader who happens to be blind!), Justin Powell, Charles Shingledecker, and Victor Stenger for their review of part or all of the manuscript, and their encouragement, corrections, and suggestions. They of course bear no responsibility for any errors that remain. Also thanks to Brian Suominen for his assistance with the index.

To the fossil hunters—scientists and assistants alike—who have dedicated so many patient hours to finding the evidence that so many choose to ignore.

Table of Contents

Introductions 1
   In the Beginning 3
      Branches of the Tree 4
         Branch I: The Word 6
         Branch II: The Creature 7
         Branch III: The Creator 8
         Damage Control 9
         Conclusions 11
   Cast of Characters 13
      Your Escorts Out of Eden 13
         A Bit of Biography 14
         A Fundamentalist Bumps into Darwin 15
         An Engineering Perspective 17
      A Great Cloud of Witnesses 20
         Fiat Creationists 21
         Progressive Creationists 22
         Intelligent Designers 24
         Biologians 25
         Deism in Denial 26
   Let There Be Replicators 29
      Ancestral Awakening 29
      Rise of the Mutants 33
      Replication Relics 36
      Plagues from Denial 39
      A Grand View of Life 42

Branch I: The Word 45
   The World the Biblical Writers Thought They Lived In 47
      Planet of the Bible 47
      The Strange Old World of the Bible 51
         For the World Is Hollow, and I Have Touched the Sky 52
         Features of the Firmament 56

Geocentrism Is Egocentrism Writ Large	59
Supporting the Sky, and Earth	61
God Most High	62
Moving On	65
Eden Disorder	67
Adams by the Sackful	67
The Evolution of Genesis	73
Creation by Combat	77
Yahweh versus Nehushtan	79
Deity Deception	86
In the Image of the Gods	88
Ancient Nomads' Intents	90
In-A-Gadda-Da-Vida	92
Everybody's Working for the Weekend	95
Days and Confused	95
Playing in the Creationist Sandbox	95
The Time Telescope	98
Not so Fast!	101
The Gap Theory	105
Last Thursdayism	106
Priestly Postulations	107
Let's be Kind	110
The Sliding Scale of Inerrancy	112
Deny Me Three Times	112
Something New Has Been Added!	115
The Bible as Ventriloquist Dummy	118
Limited Inerrancy	120
Scripture's "Mega-purpose"	122
The Island of Doctor Lamoureux	124
Are We Not Bultmenn?	127
Backtracking	129
Branch II: The Creature	**131**
Apex or Ex-Ape?	133

Aren't We Special?	133
Monkey Business	138
Do Unto Others	140
Gloating in Genesis	148
Catholic Confusion	151
Gimme that Old Time Anthropocentrism!	155
Prehistoric Propitiation	157
Justifying Jesus	160
Is Jesus Safe from Adam's Fate?	160
Jesus Christ Superchimp?	166
The Soggy Foundation of Original Sin	172
Visiting the Iniquity	172
Paul and Adam	178
Natural Election	181
Gardener's Guilt	183
The Only Expiation I Can See	186
Sinful Selection	190
Adam Made Me Do It	190
The Man of Sin	191
Lustful Angels	194
Branch III: The Creator	**197**
Peekaboo Deity	199
Designus Absconditus	199
Dithering Design	201
A Two-Edged Standard	206
Omphalos Again	209
Quantum Apologetics	217
Deconstructing Divine Action	217
Ways Past Finding Out	218
For My Mutation They Cast Lots	225
Elohist Evolution	228
Intelligent DeSade	231
A Creator Red in Tooth and Claw	231

    Pilgrim's Progress..233
    Justification by Faith...234
Damage Control..241
    Let Not Your Left Brain Know What Your Right Brain Is Doing.............243
        Desperate Dissonance...243
        Rammifications of Science..245
        Reductionism or Mystification?...247
        The Experiential "Evidence"...249
        Demonic Delusion and Noetic Dysfunction................................251
        Shoveling After the Parade...253
    Groping for Gaps..256
        Gasp! Gaps!..256
        A First Cause to Believe?..259
        Who's on First?...263
        God of the Quantum Gaps...268
        The Invisible Gardener of Eden..269
        Giving God a Place to Hide 273
        Peaceful Coexistence?...273
        Back to Bultmann and Bonhoeffer...274
    Passive-Aggressive Creationism...278
        Creationism's Last Stand..278
        Whiteheaded or Wrong-headed?...283
        Haught Culture...286
        Refined versus Piecemeal Supernaturalism................................289
Conclusions...293
    The Memes Shall Inherit the Earth..295
        Doctrinal Darwinism..295
            Revival value..295
            Origins..299
        Go Ye Therefore, and Replicate in All Nations..........................300
            Memes..300
            The Meme's Eye View..303
            Defending the Faith...305

Taming the Beast	306
Paradise Lost	310
No Turning Back	310
Magnanimous Materialism	314
Walking Upright	315
References	317
Illustration Credits	331
Index	333

Introductions

Conservative Christianity is caught between the embarrassments of simple fiat creationism which is indigestible to modern science, and evolutionism which is indigestible to much of hyper-orthodoxy.

—Bernard Ramm,
The Christian View of Science and Scripture

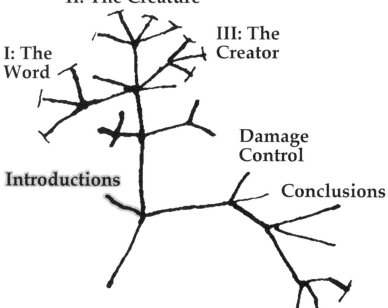

In the Beginning

Can theology after Darwin, instead of retreating from or only reluctantly accommodating Darwinian ideas, actually embrace them with enthusiasm?

—John F. Haught, *God after Darwin*

The now-indisputable reality of life's evolutionary origins—especially human life—challenges some foundational doctrines of the Christian faith. The issues range from the nature of God as a creator and guiding hand in the lives of mankind to the Fall of Man as the impetus for the sending of a Savior. With the evidence for evolution continuing to pile up in the fields of paleontology, biology, and genetics, the responses of Christian leaders have themselves evolved into different forms. The biblical literalists remain a loud and stubborn voice of denial, yet they have found themselves in a strange and unwitting alliance with the New Atheists by denying that evolution is in any way compatible with Christian doctrine. Many concerned believers are left walking a troubled middle path between Genesis and genetics, wary of the perils of losing their cherished faith on the one hand or their intellectual integrity on the other. Numerous science-savvy theologians have emerged to help them on their way.

Evolving out of Eden is a survey of the resulting theological attempts to grapple with the significance of the theory of evolution for Christian belief. There is now a cottage industry of guides working to establish their own different trails through the hostile territory outside the comforting faith-fairyland of Eden. Some remain so close to biblical literalism that we wonder if they really understand the realities of evolution, despite their often impressive scientific credentials. Others eloquently and accurately expound on the science and their acceptance of it, but seem to have hollowed out their scriptural and credal commitments in the process, to the point where we wonder what they mean by the claim of being "Christian" at all. Between the two extremes, there are now many paths on this well-trod ground.

We'll try to map out the lay of the land as we see it. What we don't seek to do is blaze any new trails of our own. Nor do we find any of the existing ones compelling, despite having begun with a conscious effort at maintaining an open mind about the whole thing. As we'll discuss shortly when introducing ourselves as your "Escorts out of Eden" in the next chapter, one of us initiated his informal study of evolution as a believing but troubled Christian, and we got to working together on this book as a way of evaluating the options left for belief in the face of scientific reality. Despite the sincere and fervent efforts of some writers who are highly qualified in both the theological and scientific realms, what we found just didn't add up for either of us. Nonetheless, we will seek to evaluate their attempts to accommodate evolution as fairly as we can, though you now know where we both came out.

In recent years, the New Atheists have put Christians on the defensive regarding their beliefs, with books full of scientific objections like *The God Delusion* by Richard Dawkins and *Breaking the Spell* by Daniel Dennett. Just as influential, perhaps even more so, are atheist voices from newer media: blogs, social networking, podcasts, and video. Predictably enough, there's been no shortage of literature, both online and in published books, that seeks to reassure the faithful that they can safely disregard (or contort beyond recognition) scientific findings and cling to the ancient myths of Genesis.

We'll make note of the claims of the creationists on one hand and try to convey some of the amazing science behind evolution on the other. The chapter "Let There Be Replicators" is our laymen's introduction to evolutionary science and what Darwin aptly called the "grandeur of this view of life." But our major focus will be on those trying to make it all fit together somehow.

Branches of the Tree

The layout of this book centers around three main "branches" of theological conception that are impacted by evolution: (1) *The Word*, i.e., the Bible that was produced by human beings and—ostensibly—their God; (2) *The Creature*, the human writers and expositors of that Bible who themselves claim to be the creatures of that God; and (3) *The Creator* whose recognition and appeasement is the ultimate object of Christian theology. It seems to us that all the theological issues posed by evolution can be classified into one of these three branches, though of course some overlap is inevitable.

The tree-branch metaphor is inspired by the idea of an evolutionary "tree of life," which appeared—perhaps for the first time in any extant diagram—in Darwin's 1837 first notebook on the "Transmutation of Species." Figure 1 reproduces it with some handwritten notes on the page. (The text of Darwin's "I think" note, not fully shown here, concerns the degree of relation between species and their common ancestors.)

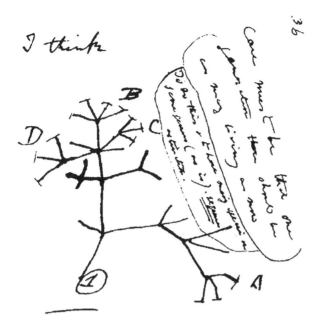

Figure 1. Darwin's famous "I think" sketch, 1837.

Our three branches go beyond the tree-branch metaphor in one important respect, however: The theological conceptions broaden in scope from one to the next. *The Word* (I) exists within the minds of the humans who wrote it, who contemplate and interpret it. Outside of human thought, the words of scripture mean nothing, whether those words are considered inspired or not. The human being is in turn *The Creature* (II) that Christian theology treats as the pinnacle of God's own creative contemplation. Finally, at the outermost level of Christian conceptualization, there is God, *The Creator* (III), who is supposed to have existed even when humans did not.

Each of these conceptions is impacted by evolution on its own successive level. Most theistic evolutionists and even some creationists try to dispense with the issues concerning the Bible by disavowing it as a science textbook, casually leapfrogging the lowest level of concern. But they are still left to answer for their own place in the universe as the product of random mutations and natural selection, rather than of any apparent divine act of creation. That's a significant issue. Yet it is not the biggest one. Even if you can move beyond anthropocentrism, accepting your place as a mere evolutionary byproduct (as the more sophisticated theologians seem almost willing to do), Christian theology is still left with some tough questions to answer about God himself.[1]

Branch I: The Word

As the foundational text of Christianity, the Bible is a source of inspiration but also vexation for the various schools of evolution-accommodating Christian thought. It will be a frequent point of reference, as we attempt to understand not just the beliefs based on the Bible but also the scriptures themselves. Somehow, certain matters usually wind up unnoticed and unmentioned in the standard treatments of the subject. We'll put a spotlight on these issues, on the passages and origins of the Bible that many Christian writers seem to prefer not dwelling on as they try to embrace modern science unencumbered by ancient misconceptions.

It is refreshing to see some of those writers at least acknowledging the presence of two incompatible creation accounts in Genesis 1 and Genesis 2-3. They are generally less willing to acknowledge that the writers of both accounts were just plain wrong about the origin and shape of the cosmos, no closer to the truth than the Sumerians, Babylonians, and Egyptians who clearly influenced them. And it turns out that there is actually a *third* creation account woven into the Bible, a fanciful one involving God fighting a chaos dragon. Once you recognize that even *this* is in the Bible, it ought to become clear that mere presence in scripture is not enough to lend credibility to any biblical creation account.

Some hard-shell creationists still insist that the biblical writers actually pictured the world as we do today, that we just need to properly understand their language. They have an awful lot of explaining to do. It is just this sort of exegetical fancy footwork that has persuaded younger Christian writers that the jig is up.

Another approach must be found. One is the argument that the Bible speaks of the heavenly bodies and their motions in a "phenomenal" and "non-postulational" manner. We do so when we casually speak of "sunrise" and "sunset," terms whose use by our ancestors reflected their belief that the earth was stationary and orbited by a tiny sun, moon, planets and stars. We still speak this way, even knowing the vast size and distance of these objects and that they do not in fact move around us. Neither use was really

---

1. "Sophisticated theology" has become a term of disparagement among the New Atheists, with Jerry Coyne appending a sarcastic trademark symbol when he references it on the *Why Evolution is True* website. We appreciate the ironic usage. But it is still true that the "sophisticated" theologians whose writings we discuss are at least trying to confront issues that those with "childlike" (i.e., childish) faith simply deny and ignore, and in that sense they deserve the term.

intended as an accurate charting of the heavens, the argument goes; the terms merely describe our perception of things. Isn't that just special pleading, though? Doesn't it imply the old Protestant Rationalism of the eighteenth century?

More and more Christian writers are daring to cut themselves loose from some of the classic Genesis-defending arguments. The most intellectually embarrassing, of course, are those of Young-Earth creationism. One notorious example is the attempt to reconcile the abundant evidence for an ancient earth with the plain young-earth sense of the Genesis creation account by suggesting that, yes, God only recently made the earth, but with false signs of great age and of having evolved. While more and more Christian thinkers repudiate this kind of special pleading, the logic of such arguments continues to influence anti-evolution thinking in unsuspected ways. We will trace these surprising connections.

Others admit the Bible writers were referring to a world as they pictured it, and that they were mistaken, presupposing the ancient "science" of their day. To admit this, these scholars must move strategically to redefine inspiration and inerrancy. They restrict these virtues either to the assertions being made (not the terms in which they are couched), or to the salvific or theological aspects of the text, freely disregarding the falsehoods in biblical descriptions of nature. Still others appeal to the old stand-by of "progressive revelation," not noticing how very inimical the whole idea is to any doctrine of revelation at all. If God limited himself to what the people of a given era could grasp and absorb, then by definition, how could he ever have had occasion to reveal *anything to anyone?*

Inevitably and quite properly, Christian attempts to grapple with evolution (once one admits evolution is the truth) bring up huge, major questions not only of exegesis (what were the ancient authors trying to tell their readers?) but also of hermeneutics (what shall we make of their messages today?). When they despair of refuting modern science in favor of scripture, fundamentalists and neo-evangelicals alike tend instead to pretend that the Bible writers really meant to describe nature as we now understand it, as if it has taken us this long to catch up with them, to understand what they were really getting at! As if whatever science tells us is the "real" meaning of those old texts. This strategy might be called "the sliding scale of biblical inerrancy," and it remains alive and well, if intellectually dishonest, among many of today's Theistic Evolutionists.

Branch II: The Creature

We are all mere human beings pondering these deep matters with a cerebral cortex that is itself the product of evolution, and recent evolution at that.

Our current capabilities of abstract thought do not seem to extend far back in our ancestral family tree. Most of our time spent even as *Homo sapiens* was more concerned with daily survival and reproduction in a harsh environment than contemplation of our place in the universe. So it is understandable that we have constructed a worldview—not just in Christianity, but in the ancient religions that preceded it—that puts us on the pinnacle of God's creation. Now, seeing a chimpanzee's cousin when we look in the unflattering mirror that evolution hands to us, we must step down from this exalted position. But we are very reluctant to do so, not just because of our theological commitments, but due to our very human nature.

Our second branch of Christian conception is all about this self-examination. What is man's role in a universe in which he is at best an afterthought? Where does Jesus fit into all this, with at least half of his DNA being the same earthy stuff that we have to wrestle with? There's no point calling oneself a Christian without some commitment to the divinity of Christ, after all. Yet Jesus *was* a human, his nature evolved from the same naturalistic processes—sex, slaughter, and selfishness—as our own.

Human nature turns out to be the subject of significant disagreement between traditional Christian theology and evolutionary science. The main issue, perhaps foremost of all the problems that evolution poses for Christianity, is the idea of Original Sin without any original sinner. There was no Adam from whom we could have inherited our supposed taint of sin-corruption, nor any Eden in which a Fall might have occurred. So what are we to make of a doctrine that goes all the way back to Paul as *the* reason God had to send a savior on our behalf?

And what is really so sinful and worthy of eternal condemnation about our sexual, selfish, and aggressive urges, anyhow? Sure, we have collectively agreed on certain moral ideas now that we have come to live in large, fixed societies rather than just roaming bands of kin. But it is *those very traits* that allowed our ancestors to pass them on, with all the rest of our DNA heritage, in a harsh and competitive world. Jesus said the meek inherit the earth, but we haven't inherited much from them, genetically. Rather, it is the "disproportionate replicators" who left their mark on us, our forebears whose drive and passion got their DNA immortalized into children who would, with enough luck as well as drive and passion of their own, continue down the line. You won't find many celibate shrinking violets in your ancestry.

Branch III: The Creator

Expanding to the outermost level of evolutionary concern for Christian theology, we finally come to the issue of God. Where's he hiding, now that we can explain seemingly everything about our origins without recourse to any miraculous actions on his part? And if a creator *is* behind it all, despite all indications to the contrary, why did he choose a method that entails so much suffering, so much waste, so much time?

The loss of the design argument is right up there with Original Sin as a deal-breaker for evolutionary Christianity. Evolution is driven by random mutations, and there is no allowance in the theory for any divine foresight. The best that our theologians can do in the face of that reality is to claim that God actually twiddles things at the quantum level, an observational curtain that we (conveniently) can never peer behind.

Perhaps God has the same agenda as Jesus in Mark's portrayal, with the parables being mystifying rather than edifying. Those (mostly) atheist scientists? Too bad for them! Let's make everything look random and happenstance, "For that seeing they may see, and not perceive; and hearing they may hear, and not understand; lest at any time they should be converted, and their sins should be forgiven them" (Mark 4:11-12). Pass the test tube, and we'll see you in hell.

Damage Control

Almost all of the issues we identify have been addressed by our cadre of evolution-savvy theologians, and we try to review their efforts as we discuss each issue in turn. There are some general themes to it all, though. We address those in a section of the book that is devoted to the overall theological effort at damage control in the face of evolution.

It is appropriate that this section follow our discussion of the issues involving the creator, as the theologians pull out all the stops to defend the ultimate object of their worship. Their efforts can be remarkably creative. Onward go the Christian soldiers, marching under God's banner. Evolution provides them with many battles to fight in his name: It is *his* written Word that has turned out to be riddled with scientific errors, *his* "pinnacle of creation" who has been exposed as a hairless primate, and *his* ill humor about innate human traits and actions that now trouble us.

The Bible, of course, is the beginning of sorrows for the theistic evolutionist. In many cases (indeed it is fast becoming the rule) Christian evolutionists are not merely accommodating the reading of the Bible to the facts of science. They seem ready to accept Bultmannian demythologizing of the Old Testament, admitting it is marked by obsolete cosmology and mythical tales from the ancient world. (Rest assured, they comfort their readers, who can see what ought to come next; the gospels are in no such danger. But aren't they?) In this way they claim (as Rudolf Bultmann did) not to be *rejecting* scripture but rather to be *reinterpreting* it. This distinction, they hope, will enable them to "sell" evolution to their evangelical brethren, suspicious as they are of the product. But one must suspect also that they are trying to cover their own posteriors, rightly sensing that their evangelical membership cards are in imminent danger of being canceled.

Many appear to be cozying up to Process theism, the brainchild of Alfred North Whitehead and Charles Hartshorne. As we will see, this radical redefinition of God is useful as a kind of sleight-of-hand; it allows them to think they are making room for God to somehow influence the evolutionary process without influencing it. There is something very slippery about this, as we hope to show. There is an implicit back-and-forth, the contradictory goals simultaneously pursued being to get God out of the way of autonomous natural processes and to find some way to say he is nonetheless guiding them. One might call the result "passive-aggressive creationism." Worse still, this "refined supernaturalism" (William James) again and again gives way to "piecemeal supernaturalism" when our evangelical authors wish to reassure their readers that God is a person after all, and that he works occasional miracles and answers prayer—something that seems downright inconsistent with the implicit deism of Process Theology.

All our new Christian evolutionists hasten to assure us that they have learned the cardinal lesson taught by Dietrich Bonhoeffer (not usually thought of as a participant in evolution debates): Don't try to locate God and his creative action in perceived gaps in natural processes, places where we cannot yet discern the causal link between A and B. Inevitably, further research fills in those gaps as we learn the hitherto-hidden causes. And then poor God gets the pink slip, each of his tasks having been outsourced to one lousy act of nature after another. The evangelical evolutionists do not want to make this mistake, partly because they recognize that it is shoddy methodology, partly because they don't want to install their deity in a new shelter only to have him evicted again. Again, Process Theology is a favorite means to this end. But we will show how these thinkers time and again default to their own version of the "God of the gaps," e.g., the Intelligent Design and fine-tuning arguments. The relevant mystery-gap in our knowledge is how they manage to remain oblivious of what they are doing.

What motivates any and all of the Christian approaches to the Bible and evolution? It is not a question of trying to reconcile disparate results from independent branches of scientific inquiry. It is rather a case of individuals with one foot in the religious camp and the other in the camp of science trying to resolve their acute cognitive dissonance. They don't enjoy switching back and forth between isolated compartments in their thinking. For integrity's sake they feel compelled to harmonize their conflicting beliefs, perspectives, allegiances. Of course, they could always resolve the tension by going into denial about evolution, pretending it's not true after all. But it's too late for that, and they know it. They could drop their belief in theism, but there are too many bridges to burn. The result is that they are hell-bent on bringing their two beliefs together by hook or by crook.

But are they perhaps betraying science right at the start by even trying to do this? Can they really get away with placing their inherited faith and their subjective religious sentiments on a par with the data of anthropology, genetics, biology, and geology? One often hears the note of indignation from Christian thinkers that atheistic scientists like Richard Dawkins and Jerry Coyne suffer from a "naturalistic bias," a willful prejudice in favor of "scientism." They supposedly exclude the upper story of reality, the very real spiritual dimension, which Christians seek to restore to the equation with their talk of divine creation and providence.

Well, who is right? Are naturalistic scientists guilty of *reductionism*? Or are their Christian counterparts guilty of *mystification*? We believe the burden of proof rests with Christians to demonstrate any good reason for factoring the supernatural or metaphysical into the total picture. The writer to the Hebrews understood that "faith is the substance of things hoped for, the evidence of things not seen" (Heb 11:1). But surely faith is not evidence so much as a substitute for it: a very different currency used in a very different realm. It is what satisfies you when you do not feel you need (or have) evidence, isn't it? And if that is the way of things, can it be genuinely scientific for believers to require scientists to factor God into the equation? To condemn them for not doing so? Are secular scientists removing one of the wheels from the automobile, or are Christians trying to squeeze a fifth wheel onto it?

The bottom line regarding creation theology and evolutionary reality, we think, is this: You wind up sacrificing as much of one as needed to preserve the other. Inevitably, the whole God concept is compromised. But that has been happening ever since Elijah had to settle for a "still small voice" (1 Kings 19:11-13) in place of the talkative, smiting warrior deity who had centuries earlier led his people's troops into battle. The de-deification process began long before Darwin or even Copernicus.

Conclusions

It is all quite disappointing in a way. It appears that the whole apologetical enterprise has once again run aground and that it is not going to be easy, probably not even possible, to secure a safety zone for one's faith or one's God.

We take no pleasure in the failure we recount, the dead end at which attempts to reconcile faith and science seem to have arrived. Christianity was an important part of both of our lives at one time. We know what comfort those transcendent promises bring to many, as well as the accompanying social benefits. One of us (Ed) is still in the process of mourning the loss of a cherished faith and church community. We get it.

We probably need not be too concerned about the fragile faithful, since few of them would ever read a book like this anyhow. It is amazing how blissfully unaware they manage to remain of evolution itself, beyond the caricature of creationist talking points. Our real misgivings are more about the necessity to criticize the theistic evolutionists whom we view as misguided allies. Many have the best of intentions (though the appeal of Templeton cash certainly doesn't hurt in the motivation department), trying to get Christians to accept the medicine of evolution by coating it in the sugar of some residual Christianity.

To thoughtful believers who are warily beginning their own dangerous steps outside Eden, we invite you to keep reading regardless. Avoiding the facts will not make them go away; the church has tried that approach over and over again for centuries, with nothing more than a series of embarrassing recapitulations to show for it. And maybe there is some way for *you* to perform the feat of accommodation that neither we nor those we discuss have yet found. If you want to begin working on a more effective strategy, you'd be well advised to assess the limitations of the failed options first, so as to avoid their mistakes as you go forward.

But in doing so, be sure you are not manufacturing a reality that simply conforms with your deepest yearnings. Yes, "Darwin doesn't provide much consolation at a funeral" (Kitcher 2007, 155). The silent, grinding design-by-death mechanism of evolution can seem nihilistic, to be sure. Then again, so is Ecclesiastes. We try to leave you with a few bits of illumination to fill the void: an overview of the fascinating (though still somewhat contentious) theory about religion being the product of cultural evolution—"memes" rather than genes—and some reflection on the wonder of being alive, if only briefly, in an amazing universe.

Cast of Characters

> *Darwin's* Origin of Species *had come into the theological world like a plough into an ant-hill. Everywhere those thus rudely awakened from their old comfort and repose had swarmed forth angry and confused.*
>
> —Andrew Dickson White,
> History of the Warfare of Science with Theology

We are writing primarily about the ideas of others: first the Bible writers who took their best shot at origins and a theology based on it, then those who found themselves having to wrestle with what that Bible had imposed on them. There's quite a cast of characters here, and it's only fair that we begin by introducing ourselves.

Your Escorts Out of Eden

This book began as a collaboration between an atheist, Robert M. Price, and a believing but troubled Christian, Edwin A. Suominen, who was wrestling with what he perceived as a grave conflict between evolution and his inherited faith. Having both accepted the reality of evolution (with considerable reluctance on Ed's part), we agreed to research its theological implications and the various ways that Christian writers have tried to smooth things over.

We set out to evaluate the various approaches with an open mind, and we like to think we were in a pretty good position to be objective about it all. Bob has long since left Christianity (except for occasional participation in the aesthetics of Episcopalian services) yet still retains a love and respect for it as well as the Bible on which it is based. The perceived failure of Christian theology to deal realistically with evolution was not a factor in his respectful departure from Christian belief. Indeed, he did not then see it as a major difficulty.

On the other hand, Ed began the project in the middle of a faith crisis that had been sparked by the realization that evolution was true and Genesis was not. He genuinely wanted to know if there was any convincing solution to what he suspected was an intractable problem for the Christianity to which he still clung, ever so tenuously. That crisis has ended, and Ed has departed his fundamentalist Christianity for reasons that include but are not limited to the evolution issue (see Suominen 2012). So, apart from some remaining tendencies toward Deism on Ed's part, neither of your co-authors retains any dogmatic commitments to the success of any of the proposed theological solutions we review.

Could that also make us predisposed to criticize and reject those solutions? We hope not, but nobody is really free of bias. Robert Burton has made a fascinating study of the brain's seemingly unavoidable over-reliance on a subjective "feeling of knowing" the truth of matters not really known. He memorably observes, "Our mental limitations prevent us from accepting our mental limitations" (2008, 1490).

The best any of us can hope for, Burton says, is "partial objectivity" (loc. 1478). Well, so be it. At the risk of sounding like the Pharisee who looked down his nose at the poor publican alongside him in the temple, we note that there is an awful lot of presuppositional thinking out there driving the discussions about evolution and human origins. Creationists relentlessly defend Genesis with viewpoints and arguments that range all across the spectrum of absurdity, seemingly oblivious of modern science. Meanwhile, the sophisticated theologians blaze their various trails outside Eden with a shared motivation to wind up at the foot of the old rugged cross, no matter what. Divided as they all are in their approaches, even between creationism and evolution itself, they still unite in *just knowing* that their beloved Christianity is true. It's hard for any attempt at objectivity to compete with "felt knowledge," whose power "cannot be underestimated, even when it exists independently of reason or any confirming evidence" (loc. 1288).

A Bit of Biography

From 1965-1972, Bob was involved in a fundamentalist Baptist church. He went on to become a leader in the Montclair State College chapter of the Inter-Varsity Christian Fellowship. Having developed a keen interest in apologetics (the defense of the faith on intellectual grounds), Bob went on to enroll at Gordon-Conwell Theological Seminary, where he received an MTS degree in New Testament. Billy Graham was the commencement speaker.

It was during this period, 1977-78, however, that Bob began to reassess his faith, deciding at length that traditional Christianity simply did not have either the historical credentials or the intellectual cogency its defenders claimed for it. Embarking on a wide program of reading religious thinkers and theologians from other traditions, as well as the sociology, anthropology, and psychology of religion, he soon considered himself a theological liberal in the camp of Paul Tillich. He received the Ph.D. degree in systematic theology from Drew University in 1981.

After some years teaching in the religious studies department of Mount Olive College in North Carolina, Bob returned to New Jersey to pastor First

Baptist Church of Montclair, the first pastorate, many years before, of liberal preacher Harry Emerson Fosdick. Bob soon enrolled in a second doctoral program at Drew, receiving the Ph.D. in New Testament in 1993. These studies, together with his encounter with the writings of Don Cupitt, Jacques Derrida, and the New Testament critics of the Nineteenth Century, rapidly eroded his liberal Christian stance, and Price resigned his pastorate in 1994. A brief flirtation with Unitarian Universalism disenchanted him even with that liberal extreme of institutional religion.[2]

Evolution hasn't been a major focus in Bob's studies, but a few of his published works do address the issue: "The Return of the Navel: The 'Omphalos' Argument in Contemporary Creationism," *Creation Evolution Journal* 1, no. 2, 26-33 (Price 1980); *Inerrant the Wind* (Price 2009, 24-28); "Apex or Ex-Ape?" *The Humanist*, Jan./Feb. (Price 2010). His New Testament work can be found in books including *The Widow Traditions in Luke-Acts* (1997), *Deconstructing Jesus* (2000), *The Incredible Shrinking Son of Man* (2003), *The Pre-Nicene New Testament* (2006), and *The Amazing Colossal Apostle* (2013). His theological works include *The Reason Driven Life* (2006) and *Beyond Born Again* (2008).

For his part, Ed makes no pretense of having any professional credentials regarding either theology or evolution. (His education is a Bachelor's in Electrical Engineering—University of Washington, 1995—and the unfortunate history of engineers advocating for creationism is duly noted.) Ed's major qualification for this project was simply a passion for it, after experiencing the sting of evolution's impact on a cherished Christian faith. He is an amateur—a person whose pursuit of a field arises solely from "the love of it," *amatorem*—and proud of it.

A Fundamentalist Bumps into Darwin

A few years back, after spending his whole life up to that point in a fundamentalist Christianity where even theistic evolution had been called "an outrage to the word of God" (Reinikainen 1986), Ed got introduced to Darwin in a most unexpected way. He certainly wouldn't have gone out looking to meet him, having been indoctrinated against evolution to the point where it was almost difficult to say the word without negative connotations. It all began from some research about an intriguing way of optimizing design parameters without the engineer having to explicitly specify those parameters: genetic algorithms.[3]

---

2. These three paragraphs are adapted from the "Biography" page of Bob's website, robertmprice.mindvendor.com/bio.htm.

3. The material under this heading and the next was adapted and expanded from Suominen 2012, §1.1 and §4.3.1.

It's one form of evolutionary computation, which, as Daniel Ashlock describes in his textbook on the subject,

> operates on populations of data structures. It accomplishes variation by making random changes in the data structures and by blending parts of different structures. These two processes are called *mutation* and *crossover*, and together are referred to as *variation operators*. Selection is accomplished with any algorithm that favors data structures with a higher fitness score. [Ashlock 2006, 13]

Sounds innocuous enough, right? But it's evolution, plain and simple: the "ability to produce new forms, in essence to innovate without outside direction other than the imperative to have children that live long enough to have children themselves." That, says Ashlock, is the key feature that evolutionary computation tries to reproduce in software (p. 13), and he spends another 500 or so detailed pages showing the various ways it's been done.

After reading about evolutionary computation and playing around with demonstrations of it for a while, using readily available open-source software, Ed was soon hooked on its elegance and power. The software sets up an artificial chromosome with each "gene" determining a parameter for some widget you want to design. Then you run a simulation of your widget a few hundred different times, with different sets of parameters specified by random numbers in the genes of each chromosome. Each simulation produces a "fitness" metric, a value that shows how well the widget works in its simulated environment with the particular "DNA" that it was randomly assigned as a starting point.

Then the fun starts: The widgets mate with each other, crossing over their chromosomes in the same way that those from your parents do in real life, during the production of sperm or eggs in your body. Each widget in the next generation has a randomly shuffled combination of the genes from two widgets in the first population, plus a few mutations sprinkled in. Things are set up so that only the "fittest" widgets from the first generation are likely to be parents of those in the next.

The result: evolution by a simulated form of natural selection. Ed had been raised believing that Adam and Eve were his ancestors and Darwin was of the devil, but now Darwin had come to his computer. What was happening on the screen before his eyes not only worked but made a lot of sense. He could understand exactly what was happening, because it was computer code, and pretty simple code at that.

Ed decided that he should learn a little bit about this evolution business to help give him some perspective about how to use this new engineering tool.

As things turned out, it wasn't needed for the parameter-optimization project he had been contemplating, but the hook was set in his mind regardless: It was fascinating stuff, and made so much sense out of everything! Could there really be something to this after all? He started cautiously reading, initially feeling guilt and anxiety about leafing through evolution books as if he were over at the rack of porn magazines instead of the Natural Sciences section of the bookstore. But read he did, and, after a few hundred hours of study, came to the conclusion that evolution was true and Genesis 1-3 was not. It was not an easy or welcome discovery for a fundamentalist Christian to make.

An Engineering Perspective

The realization that evolution might possibly have some truth to it was the most disturbing event in Ed's Christian life. It took quite a while to really understand the scientific issues, and to accept their profound implications both theologically and for his own place in the universe as a conscious, self-aware organism. What he saw right away, however, was that an *unguided, natural* process of evolution threatened to remove the strongest intellectual prop that had been shoring up his faith. It provided a simple, elegant, and tangible answer to the question for which the *guided, supernatural* process of creation was previously his only answer: "How could all of these amazing forms of life, myself included, have just happened to arise?"

Creationism has so profoundly poisoned the discourse about evolution in the United States that Ashlock felt compelled to defend the basic concept at the beginning of what he notes "is, essentially, an interdisciplinary computer science text." Most students, he says, "come into the field of evolutionary computation in a state much worse than ignorance" due to "quite vigorous opposition to the teaching of evolution," having "heard only myths, falsehoods, and wildly inaccurate claims" about it (Ashlock 2006, 12). It is a sad state of affairs when he must make this appeal to future engineers and computer scientists about the scientific basis for a design tool having proven effectiveness:

> Within the scientific community, the theory of evolution is viewed as well supported and universally accepted. However, you do not need to accept the theory of evolution in biology to do evolutionary computation. Evolutionary computation uses the ideas in the theory of evolution, asserting nothing about their validity in biology. If you find some of the proceeding material distressing, for whatever reason, I offer the following thought. The concept of evolution exists entirely apart from the reality of evolution. Even if biological evolution is a complete fantasy, it is still the source from which the demonstrably useful techniques of evolutionary computation spring. We may set

aside controversy, or at least wait and discuss it over a mug of coffee, later. [p. 12]

The "controversy," of course, is all about religion. Jerry Coyne, author of *Why Evolution Is True* and veteran of many battles with creationism, finds it "palpably obvious that, despite the presence of the few atheists or agnostics who deny evolution, virtually all opposition to evolution in America, and other countries as well, has religious roots" (Coyne 2012, 2). Most everybody accepts the benefits that evolutionary science has provided in medicine, genetics, agriculture, and, yes, even electrical engineering. Imagine the irony of a scene that has probably played out all too many times in recent years: some creationist lecturer asking for directions to the church where he is planning to spew his nonsense, calling the pastor with a cell phone whose antenna was designed by an evolutionary algorithm.[4] Thanks to an advanced new antibiotic, he has just fought off a nasty infection of bacteria that long since evolved resistance to penicillin. The grain in his breakfast cereal is the result of seeds, pesticides, and herbicides that were all developed with evolutionary science in mind.

Religion aside, there is a mental roadblock that seems to stand in the way of many thoughtful, educated people accepting that, as Darwin famously put it, "from so simple a beginning endless forms most beautiful and most wonderful have been, and are being, evolved" (1859, 492). How could such complexity *just happen to* arise from a single first fragment of life, not yet even a complete cell? But Ed's electrical engineering work in signal processing allowed him to quickly bypass that roadblock. He understood that filtering out random noise is the key to selecting weak bits of intelligence from noisy communication channels. It didn't take long to appreciate how evolution uses its own type of filtering to select those few useful and beneficial variations that occasionally appear in the midst of the noise of random genetic mutation. Eventually, the complexity and *apparent* design of cells, organs, organisms, and ecosystems emerge much as the faint tones of Morse code signals became perceptible (with appropriate audio filtering) amidst the static of his ham radio receivers decades ago.

Cells do an amazingly accurate job of DNA replication, but on such a massive scale that some errors are guaranteed. That's mutation. Our DNA is made of paired-up building blocks called nucleotides: adenine with

---

4. Mobile phone antenna design is one of the major engineering success stories for genetic algorithms. Google scholar (scholar.google.com) reports over 5000 hits for the search query "genetic algorithm cellular antenna." Despite a hundred years of engineering work on radio antennas, it turns out that unguided evolution does a better job of meeting the challenges of hiding them inside the tiny, sleek cases of mobile phones than "intelligent designers" do.

thymine, or cytosine with guanine. The order of pairing is significant, so there are four information-bearing combinations: AT, TA, CG, and GC, letters of a four-symbol alphabet. There are about three billion of these letters in our DNA, and somewhere around 40-100 of them are copying errors, producing novelties not found in the DNA of either parent (Roach et al. 2010, 637; Wells 2006, 16). An AT pair might turn into a CG or TA. One or two base pairs might be inserted or deleted. Sometimes, more drastically, there is a chromosomal mutation, a copying error—duplication, deletion, rearrangement—that affects a long string of DNA base pairs *en masse*.[5]

Another source of randomness is "shuffling of the chromosomes and genetic recombination" that occurs in sexual organisms during the production of sperm or eggs (Haarsma and Grey 2003, 302). Your chromosomes come in pairs (this is totally separate from the AT and GC pairing of nucleotides), with one chromosome from each parent. When a new sperm cell or egg is formed, part of each paternal chromosome is stitched together with part of its maternal counterpart. At one or more points along the length of each chromosome, there is a "crossover" from using the father's version to the mother's, or vice versa.

It's like walking along a street sampling the goodies from food vendors on one side, and then crossing to the other side (without doubling back) to get what's being offered there instead. Ms. Egg might get a North Side hot dog and fresh pretzel but a South Side ice cream cone. Over in a different neighborhood (he hopefully has different parents), Mr. Sperm might get a North Side hot dog and a South Side pretzel and ice cream cone. One study of complete parent and child genomes showed that this crossover happened, on average, a little more than twice for each egg chromosome and once for each in the sperm (Roach et al. 2010, 637). During conception, the new chromosomes will pair up again; the child will wind up with the North Side hot dogs and the South Side ice cream cones from each street, but a pretzel from each side of each street. It could have gone several different ways, by chance.[6]

---

5. See "Mutation," cod.edu/people/faculty/fancher/genetics/mutation.htm (accessed November 2012), and the Wikipedia entries for DNA, Nucleotide, and Mutation.

6. The biologist Paul Ehrlich provides a technical but succinct description of this "reshuffling of genes that occurs during the process of sexual reproduction" in the testes and ovaries, called *meiosis*. This cell division process, as opposed to the usual *mitosis*, cuts the number of chromosomes in half and mixes "the paternal and maternal chromosomes carried by both males and females." The resulting sperm contains "chromosomes from both parents of the male," the egg containing "chromosomes from both parents of the female. As the chromosome number is halved, DNA is transferred physically from chromosome to chromosome, increasing

All this variation is the raw material of evolution, which is an unguided process that results in the "preservation of favourable variations and the rejection of injurious variations," a phenomenon that Darwin called natural selection (1859, 81). The randomness of the mutations and crossover is the noise, and natural selection is the filter. Every "slight modification which in the course of ages chanced to arise, and which in any way favoured the individuals of any of the species, by better adapting them to their altered conditions, would tend to be preserved; and natural selection would thus have free scope for the work of improvement" (p. 81). Natural selection "is just differential reproduction. Some organisms because of their features do better at reproduction than others. That is all there is to it. Nothing more" (Ruse 2010, 164).

A Great Cloud of Witnesses

From its earliest years, Christianity has been fractured into numerous competing ideologies. Even in the Bible we see the discord: Paul vs. Peter, the temperamental human Jesus of Mark vs. the lofty Father-Son Christology of John, Pauline justification by faith vs. James' "faith without works is dead." So it shouldn't be any surprise to see many fault lines appearing in origins theology now that science has put the traditional Genesis story under pressure.

There is a range of willingness to cede ground from Scripture to science. How much of the store are you going to give away to make peace with science? How much reality will the craving for intellectual integrity and scientific respectability force a creationist to assimilate, as he seeks a new identity? All of those wrestling with the origins problem while trying to defend some form of Christianity seem to think they can draw a line somewhere. But science is not about establishing some contrived position that everyone can be happy with. It's about a search for the truth, based on data rather than dogma.

We might picture the various schools of thought in the Christian vs. evolution discussion as Israelite tribesmen, wandering through the wilderness of science. In the evening firelight we can see silhouettes. The first tent in the row belongs, let's say, to Lemuel, the consistent literalist, a rare breed even in this company. He represents today's flat earthers and geocentrists. (Though marginalized, these groups are still quite real, and mean what they say.) Think of his camel as representing science; it is

---

genetic mixing by making new combinations of genes on the chromosomes" of the sperm (in the testes), and the egg (separately, in the ovaries). That mixing process is recombination" (Erhlich 2000, 22). Remember, all this occurs separately in the male and female, entirely independent of sex and conception.

hitched outside, snoring peacefully. Lemuel has the whole tent to himself, and he, too, snoozes without a care.

Next to him, we find Yehudah, the young-earth creationist. His camel has managed to insinuate his nose under the edge of the tent flap. Yehudah does not doubt that the earth is a ball revolving about a central sun, and it does not even occur to him that the Bible says otherwise. He hardly knows Lemuel is there, but he has occasional arguments with his next neighbor, Obadiah, the old-earth creationist.

Obadiah's tent is a bit crowded, because he has allowed his camel to get farther into his tent. One hump now occupies half the tent, but Obadiah can't figure how to get him back out! The evidence for an earth billions of years old is just too cogent. But he still thinks God created the various species one by one, albeit at intervals of many millions of years. Sure, sure, he is glad to admit, there is "micro-evolution" within species ("kinds"), but of course there's no "macroevolution" from one discrete species to another. No, Obadiah thinks, the Bible does not leave room for that.

Next down the line is Thammuz, the theistic evolutionist, who believes species did indeed evolve from one another, but that God somehow employed this process as the instrument of his "creation." This poor guy can't even find enough space in the tent to lie down for a peaceful nap, because by now both his camel's humps are inside the tent, with only the tail protruding. But poor Thammuz is by no means the worst off.

His pal Perechiah, the Process Theologian, tries to grab a few winks as he shudders from the desert chill. His camel is hogging the whole tent, and all of Perechiah's anatomy that enjoys the warmth of the tent is *his* nose! The rest of him camps out on the sand.

The next tent, at the extreme left of the camp, is also filled with camel, but his former rider, Seth, got up and left hours ago. His faith, having been more and more marginalized, has yielded completely to science.

Fiat Creationists

The earliest creationist we refer to as such is **Martin Luther**, because he did his writing at a time when the alternatives to Genesis literalism were just beginning to become apparent. He's an important figure because of his outsize influence on Christianity. Most Protestants (even Lutherans) know little of what Luther actually taught, but those teachings have nonetheless laid the foundation for many of their beliefs. His *Lectures on Genesis* from around 1535-1536 provide insight into Luther's thinking about origins (or at least contemporary *Lutheran* thought–the *Lectures* were possibly subjected to third-party editing), and we cite from that work quite a bit.

Luther accepted—and likewise insisted that others accept—the idea of an earth only thousands of years old, even the heaven being a "firmament" holding up an overhead sea of waters. But at least he had the excuse of access only to a primitive and incomplete science. Now the positions he held are maintained only by an extreme lunatic fringe. They *know* what the Bible says, and they don't care about your fancy modern science. True Genesis literalism is very much an endangered species at this point, but young-earth creationists who hold to the creation of "every *kind* of living thing in six days—only a few thousand years ago" (Kurt Wise) and envision dinosaurs cavorting with humans (Ken Ham and his "Creation Museum") are not far to the left.[7]

Progressive Creationists

As it became increasingly ridiculous to claim that science somehow was getting it all wrong about the age of the earth, there emerged a more nuanced view of Genesis and origins.[8] An early and influential advocate was **Bernard Ramm**, a prominent neo-evangelical progressive creationist whose classic book *The Christian View of Science and Scripture* we cite frequently.

Ramm saw evangelicals as being on the losing side in a battle fought in the previous century between the Bible and science. Fiat creationism (God spoke, and it was so) shared in the blame for giving the Bible a bad name. Ramm lamented, "For all practical purposes science is developed and controlled by men who do not believe in the scientific credibility of Holy Writ. Evangelicals in science are considered by scientists as anachronisms or unnecessary perpetuations of the medieval mentality into the modern period." To him, it was a tragic outcome: "The detrimental influence of this on Christianity is beyond any possible calculation" (Ramm 1954, 19). There were errors on both extremes, he thought, with the "modernist" writing off "the supernatural character of the Bible by a destructive theory of accommodation," and the hyper-orthodox failing "to see that there is a measure of accommodation" (p. 49).

In a sense, Ramm was a "transitional form" in the development from literalist creationism to theistic evolutionism. While he personally preferred

---

7. Kurt Wise, "Noah's World—Same Time, Different Place" (answersingenesis.org/articles/am/v6/n4/noahs-world); Creation Museum homepage, "Children play and dinosaurs roam near Eden's Rivers" (creationmuseum.org). Both accessed August 2012.

8. It might be better to say "re-emerged," since allegorical views of Genesis, including the length of the days, have been tossed around since the first centuries CE, as we discuss in "Everybody's Working for the Weekend."

progressive creationism (God created discrete species one by one but at great intervals of time), he defended fellow evangelicals who opted for theistic evolutionism. Their view was theoretically viable and by no means heretical. And he shows some world-wisdom in warning against the "brittle thinking" of the hyper-orthodox creationist who

> makes his entire theological system—the Deity of Christ, original sin, atonement, resurrection—hang on sudden creation, and one bone from a fossil pit can potentially bring the whole edifice down. Surely, Christianity cannot live in constant dread as to what some palaeontologist or archaeologist is going to bring to light—so that one fossil can spell the doom of orthodoxy. [p. 180, emphasis omitted]

The lesson was partly learned by old-earth creationists that would follow, like Hugh Ross of Reasons to Believe. They avoid the foolishness of denying the earth's age but somehow remain impervious to the equally irrefutable evidence for evolution. We don't devote much space to their arguments because, frankly, we find them as unworthy of serious discussion as the rest of creationism. No matter how much they may crave the respectability of science, they cherry-pick its findings just enough to distance themselves from the lunatic asylum of young-earth creationism. A six-thousand year old universe? Don't be silly. But *of course* the Bible is right about the first human pair!

Let us mention Kenneth Miller now, out of turn. We introduce him below as a theistic evolutionist, and as such, he gets his share of our criticism for various problems we see in attempts to defuse the evolution dilemma. But we have nothing but admiration for the way Miller advocates for the science of evolution. In *Finding Darwin's God*, he spends page after page indicting the creationists who stubbornly refuse to acknowledge the evidence. He bemoans

> the view of the Creator that their intellectual contortions force them to hold. In order to defend God against the challenge they see from evolution, they have had to make Him into a schemer, a trickster, even a charlatan. Their version of God is one who intentionally plants misleading clues beneath our feet and in the heavens themselves. Their version of God is one who has filled the universe with so much bogus evidence that the tools of science can give us nothing more than a phony version of reality. In other words, their God has negated science by rigging the universe with fiction and deception. To embrace that God, we must reject science and worship deception itself. [Miller 2007, 80]

Well said! And as a Christian (Roman Catholic) and professor of biology, he is well positioned to say it.

Intelligent Designers

Michael Behe and William Dembski are well known as vanguards of the "Intelligent Design" effort to dress creationism up "in a cheap tuxedo," as the paleontologist Leonard Krishtalka memorably put it (Slevin 2005). They're not fooling anyone, at least not if the decision in *Dover vs. Kitzmiller* and the subsequent electoral housecleaning of the Dover school board is any indication. We don't bother discussing their agenda-driven statements much, either, focusing more on the writings of a supposed theistic evolutionist who can't quite seem to leave the idea of ID behind: **Denis Lamoureux**. "Scripture affirms that intelligent design is real and that humans have the ability to understand this revelation inscribed deeply into the Book of Nature" (2008, 65). He finds it not surprising "that the reflection of intelligent design revealed by science would be disregarded, perverted, or rationalized away by those not wanting to acknowledge the Creator's existence and their accountability to Him" (p. 72).

Now, Lamoureux acknowledges life's evolution through natural processes (p. 8) and describes evidence for it, including some from his dentistry background (pp. 355-62). He also saw early on that Behe's "irreducible complexity" amounted to a God-of-the-gaps argument (p. 360). And it is with refreshing clarity that he points out one big elephant in the church sanctuary: "Adam never existed" (p. 367). But his immediate follow-up to that statement, "this fact has no impact whatsoever on the foundational beliefs of Christianity," we find exemplary of a duplicitous wishful thinking that pervades his writing. Yes, regardless of what Genesis and your pastor say, these uncomfortable scientific facts are real. But guess what! You can still be a Christian, he tells his readers. In fact, somehow—despite, we would add, all the naturalistic explanations, the double standard about biblical revelation, the disappearance of Paul's rationale for Christ, etc.—you can celebrate a stronger faith in Christianity!

The anthropologist **James P. Hurd** is another scholar with Christian commitments (Bethel University) who seems torn between the science of human origins and the faith of the creationist. "Whichever scenario of origins Christians embrace, we can agree that God is the creator of all and that humans are unique because they partake of God's spirit" (2003, 230). And he seems to find Adam and Eve as real as the evidence he cites for *Ardipithecus* and *Australopithecus*: "Adam and Eve demonstrated complementary roles and both had a moral responsibility to God. We learn from them that we must take responsibility for our own actions. Adam and Eve's tragic story explains how humans transgress God's commands, become conscious of their own sin, and despair of measuring up to God's standards" (p. 230).

Biologians

Francis S. Collins, author of the *New York Times* bestseller *The Language of God*, founded the BioLogos Foundation with funding by the John Templeton Foundation. According to the biologos.org website, their goal is "exploring and celebrating the compatibility of evolutionary creation and biblical faith, guided by the truth that 'all things hold together in Christ.'"[9] Jerry Coyne summarizes their agenda more bluntly: "convincing evangelical Christians that they can accept both Jesus and Darwin," with the hope "that by showing the faithful that science and evolution do not automatically lead to atheism, many of them will retain their faith but abandon creationism" (2012, 5).

Collins was joined early on in the effort by **Karl W. Giberson**, and the two of them have co-authored one of the books we cite, *The Language of Science and Faith*. Their scientific capabilities are certainly not in question; both have PhDs, Giberson in physics and Collins in biology. But no one should assume these credentials guarantee the cogency of their theological ruminations. Reading C.S. Lewis does not make one a theologian, though it may be a good start.

As we mentioned, **Kenneth R. Miller** is also a scientist and defender of evolution. He dismantles creationism and ID with the same rigor and clarity as his largely atheist colleagues. Yet he maintains belief in a God whose "magic lies in the fabric of the universe itself," as opposed to one who fabricated the fossil record as "a series of sequential tricks ... for no purpose other than to mislead" (Miller 2007, 128). He admits the theological difficulties, but does not seem overly troubled by them. He takes comfort in quantum indeterminacy and what he sees as the result: "the breaks in causality at the atomic level make it fundamentally *impossible* to exclude the idea that what we have really caught a glimpse of might indeed reflect the mind of God" (p. 214). Genesis is not natural history but "a true account of the way in which God's relationship with the world was formed" (p. 257).

Peter Enns has been associated with both BioLogos and the conservative Westminster Theological Seminary. His book *Inspiration and Incarnation* earned him the boot from Westminster, not surprising given his near-Bultmannian approach to the Old Testament. He embraces the demythologizing hermeneutic, as Bultmann did, in order to accommodate the Christian message to modern science. Whether such a lopsided hybrid possesses any survival value remains to be seen. We cite from his most recent book, *The Evolution of Adam*.

---

9. "About the BioLogos Foundation," biologos.org/about (accessed January 2013), citing Colossians 1:17.

Joan Roughgarden is a transgender theistic evolutionist, much of whose work seems aimed at correcting what she regards as male chauvinist elements in Darwin's sexual selection theories, as well as to demonstrate the unsuspected amount of gender and sex role diversity throughout the animal kingdom. She affirms the divine involvement in the evolutionary process, while rejecting both traditional creationism and its latter-day guise of Intelligent Design. (At least those latter perspectives can explain how they picture God's involvement, even if they are scientifically indefensible.)

One might interpret all this as a mode of apologetics, though that hardly vitiates Roughgarden's scholarship. Her apologetical approach to the scripture vs. science dilemma is based on the notion that Jesus, as depicted in the gospels, appeals to the text using non-literalist methods. If it was good enough for him, she asks, then why not also for modern, science-minded Christians?

Another Christian who accepts evolution without thinking she is doing violence to her faith is **Tatha Wiley**, author of *Original Sin: Origins, Developments, Contemporary Meanings*. It is a comprehensive study of the topic that is refreshing in its honest approach:

> If the doctrine of original sin lies at the heart of Christian belief, as Christian theology has long asserted, it has to be coherent and speak to our experience and situation as human beings. Blind faith renders religious beliefs incoherent. Faith, in its journey toward authenticity, embraces the search for the intelligibility of the religious truths we affirm. [Wiley 2002, loc. 44]

Wiley focuses on her specific topic, spending little time discussing the science of evolution, but she acknowledges that the doctrine of Original Sin cannot correspond to any historical Adam and Eve. Her recourse is to an understanding of it "as the sustained unauthenticity of human beings, their alienation from the divine source of their existence, and the personal and systemic evils that issue forth from individual and collective unauthenticity" (loc. 2525). Whatever that means.

Deism in Denial

Almost in a class by himself, **John F. Haught** stands at an extreme end of the range of origins theology, opposite the flat-earthers. It seems to us that there is little he does not understand about the scientific concept of evolution, and as little about Christian doctrine he is unwilling to cede in order to accommodate it.

Haught is one of the best writers out there at articulating the problems that evolution poses for Christian theology. In the introduction to his *Making*

Sense of Evolution, he invites the reader to just ponder, for a moment, "Darwin's claim that all life on earth has descended from a single common ancestor that lived ages ago." Then he manages to pack most of the major issues raised by that idea into a single succinct paragraph:

> What does [Darwin's] idea of common ancestry mean for our understanding of life, of who we are, and of what our relationship with the rest of nature should be? Or consider Darwin's idea of "natural selection," the impersonal winnowing mechanism responsible for the emergence of new species over an unimaginably immense span of time. If all the diverse species arose gradually by way of a blind natural process, in what sense can God still be called the author of life, if at all? And if our own species is a product of natural selection, can Christian theologians still responsibly pass on the news that we are created in the "image" and "likeness" of God (Gen. 1:26)? Since human beings apparently evolved as one species among others, what does this imply for our ideas about the soul, original sin, and salvation? And what does "Christ" mean if Jesus also is a product of evolution? [Haught 2010b, xii]

This is evidence of a highly capable mind with a firm grasp of the science behind evolution as well as the conflicts that arise when it intersects with Christian theology. Our criticisms are reserved for Haught's attempts at resolving those problems, attempts that seem as maddeningly ethereal as the God he describes: "Always beyond reach, God abides in the depths of an elusive but forever-faithful future, which keeps opening the world to unpredictable outcomes" (p. 135).

Haught draws much inspiration from the writings of **Pierre Teilhard de Chardin**, a heterodox Jesuit paleontologist who devoted thirty years "to the pursuit of interior unity" between "two domains of life which are commonly regarded as antagonistic" (Teilhard de Chardin 1934, loc. 1238). His labors left him with the feeling that a "synthesis has been effected naturally between the two currents that claim my allegiance," with the one reinforcing, rather than destroying, the other (loc. 1242). We suspect the supposed synthesis was really the product of wishful thinking, after too much time spent allowing "two apparently conflicting influences full freedom to react upon one another deep within" (loc. 1241) a mind greatly troubled by cognitive dissonance and ecclesiastical rejection of his writings.

Teilhard conjured a breath-taking vista in which all life-forms are being swept along (albeit via a glacially slow process!) in a grand transformation of the Biosphere into a "Noosphere" (from the Greek *nous*, mind) of universal, uniative consciousness. The final attainment of this culmination

Teilhard called "the Omega Point," that which will fulfill the prophetic hope for the Second Coming of Christ.

It is a powerful vision, strongly recalling both Gnosticism and Mahayana Buddhism. But how does it fare as a description of reality? Is this vision a scientific hypothesis? Jacques Monod excoriates it as a gussied-up revival of primitive Animism, the belief that even inanimate material objects possess souls and that nature is propelled by some guiding "Life Force" (Monod 1974, 39-40). Indeed, Teilhard's schema, really more theology than biology, is *Exhibit A* for Don Cupitt's claim that theological and philosophical systems are more like self-referential tapestries than models of external reality (Cupitt 1988, 245-46).

Let There Be Replicators

> *Four thousand million years on, what was to be the fate of the ancient replicators? They did not die out, for they are past masters of the survival arts. But do not look for them floating loose in the sea; they gave up that cavalier freedom long ago. Now they swarm in huge colonies, safe inside gigantic lumbering robots, sealed off from the outside world, communicating with it by tortuous indirect routes, manipulating it by remote control. They are in you and in me; they created us, body and mind; and their preservation is the ultimate rationale for our existence. They have come a long way, those replicators. Now they go by the name of genes, and we are their survival machines.*
>
> —Richard Dawkins, *The Selfish Gene*

In this chapter we attempt to provide a brief glimpse of the grandeur of evolutionary science. We've already seen how breathtakingly simple and elegant the idea is. The evidence for it is so overwhelming, so carefully documented and attested by the entire scientific community (with the inevitable exception of a few dogma-driven cranks) that we feel a bit lost trying to convey its import in a few pages. We are not experts in the field by any stretch of the imagination, so that makes our task a bit more difficult. But our passion for the topic and its importance for this book leaves us determined to give it our best shot. We owe you that.

There is no appeal to faith here. Every one of these assertions is backed up by a wealth of evidence and expert knowledge, not our own but those of the scientific giants on whose shoulders we humbly crouch rather than stand.

Ancestral Awakening

Many tens of thousand of years ago, there lived a man who was the ancestor to every male human who now walks the earth.[10] This individual is often referred to as "Y-chromosome Adam," but he wasn't the guy we read about

---

[10]. Wells puts the date at 60,000 years ago (2006, loc. 1720), while Wilcox would have it a bit more recent at 50,000 years ago (2003, 250). The anthropologist John Hawks says that most "estimates put it within the last 70,000 years," which he thinks is too young (Hawks 2011). His view recently got some backing from a study by Mendez et al. that has the most recent common Y ancestor going back more than 230,000 years. That exceptionally old estimate is the *lower* end of their 95% confidence interval, which they report as 237,000–581,000 years ago (Mendez et al. 2013).

in the opening chapters of Genesis. He wasn't even close to being the first member of *Homo sapiens*: Paleontologists have dated fossil skulls and skeletons with essentially modern characteristics to a hundred thousand years ago or earlier (Robson-Brown 2011, 164; Fagan 2010, loc. 327).

One of the locations where these fossils were found—Ethiopia, Sudan, Tanzania, South Africa—was probably this genetic Adam's home. That's also where the closest copies of his Y chromosome live on today, in the Ethiopian, Sudanese, and South African men who carry haplogroup A of that chromosome (Wells 2006, loc. 2430, but see Mendez et al. 2013). It's a stunning concordance of conclusions about human evolution by two entirely separate scientific disciplines. And it's confirmed by yet another discipline, anthropology. Individuals with haplogroup A who live in these areas

> often practice cultural traditions that are representative of the ways of life of their distant ancestors. For example, some live in traditional hunter-gatherer societies once common to all humans. They also may still speak ancient click languages, like those of the San Bushmen of the Kalahari and the Hadzabe of Tanzania. [loc. 2434]

Genetic Adam was not the loner described in Genesis. He had companions, including male rivals who carried Y chromosomes of their own. The fact that all living men share variants of his Y chromosome is simply the result of *genetic drift*, where one gene (or chromosome) winds up doing a bit better than its rivals due to nothing more than sampling error, and eventually comes to dominate (loc. 1124).[11]

Figure 2 is a cartoon of how that can happen with the Y-chromosome, using an obviously simplified example of a four-man population that is stable in size over three generations. The guy with the grey Y-chromosome (that's really what it and its X counterpart look like, except for the shading) has two sons, whereas the one to the left of him and the one to the extreme right have none. His neighbor to the right has two sons, too, but you can see what happens in the next generation. Only Mr. Grey (no, not *that* one!) is a grandfather, and his grandsons wind up making up the entire population. Given ample mutations and thousands of generations, it's no surprise to see this sort of dominance, even without someone like Mr. Grey having any particular fitness advantage over the others (Levine and Miller 1991, 241).

---

11. Genetic drift can even establish, i.e., "fix," negative mutations in the population, as long as the "deleterious effect . . . is small enough to escape efficient elimination by purifying selection" (Koonin 2011, loc. 676).

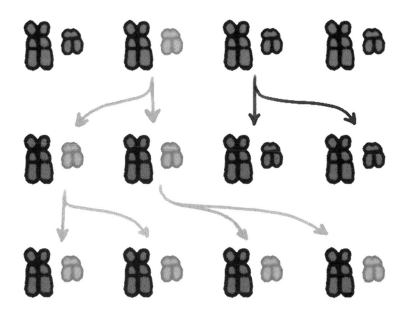

FIGURE 2: Eventually, all men have versions of a single Y chromosome.

There is another important side to this story: women. We refer to Adam first because that's what the Bible does, but Genetic Eve goes back much further. Genetic Adam and all of the humans alive with him likely carried mitochondria with vestiges of the DNA that *another* common ancestor had conveyed to them from a much longer time ago. She lived somewhere around 170,000 years ago (loc. 1728). Today, all humans, male and female, have mutated versions of this ancient African woman's mitochondrial DNA.[12]

So, with the Tigris and Euphrates nowhere in sight and a hundred thousand years separating the male and female genetic ancestors, this is no scientific substitute for the Garden of Eden story. Certainly not when you consider that "the mitochondrion was once a free-living bacterium that was absorbed into a cell, probably over a billion years ago, and gradually became part of

---

12. Ehrlich notes a distinction between mitochondrial ancestry and "the rest of our genetic endowment," the DNA in the nucleus (2000, 99). Women contribute as much to the latter, much more extensive type of DNA as men—actually more, since the X chromosome is bigger than the Y. But it's not as easy to trace ancestry with the chromosomes other than the Y because their formation includes crossover between each parent's paternal and maternal lines, as discussed in the previous chapter. The Y chromosome and mitochrondrial DNA, however, stand alone, changing only by mutation. "All modern human beings do not share just a single female ancestor" (p. 99), or a male one either. Genetic Adam's rivals contributed lots of DNA to our legacy, even if their Y chromosomes didn't make the cut.

the cellular machinery." It "has its own membrane and DNA, which is circular, unlike nuclear DNA, which is linear. This gives a clue to the origin of the m[itochondrial] DNA, since in nature only bacterial DNA is circular" (loc. 680). Not only is the mitochondrial DNA separate and distinct from the DNA packed into the cell nucleus, but it is evolving at a different rate, too (Prothero 2007, 157).

This Adam did not suddenly appear *ex nihilo*, fully formed. The absurdity of the idea is readily apparent upon a moment's thought. Here he is, with that mitochondria in his every cell, living fossils of captive bacteria that found a better living cranking out chemical energy inside other cells than trying to make it on their own. Shortly we will see how Adam had fossil viruses captive in his DNA, fossil genes that no longer code for anything, and vestigial features that no longer do anything useful. All this showed up at once, for no purpose than to deceive us into suspecting that evolution, even *human* evolution, might be true?

Genetic Adam's parents were pretty much just as human as he was (Dawkins 2009, 203). His brain had evolved to a point where it was about as big and sophisticated as our own (Ramachandran 2011, loc. 2294). The hardware was in place, but the software development would continue, via *cultural* evolution, for thousands of years to come. That's not to say there wasn't already complex thought going on. Signs of that go back a long ways indeed. Some sort of ceremony accompanied a human burial around 90,000 years ago, arms bent upward and deer antlers laid across the chest (Robson-Brown 2011, 167). About 70,000 years ago, for reasons we'll probably never know, someone took a piece of ocher and decided to etch three parallel lines with short diagonal lines crisscrossing back and forth between them (p. 166). It was around this time that people may have started wearing clothes; that's when DNA evidence shows that the body louse (capable of living in clothing) evolved from the head louse (Wade 2006, 4-5).

Some forty thousand years later, a Cro-Magnon would sit in a cave in what is now Southwestern Germany, playing a flute that had been constructed from the radius bone of a swan (Münzel et al. 2002, 107-108). He may have been a son of Adam like your co-authors and all the other billions of men now living, but perhaps not. It would have taken many generations for the Y chromosomes of Adam's male counterparts to fade into evolutionary oblivion.

This prehistoric flutist had no idea what people would be claiming thousands of years later about their origins, and his. Did he look up at the night sky and wonder about it himself? He certainly could have had the imagination for it; the artifacts left from his era in that cave include "four carved ivory figurines depicting human, mammoth, (cave)bear and bison, a limestone pebble painted in three colours, as well as ivory beads, perforated and dyed fish vertebrae and ornamented objects of antler and ivory" (p. 107). There is some beautiful cave art from this period, but nothing that tells us how these people thought they got there. Perhaps that speaks to a lack of interest in the topic, certainly as compared to the animals they hunted and feared on a day-to-day basis. It would take another thirty thousand years or so before the appearance of our first records of humans attempting to explain themselves: the Mesopotamian *Enuma Elish* and then the Genesis account that, as we will see in the next chapter, borrowed so much from it.

Rise of the Mutants

Adam was a "survival machine," patiently and mindlessly constructed by genes as a means for replicating themselves (Dawkins 2006, 19). That sounds backwards to us, because we are used to "top down" thinking about our origins rather than the "bottom up" reality of evolution being driven, at bottom, by random mutations. But it is a counterintuitive fact of nature, says Victor Stenger, that "complexity can arise naturally from simplicity." He notes that the "development of complex systems from simpler systems has been demonstrated in virtually every field of science." One example is the spontaneous development of snowflakes from water vapor (Stenger 2009, 151).

To use Daniel Dennett's terminology, we are prone to look for "skyhooks" that drop organisms into place, newly designed, when the reality is that "cranes" laboriously built things up bit by bit (Dennett 1995, 73-80). There's no grand design to any of it, just the simple calculus of evolution: Those genes that stumbled blindly into ways to replicate themselves *are the ones who survived* to evolve ever more productive mechanisms for replicating further.

At first, this may seem like a "strange inversion of reasoning," as an anonymous critic of Darwin complained in 1868 (Dennett 1995, 65). But then you give it some thought and realize it's the only way things could possibly work: "Skyhooks are miraculous lifters, unsupported and unsupportable. Cranes are no less excellent as lifters, and they have the decided advantage of being real" (p. 75). Dennett asks us to "imagine all the 'lifting' that has to get done in Design Space to create the magnificent organisms" of our world. "Vast distances must have been traversed since the dawn of life with the earliest, simplest self-replicating entities" (p. 75).

It is not done by some divine skyhook but "the crudest, most rudimentary, stupidest imaginable lifting process—the wedge of natural selection." Darwin's claim was that the process can traverse the huge distances of Design Space by "taking tiny—the tiniest possible—steps, . . . gradually, over eons" (Dennett 1995, 75). "At no point would anything miraculous—from on high—be needed. Each step has been accomplished by brute, mechanical, algorithmic climbing, from the base already built by the efforts of earlier climbing" (p. 75).

Would it be (genetically) useful to build a multicellular structure by converting energy from the sun? Parasitize the cyanobacteria that have been performing photosynthesis for a billion years, greening your cells with the chloroplast descendants of those bacteria (McFadden 2002, 18). As a fish discovering a life out of water, would some walking limbs be useful after dragging yourself around on lobe-shaped fins for millions of years? Evolve some structural bones in those fins, and eventually digits (Coyne 2009, 36-37; Prothero 2007, 224-25). The examples go on and on: penguins' flippers evolving from wings, balancing organs (halteres) of flies with two wings (*Diptera*) evolving from the hindwings of ancestors with four, baleen in toothless whales evolving from the skin of the palate (Deméré et al. 2008).

Of course, there are many intermediate steps for such transformations, and some of those details are probably lost to the ages forever. Fossils can "only provide a few fragments of life's history. In order to become a fossil, a carcass had to be properly buried in sediment, turned to rock, and then avoid destruction by volcanoes or earthquakes or erosion" (Zimmer 2001, 49). The odds are stacked heavily against the paleontologist, which makes the variety of specimens we do have all the more remarkable, including an "impressive array of hominid fossils that connect us to the apes and to all of the rest of the animal kingdom" (Prothero 2007, 332).

Life looks amazing and miraculous to us because what we see still living is only its latest stage, highly refined after billions of years of experimentation carried out in Natural Selection's ruthless design laboratory of mutation, reproduction, and death.[13] This is a laboratory that runs itself, starting a new experiment whenever a mutation occurs. Its countless instances of failure are unceremoniously discarded in the genetic trash can of reduced fertility and premature death. The only records of its success are found in the genetic code now occupying the biosphere: the current batch of experiments, which proceed unthinkingly and unrelentingly.

---

13. We humans have a natural bias toward overemphasizing the significance of what we see walking around today, in cities, forests, and zoos. We can't see cells without microscopes, and it was not that long ago in our history that we figured out that they even exist, much less the complex structures operating within them. The molecular biologist Eugene Koonin reminds us that, despite major exceptions like "the emergence of eukaryotic cells or multicellular eukaryotic organisms, ... most of the fundamental evolutionary innovations are crammed into the earliest 5% of the history of life" (2011, loc. 6510). Many "adaptations have to do with maintaining the integrity of cellular organization, preventing malfunction, and performing damage control."

The most successful product of this autonomous laboratory is the ~ one: the virus. A cold or flu virus has only ten genes (Zimmer 2011, loc. 233), a mere scrap of RNA or DNA encased within a tiny protein shell. Endogenous ("generated within") retroviruses don't even have their own physical housing; they just insert their DNA into somebody else's genome. Like ours, for example:

> Over millions of years, our genomes have picked up a vast amount of DNA from dead viruses. Each of us carries almost a hundred thousand fragments of endogenous retrovirus DNA in our genome, making up about 8 percent of our DNA. To put that figure in perspective, consider that the twenty thousand protein-coding genes in the human genome make up only 1.2 percent of our DNA. [loc. 679]

Viruses are humble replication vehicles. It's debated whether even the free-floating kind of virus should even be considered a form of life.[14] But "viruses are the most abundant biological entities on Earth" (Koonin 2011, loc. 4821). There are up to a hundred billion of them in a liter of seawater, outnumbering "all other residents of the ocean by about fifteen to one" (Zimmer 2011, loc. 530). The virus has proven to be a highly effective way for genes to make copies of themselves, and that is what evolution is all about.

Much of evolution has occurred *for evolution's sake*, an amazing "expanse and complexity of the molecular machinery that is dedicated to quality control of each of the major information transfer processes: Systems of DNA repair and protein degradation, and molecular chaperones are all cases in point. Moreover, much, if not most, of the evolution of protein-coding genes appears to be driven by selection for robustness to misfolding" (loc. 6496).

Everything else builds on this ancient biochemical foundation. Cranes, rather than skyhooks, all the way down.

14. In his *Philosophy of Death*, Steven Luper considers it more reasonable to classify viruses as "surviving examples of proto-organisms" rather than living organisms. They lack autonomy, and "can exist only by taking control of the vital processes of some organism." Many living things are not organisms, and he acknowledges "the fact that viruses are not organisms does not show they are not living things." Counter-examples are "blood and muscle, and various organelles within cells of your body," which nonetheless are alive (Luper 2009, 14). But they are parts of a greater whole, and "we can rule out viruses as living things since they are neither organisms nor components of organisms" (p. 14).

Koonin finds "the denial of the 'alive' status to viruses" unfortunate because it carries the implication "that viruses are of no substantial relevance to the evolution of cellular life forms" when the opposite is true (2011, loc. 4854). Viruses are not just an outcome of evolution, but a major player in the evolution of "living" organisms.

The most prolific undisputably *living* organism is a bacterium that thrives in the world's oceans via photosynthesis, *Pelagibacter ubique* (Koonin 2011, loc. 4054). With a mere 1300 or so genes encoded by 1.3 million DNA base pairs (compare that to our 3 billion), it has the smallest genome "of any cell known to replicate independently in nature" (Giovannoni et al. 2005, 1242). Like the viruses, it makes virtue of simplicity: "Evolution has divested it of all but the most fundamental cellular systems" (p. 1245). Its streamlined little genetic payload occupies about a third of its volume, arguably the smallest of any free-living cell studied in the lab (p. 1242).

Virus genes don't care that they encode for an inert parasite that must commandeer a host cell in order to replicate, nor do *P. ubique* genes mind occupying so much of a tiny microbial speck that is just complex enough to get the job done on its own. Even as we look beyond the microbial world, we see the antlike determination of the genes to replicate, for replication's own sake. Literally so: There are probably *ten trillion* ants on earth (Ridley 1996, 11). Whether they reproduce via the quick opportunism of a mere virus or prefer the "pulpy, throbbing colonies of tens of trillions of cells" that form human beings (Ramachandran 2011, loc. 354), the genes "march on. That is their business. They are the replicators and we are their survival machines," to be cast aside in death when we are done serving them (Dawkins 2006, 35).

Replication Relics

The former bacteria that now live inside our cells as mitochondria are *functioning* relics, like a Roman bridge that is still in use. (Yes, amazingly, some are.) They show evolution's opportunistic way of getting things done from existing materials—cranes rather than skyhooks—but perform a critical task in their captive role. There are many leftovers in our bodies that are less flattering to evolution's efficiency, and thus even more compelling as evidence of it.

A prime example is the recurrent laryngeal nerve. This thing goes from the brain down to the heart where it loops around the aorta and heads back up again to the larynx. Since "the major blood vessels of our chest are the messed-about relics and remnants of the once clearly segmental blood vessels serving the gills" of our remote fishy ancestors (Dawkins 2009, 359), this particular nerve got routed in between some of those blood vessels. That wouldn't be so remarkable—the wiring has to get put somewhere, after all. But when mammals evolved,

> the neck stretched (fish don't have necks) and the gills disappeared, some of them turning into useful things such as the thyroid and parathyroid glands, and the various other bits and pieces that

combine to form the larynx. Those other useful things, including the parts of the larynx, received their blood supply and their nerve connections from the evolutionary descendents of the blood vessels and nerves that, once upon a time, served the gills in orderly sequence. [p. 360]

As the brain moved farther away from the heart, the "vertebrate chest and neck became a mess, unlike the tidily symmetrical, serial repetitiveness of fish gills" (p. 360), and the recurrent laryngeal nerve was stuck making an increasingly ridiculous detour. The diversion could not be avoided due to evolution's inability to make a clean break with the past, like the Matthean writer sending Jesus' family on a vacation trip to Egypt during Mary's maternity leave because of something Hosea had written. In the most extreme example, the giraffe, the excess length required for the detour is a full *fifteen feet* (Coyne 2009a, 82). "The indirect path of this nerve does not reflect intelligent design but can be understood only as the product of our evolution from ancestors having very different bodies" (p. 83).

People have been unable to avoid noticing these oddities, for a long time now. Even Darwin's nemesis Richard Owen remarked that "there is no greater anomaly in nature than a bird that cannot fly" (from Darwin 1859, 136). That is of course exactly what you would expect to find in birds who evolved gradually to new environments. Their wings "are a *vestigial* trait: a feature of a species that was an adaptation in its ancestors, but that has either lost its usefulness completely or, as in the ostrich, has been co-opted for new uses" (Coyne 2009a, 57). You don't need to get airborne anymore if you've wound up in a place where you're not hunting or being hunted. Saving energy and reducing risk of injury to delicate structures becomes an advantage, and "selection would directly favor mutations that led to progressively smaller wings, resulting in an inability to fly" (p. 59).

Too bad Owen in his obtuseness couldn't appreciate how his eye for science had mutated into a useless vestige, just like those of blind mole rats. Nor could he see the utility of natural selection to explain why they had eyes left at all. What sort of theory is it that makes you fear the evidence rather than accounting for it? "The light shineth in darkness, and the darkness comprehended it not" (John 1:5).

In case you're wondering, the eastern Mediterranean blind mole rat lives underground for its whole life, yet "retains a vestige of an eye—a tiny organ only one millimeter across and completely hidden between a protective layer of skin." The buried vestigial eye still has light-sensitive pigment, which can't form any images but does help the mole rat keep track of its daily cycle of activity (Coyne 2009, 59-60). True moles don't have any sort of eyes left at all, "retaining only a vestigial, skin-covered organ that you can see by pushing aside the fur on its head" (p. 60).

The examples abound. Whales have small hindlimb and pelvis bones, buried inside the body, disconnected from the spine and useless (Prothero 2007, 108-109), while fossils of earlier extinct proto-whales show vestigial limbs that still protruded from the body, and still earlier species had functional limbs (pp. 318-21). You can see a vestigial digit just by looking at the legs of a horse, dog, or cow: Those little nubs and dewclaws that hang absurdly above the ground aren't accomplishing anything anymore, but once they were fully functioning toes.

Well, perhaps *those* critters evolved, but not us! Right? Not so fast. We have recurrent laryngeal nerves, and they take that circuitous route in our necks, too, just not as extreme as for the giraffe. We have an appendix that causes far more trouble than it's worth, now that we're no longer eating leaves (Coyne 2009a, 60-62). Tiny muscles remain under our skin to fluff up body hair that is now too sparse for warmth or conveying threats (p. 62). Other muscular vestiges are the rudimentary extensor that some people have for a now-rigid tailbone, and the three scalp muscles attached to ears that no longer move around for sensing direction of sounds, though some people (like one of us, Bob) can still wiggle them a bit (pp. 62-64).

There is no good reason for retaining any of these features except evolution's housekeeping laziness. The natural selection custodian passes over these dark corners of our genome because there's no payoff in tidying them up. You get goosebumps when you're cold or scared? So what? It's not costing anything in terms of your survival or reproductive fitness. There is only selection pressure to change what *isn't* working. When something does, evolution has no incentive to mess with it; that's why *Triops cancriformis* is still the same species of shrimp it was 180 million years ago (Levine and Miller 1991, 259).

Evolution is a slacker even when it comes to canceling its previous work. For whatever reason (or *no* reason, if genetic drift is at play), a particular feature may find itself fallen out of favor in a species, like tails in humans and our more recent ancestors. There is no abrupt, drastic redesign in which the gene or genes involved with that feature are cut out of the genome. They are just left there and disabled, by either a fatal mutation or an explicit "off" setting in one of the many "genetic switches" (Carroll 2005, 111) that regulate gene expression. Sometimes, the off switch gets knocked back into the "on" position, and we see atavistic curiosities like a tail extending above a person's buttocks (Prothero 2007, 345) from our vestigial human tailbone, or actual toes (not just nubs) on either side of a horse's hoof (p. 98-99).

Plagues from Denial

We, along with many of the theistic evolutionists whose writings we criticize, find it infuriating how creationists deny the reality—and wonder—of evolutionary theory. Its explanatory power is stunning; all the scientific puzzle pieces fit into place, from anthropology to zoology. It is difficult to emphasize enough just how strong the evidence is, yet most Americans doggedly persist in remaining ignorant of it. Worse, the most devout among them view it as almost a holy calling to enforce that same ignorance on their children and churches.

For those not blinkered by an outdated dogma, there is no longer any debate about the truth of evolution. The debate has been over for a hundred years. The evidence has continued to pile up, in new fields like molecular genetics that Darwin couldn't have dreamed of. We have long since reached the point where evolution—micro, macro, human—is no more productive a topic for argument than computing epicycles in case that "theory" of Copernicus turns out to be wrong, after all. Two of our pious scientists put it frankly to their Christian brethren: "When there is a near-universal consensus among scientists that something is true, we have to take that seriously, even if we don't like the conclusion" (Giberson and Collins 2011, 29). Sure, there are some crackpots who reject what is squarely in front of their faces, even a few educated ones who ought to know better. But the "percentage of scientists who reject evolution is very small—so small that in most large gatherings of scientists you would not find even one person who rejects the theory of evolution" (p. 30).

The fact is that creationists are just parasites, living off the intellectual metabolism generated by the hard work of real thinkers while contributing nothing but the fever of the camp revival. They crave respectability for their faith, but they show nothing but contempt for the careful research of the scientists who find the hard data that they persistently ignore and deny. "There are no transitional forms!", they whine, despite a wealth of fossils showing various types of intermediates (Prothero 2007). Meanwhile Neil Shubin goes to Greenland year after year to dig for a fossil evidencing the transition from water to land. And he finds it, too: *Tiktaalik roseae*.[15] Did the creationists gather at the Field Museum to examine this extraordinary find that plugs the evidenciary hole they had been complaining about? Of course not. They just go on talking, about smaller holes.

---

15. See, e.g., tiktaalik.uchicago.edu (accessed January 2013). "*Tiktaalik roseae*, better known as the 'fishapod,' is a 375 million year old fossil fish which was discovered in the Canadian Arctic in 2004. Its discovery sheds light on a pivotal point in the history of life on Earth: when the very first fish ventured out onto land."

It just takes a little knowledge to leaven a lump of ignorance into something far less benign: deceit. The great nineteenth-century writer and orator Robert G. Ingersoll compared the church pulpits of his day to pillories where each straight-and-narrow minister stood as a "hired culprit, defending the justice of his own imprisonment." Every one of those poor preachers "knows that every member of his church stands guard over his brain with a creed, like a club, in his hand. He knows that he is not expected to search after the truth, but that he is employed to defend the creed" (*Lecture on Individuality*). We wonder if the situation is really any different for many of today's creationist mouthpieces.

Even the more sophisticated wanderers through the wilderness of science, kicking at the camels who are crowding into their tents of threatened faith, seem to need some denial to get through the night. Like evangelical New Testament critics (really, half-apologists) who are tempted by the explanatory power of biblical criticism, they can't deny that evolution makes more sense out of things. But they have to sanitize its findings with dogma. They just can't bear to go all the way with it and leave the innocence of their faith, so they wind up being inconsistent with both.

Denial and deceit can only take you so far, though. Every creationist who has taken a new antibiotic to get rid of penicillin-resistant bacteria knows that some sort of evolution is going on. They try to compartmentalize these seemingly modest changes as "microevolution" while dismissing the idea of new features or species forming as "macroevolution." An old favorite is, "If we came from apes, why are there still apes around?" to which one might well ask why there are still Jews if Christianity came from Judaism.[16] The creationists' sneering skepticism would never be applied to, say, a poorly attested (even in scripture) story about a crucifixion victim rising from the grave after undergoing the irreversible brain death of asphyxiation and the breakdown of cellular structure that commences quickly after death.

Why? Because the creationist doubting Thomas didn't ever see it happen! Well, the scientists invite him to put his hand into the print of the fossils and see for himself. Don't take our word for it. The evidence for evolutionary alterations and additions—not minor tweaks here and there but significant changes occurring over time—is overwhelming, and readily

---

16. The more technically correct answer is that chimps are still mostly forest-dwellers, like our common ancestor, and thus haven't experienced the selection pressure that causes evolutionary change. The human lineage, on the other hand, "left its forest home and took its chances in the open woodland, adapting to a quite different set of challenges" (Wade 2006, 14). As we have seen, there is a dithering and clumsy partnership between randomness and natural selection that leaves vestigial features around long after they are no longer useful. It is in no hurry to change designs that still work just fine after millions of years.

available to the public in any number of informative, engaging books. As Michael Shermer says in his foreword to one fine example, Prothero's beautifully written and illustrated *Evolution: What the Fossils Say and Why It Matters*, the "visual presentation of the fossil and genetic evidence for evolution is so unmistakably powerful that I venture to say that no one could read this book and still deny the reality of evolution. It happened. Deal with it" (from Prothero 2007, xiii).

"New species can arise even faster than people once thought," Prothero says (p. 115), describing some cases of significant evolution occurring in just decades. Two populations of salmon that colonized different types of waters in Western Washington have, in less than forty years, diverged to the point where they are genetically isolated and "would be recognized as separate species in most organisms." One group has males with more slender bodies to fight strong currents and bigger females "that can dig deeper holes for their eggs to prevent the river from eroding them away," while the other group's males have "deeper, rounder bodies, which are better at fending off rivals for mating privileges" (p. 115). Another example is the three-spined stickleback, whose body armor and spines have evolved significantly in a dozen years at one location, around thirty at another (p. 116). Then there

> is the classic case of industrial melanism, so familiar from every textbook. The peppered moth, *Biston betularia*, normally has a speckled appearance that blends in well with [the] mottled appearance of trunks and branches of trees. During the Industrial Revolution, soot in the air made the tree trunks black, and the normal form was conspicuous. Instead, a dark-colored mutant became dominant, because they were well camouflaged against the dark tree trunks, while the birds picked off the normal speckled varieties. When environmental regulations cleaned up the air and eliminated the sooty tree trunks, the normal speckled varieties returned, and the dark mutants were again selected against. [pp. 116-17]

Unlike the disciples staring at the walls of their locked room and wondering how a brutally executed Jesus got there, alive, we have a very plausible story behind these dramatic transformations. What would be miraculous is if they did *not* exist, because "evolution by natural selection and drift is an inevitable consequence of error-prone replication of digitally encoded genetic information" (Koonin 2011, loc. 905). And, eventually, speciation is also inevitable, the "result of genetic barriers that arise when spatially isolated populations evolve in different directions" (Coyne 2009a, 176).

It is a demonstration of ignorance and crippled imagination to object that we don't see apes turning into men, as if the multi-generational process of evolution were something that could occur in a single lifespan of the creature involved. Coyne estimates, along with Allen Orr, that somewhere between 100,000 and five million years of evolution are required to go from one ancestor to a pair of reproductively isolated descendants (p. 176). Saying that new species don't evolve because you haven't seen it happen is like sitting on a lawn for an afternoon and concluding that the grass isn't growing.

The doubting creationists turn out to be more in the position of Peter than Thomas, anyhow. We have direct, observational evidence of full speciation occurring in recent memory. The Welsh groundsel, first observed in 1958, is a natural, reproductively viable hybrid of two other species. Its set of sixty chromosomes is a doubled combination of those found in the two species that gave rise to it. Its hybridization has been duplicated in the laboratory, with exactly the same result as what evolved, about a century ago, in the wild (Coyne 2009a, 187-88; Young 1989, 50). We have plenty of evidence for speciation, with "lineages splitting in the fossil record" and "closely related species separated by geographic barriers" exactly as evolutionary theory would predict. But, if nothing else, "those critics who won't accept evolution unless it happens before their eyes" ought to be satisfied with the Welsh groundsel and four other instances of real-time plant speciation (Coyne 2009a, 189).

Of course you know they won't be, though. Not when nearly two-thirds of Americans admit they would simply reject *any* scientific finding disproving one of their religious beliefs (Coyne 2012, 3). The scientist walks by sadly as they warm themselves around the creationist campfire and utter oaths of denial to those standing nearby.

A Grand View of Life

In the words of Psalm 139, we are fearfully and wonderfully made. It's still true, even if we weren't assembled by a creator in the darkness of the womb with our entire lives and thoughts mapped out ahead of time. The reality is even grander than that. You are "little more than an eddy in a great river of life," an evolved "collection of self-regulating and continually dividing cells" that is "continuous with your genetic precursors: your parents, their parents, and backward through tens of millions of generations" (Harris 2005, 210). You are not separate from nature, but a living, breathing example of its wonders. Atoms with curiosity, as Carl Sagan put it, nature contemplating itself.

Yes, there is grandeur in this view of life, as Darwin famously concluded in his masterpiece. It is a grandeur that we can only see if we finally leave the cave—not just the smoky recess in the rock in which that Cro-Magnon played his flute, but Plato's cave of cozy ignorance. Those lingering inside only see the shifting shadows, without any knowledge of the realities behind them. The great enlightenment is when man manages to slip outside and see his true likeness for the first time, in the full flood tide of the sun. That's what Darwin gave us, and there was no shortage of smug scoffers like Richard Owens and Bishop Wilberforce wondering why he was going to the trouble and letting in a draft.

But we understand why, and you probably do, too. It is only when we emerge from the shadows, away from the suffocating warmth of our self-satisfaction provided by stale ancient myths, that we see an explanation for ourselves and all other life. That is the breathtaking power and elegance of evolution.

Branch I: The Word

The battle to keep the Bible as a respected book among the learned scholars and the academic world was fought and lost in the nineteenth century.

—Bernard Ramm,
The Christian View of Science and Scripture

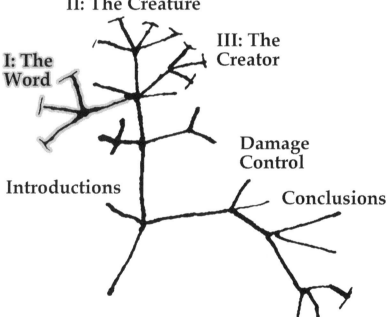

The World the Biblical Writers Thought They Lived In

> *Moses says in plain words that the waters were above and below the firmament. Here I, therefore, take my reason captive and subscribe to the Word even though I do not understand it.*
>
> —Martin Luther, *Lectures on Genesis*

It is remarkable how so many Christians imagine themselves to take the Bible literally and reject evolutionary science on that basis. But they are clinging to the tip of an iceberg that is mostly invisible to them under the surface of a superficial, Sunday school reading of the Bible. All they have seen of the text that fails to match up with modern science is the notion of a rapid-fire creation of light, darkness, water and dry land, animals and humans. There is, it turns out, much more that removes the Bible from any compatibility with the universe we now know about. Even fundamentalists comfortably take for granted certain scientific findings, so much so that they no longer even recognize the astonishingly strange features of the biblical portrait of the universe.

Planet of the Bible

Various reinterpretations of the Bible and theology stem from Christian attempts to take evolution seriously, either rejecting or affirming it. To discuss these in an informed manner, we must first examine what the Bible actually says about earth and its origin.[17] The picture is nowhere near as simple as either so-called scientific creationists or today's Christian evolutionists would make it.

So-called biblical literalists find themselves in a strange double bind when dealing with the numerous Old Testament references to the common world-picture of the ancients. Their primary commitment is to believe whatever the Bible may say on any subject, which is difficult enough when it speaks of unseen and unverifiable realities. The dogma of a virgin birth requires Christians of all stripes, from Ken Ham to Francis Collins, to profess that God created an entire set of chromosomes *ex nihilo* to pair up with the ones

---

17. Some material in this chapter was adapted from a Web page authored by Robert M. Price in collaboration with Reginald V. Finley, Sr., "Heaven and Its Wonders, and Earth: The World the Biblical Writers Thought They Lived In," infidelguy.com/heaven_sky.htm.

in Mary's egg. It's a stretch, but there's no way to prove that God didn't throw that monkey wrench (just once!) into the otherwise seemly order of nature. However, believing the Bible when it says things *contrary to* massive amounts of irrefutable empirical evidence is another challenge entirely, and that's what literalists are up against with the biblical cosmological references.

Bernard Ramm tried to get around the problem by piously denying that the words of the Bible really pose any problem. The references its writers make "to natural things are popular, non-postulational, and in terms of the culture to which the writers wrote" (1954, 65). Its language "with reference to cosmological matters is in terms of the prevailing culture." The Bible, you see, uses "the language of antiquity and not of modern science" (p. 65). And so, evading all responsibility for the actual meanings of words, Ramm dispensed with "the efforts of radical critics to impose a cosmology on the Bible as an artificial, stilted and abortive effort" (p. 66). "When the biblical account is set side by side with any other cosmology," he gushed, "its purity, its chasteness, its uniqueness, its theocentricity are immediately apparent" (1954, 69). Well, let's see about that.

Put briefly, the Bible seems to describe the earth as a flat disk afloat upon a vast cosmic ocean. The sky it represents as a solid dome with windows and gates, and as resting upon great pillars thrust up from below. The sun, moon, and stars appear to be set into the heavenly vault like ceiling lights, smallish and at no great distance from the earth. H.E. Fosdick articulated this "radical view" in 1924 with a single paragraph quoted (and sniffed at) by Ramm:

> In the Scriptures the flat earth is founded on an underlying sea; it is stationary; the heavens are like an upturned bowl or canopy above it; the circumference of this vault rests on pillars; the sun, moon, and stars move within this firmament of special purpose to illumine man; there is a sea above the sky, "the waters which were above the heavens," and through the "windows of heaven" the rain comes down; within the earth is Sheol, where dwell the shadowy dead; this whole cosmic system is suspended over vacancy; and it was all made in six days with a morning and an evening, a short and measurable time before. This is the world view of the Bible. [from Ramm 1954, 66]

A great deal of ancient evidence, both textual and archaeological, makes it clear that this is simply the common world-picture yielded by ancient natural philosophy, i.e., scientific speculation as yet unaided by observational technology such as we possess. Indeed, we should think the same thing were we in their place, for the world surely appears to be flat, albeit of variable altitude. The sky appears to enclose the flat vista on all sides and to descend to meet its edges in whatever direction one looks. Rain falls from the sky and water wells up from beneath. Such a view of the world is not the product of stupidity but rather of shrewd and careful observation. The unaided eye and mind could not be blamed for thinking that was what the world was like (Figure 3).

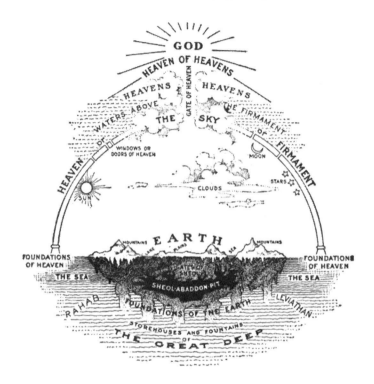

FIGURE 3: *The Ancient Hebrew Conception of the Universe.*

So where does the literalist stand? He is in the impossible position of trying to make the Bible the norm and source of his beliefs, on the one hand, and yet to keep the Bible seeming believable by the standards of modern knowledge on the other. He cannot bring himself to deny what modern instruments have shown to be the truth of cosmology, so he cannot believe the world looks as described in scripture, but neither can he bring himself to admit that the Bible is mistaken. So, in order to defend the literal truth of the Bible (the proposition that it describes things the way they are, whether things on earth or things in heaven), he must resort to non-literal reinterpretation of the cosmic-descriptive passages of the Bible.

Literalists remain in this conceptual Slough of Despond because they feel mired in it. If they admit the Bible writers pictured the world the wrong way, despite their ostensible divine inspiration, they know that *all* the encumbrances of mythology are fair game, not just the ancient three-story picture of the world. And that would be the end of supernaturalism and miracles, the virgin birth included.

But it is too late for anything else. Once you are even aware of the danger, the horse has escaped the barn. To insist on a set of beliefs that would be more comforting just because you dread the result of facing the truth is fatal to the conscience. "Faith" from that point on rests on the rotten foundation of self-deception, which requires you to "suppress the truth in unrighteousness" (Rom. 1:18, NASB). Indeed, it is to that fatal misstep that we owe the pervasive dishonesty of apologetics: anything to defend the party line.

Because the Hebrews were intense monotheists," Ramm protests, "they were free of mythology and superstition so uniformly connected with the religions of peoples with reference to the objects of nature" (1954, 55). As was then typical, he reads Rabbinic Judaism back into ancient Israelite belief, contra ample evidence of polytheism (and mythology). He claims "there is a chaste, wholesome, refreshing view of Nature in the Bible that is richly theistic," and wishes that "men like Pascal, Kierkegaard and Barth had reflected more sympathetically with the Biblical view of Nature" (p. 55). So it's *their* fault that this stuff looks crazy now!

> If the scientist insists that the Bible teaches that the earth is flat, or the heavens solid, or that the entire solar system came to rest at Joshua's command, then through his own misinterpretation he brings the Bible into conflict with science. [p. 38]

Actually, the scientists are actually better Bible interpreters than Ramm. In fact he is *reinterpreting* the crucial passages, trying to make it look like they comport with what scientists say about these matters, then pretending the scientists are falsely accusing scripture. And he has the gall to claim, "Only the man of faith has the correct perspective and motivation to harmonize Scripture and science. Men without faith cannot but clash the gears" (p. 39). Yes, because they see the firmament holding up the waters of heaven—to say nothing of the classification of rabbits as ruminants and bats as birds, the breeding of spotted cattle by gazing on a speckled rod, or demon possession as the cause of epilepsy—for what they are: mistaken ideas thought up by ancient people who had no actual idea what they were talking about.

Ramm shields his eyes from these inconvenient facts and scientific conflicts (which were all too evident even in 1954) when he notes "with pleasure the complete lack of the fantastic, absurd, mythological, or superstitious elements common to most [ancient] peoples" (p. 60). He cannot arrive at this conclusion without engaging in a text-twisting Rationalism, however. And his motive is exactly the same as those of his eighteenth century predecessors: to "save the appearances" of scripture while substituting science for the supernaturalism embraced by the ancient writers. With his allegiance to "our doctrine of the inspiration and infallability of the Scriptures" firmly in place, Ramm can find it apparently "miraculous that the writers of the Bible are free of the grotesque, the mythological, and the absurd" (p. 49).

Despite that pious awe, Ramm's "position that the Biblical writers do not teach any cosmological system or follow any cosmogony, ancient or modern" seeks to relegate the Bible to being

> pre-scientific and phenomenal or non-postulational. As far as their vocabulary and method of speaking are concerned they use the cultural terms and expressions of their time. The Spirit of God restrained them from any use of the polytheistic or fantastic. [p. 5]

We will see how wrong this last sentence is, Spirit of God or no. For the moment, suffice it to add one more voice. Conrad Hyers holds the same view as Ramm:

> Biblical references to nature were not scientific statements, which then might be said to be in conflict with scientific data, observations, and theories. The Bible uses the common, everyday, universal language of experience. [Hyers 1984, 19]

But this is all gross special pleading. It is to make a gratuitous, though convenient, assumption for "your side" that cannot be afforded to the "other side." If, instead, we want to be consistent, we should have to suppose the same is true of all the ancient writers and sculptors who portray the world in the ancient way. We would then be assuming that *all* references to "ancient" cosmology were extravagant metaphor and that all the ancients really understood the world to be arranged as modern astronomy and geology tell us. But, save for the need to extricate oneself from a tight spot, why suppose such a thing? And then what to make of the history of astronomical discovery? They must already have known everything, right? You see just how convoluted and indeed surreal it becomes. That way, as the saying goes, lies madness.

The Strange Old World of the Bible

The tireless, faith-driven efforts of apologists to prop up the Bible's relevance in a scientific age can easily leave the mistaken impression that the Genesis creation account wasn't that far off the mark after all. (Actually, it's *accounts*, plural, as we'll see in the next chapter.) Gerald Schroeder's *The Science of God* is an example. One of us (Ed) bought and read a copy as the first vestiges of doubt started to creep in about his biblicist faith, and found reassurance from conclusory statements like this one about Genesis:

> Of all the ancient accounts of creation, only that of Genesis has warranted a second reading by the scientific community. It alone records the sequence of events that approaches the scientific account of our cosmic origins. [Schroeder 1997, 80]

Now, of course, we both see this for the pious cheerleading that it is. Despite Schroeder's attempts to privilege his particular ancient text above those of the Greeks, Romans, Norse, and Babylonians (pp. 80-81), the fact is that the Bible is no further ahead in its depiction of the natural world. The earth beneath the firmament was a flat disk, supported by foundations as the firmament was by pillars. One ocean weighed down upon the dome from above, another buoyed the earth up from below. This earth was anchored immovably and formed the center of solar and planetary motion. It was a very different world than the one we live on.

If we, like our ancestors, lacked modern observational technology, we might still share these beliefs, and we would be fools no more than they. In fact, ancient, post-Genesis Jews did readily accept major or minor elements of Greek cosmology. One of those is the Ptolemaic schema whereby a still and spherical earth rests at the center of a great number of crystalline spheres in which the sun, moon, planets and stars were set, orbiting us in complex dances. The ancients were mistaken in those theories, too. But they were in all cases quite correct to accept, however provisionally, the latest results of careful thinkers on the subject. It is both ironic and tragic that today's biblicists, so eager to think the thoughts of the Bible's writers after them, refuse to budge from the same obsolete world pictures those scribes and prophets would have discarded in a moment had better knowledge been available.

Let's take a look at three aspects of this biblical universe: the heaven above, the earth below, and the structure propping it all up.

For the World Is Hollow, and I Have Touched the Sky

All the ancients, like many today, spoke of "heaven" and meant "the sky." The Bible even uses the same words for both, interchangeably. "The kingdom of heaven" is the same as "the kingdom of the sky." The Greek Titan Ouranos (Uranus) is simply the Sky, the Heaven, personified. It turns out that modern thinking on heaven as the abode of God and the location of the blessed afterlife has undergone a hasty retooling in light of scientific knowledge, namely that there is no absolute up or down, that the sky and outer space are not *up* there but *out* there.

Most Christians have reacted to this secularization of the sky and of space by redefining the religious heaven, the theological heaven. And they have done so in a vague manner drawn more or less from science fiction. Now people speak increasingly of heaven as "another dimension," whatever that means. It is surprising how little comment this great shift has occasioned. No one who says it appears to have much in mind. It is simply a way of trying to fend off the facts of science. "God turned out to be absent from the heaven of the sky? Okay, then, there must be some other heaven for him to be in!" As we will see, the Bible writers certainly drew no such distinction.

Edward Babinski notes that the usual Hebrew word for heaven in the Bible is *shamayin*, which "appears more than 400 times in the Bible and applies to a wide variety of things, from 'birds of heaven, angels of heaven, foundations of heaven, pillars of heaven, to the firmament of heaven' (Genesis 1:14). In contrast, firmament (*raqia'*) only appears 17 times in the Bible," a special cosmological creation "in the midst of primeval waters that made heaven/sky possible, as well as earth/dry land" (Babinski 2010, loc. 1479).

When they wrote, "Let there be a firmament to separate the waters from the waters" (Gen. 1:6), the biblical writers must have been thinking of something like a giant version of the Astrodome. The translation of the word preserves this idea; it contains the element "firm," which reflects the fact that the underlying Hebrew denotes a solid dome of metal or crystal. The Latin noun *firmamentum* comes from the verb *firmare*, "to make firm." It is a good word to choose to translate the Hebrew *raqiya*, which denotes "a dome beaten out of metal sheets."

FIGURE 4: The Flammarion engraving.

Though the very word underlying "firmament" implies composition out of hammered metal plates, sometimes the solid matter out of which the heavenly dome is constructed is pictured as crystalline. According to the Akkadian text in *Keilschrifttexte aus Assur religiösen Inhalts* (Ebeling 1915, 307), the cosmos has six levels: three of them celestial and three terrestrial. The following is from Wright's translation (2000, 34):

> The upper heavens are *luludanitu* stone.
> . . .
> The middle heavens are *saggilmut* stone
> . . .
> Bel sat there in a chamber on a lapis-lazuli throne.
> . . .
> The lower heavens are jasper.
> They belong to the stars.
> He drew the constellations of the gods on it.

This multi-story heaven appears to anticipate the later Ptolemaic cosmology with its numerous crystalline spheres, but that is not our concern here. The point is not to say that all the ancients, even within a single culture, had exactly matching conceptions of the universe, only that the general trend looked nothing like the universe as portrayed by modern scientific calculation.

Exodus 24:9-10 assumes precisely the same sort of a sapphire (or lapis-lazuli, same word in Hebrew) heavenly pavement. "Then Moses went up with Aaron, Nadab and Abihu, and seventy of the elders of Israel, and they saw the God of Israel; and under his feet there appeared to be a pavement of sapphire, as clear as the sky itself" (NASB). Not surprisingly, the writer of Ezekiel 10:1 takes for granted the same idea: He looked, and "in the firmament that was above the head of the cherubims there appeared over them as it were a sapphire stone, as the appearance of the likeness of a throne."

The word derives from *raqa*, "to expand by beating out," hammering out metal in order to shape it. Elihu asks Job, "Can you, like him, spread out [*raqa*] the skies, hard as a molten mirror?" (Job 37:18 RSV). Similarly, "gold leaf was hammered out (*raqa*) . . ." (Exod. 39:3); and "silver spread into plates (*raqa*)" (Jer. 10:9 RSV). Ezekiel 1:22 relates, "Over the heads of the living creatures there was the likeness of a firmament, shining like crystal." Ramm tells us, "The best meaning of *raqia* is *expanse* or *atmosphere*" (1954, 67), but he is just indulging in wishful thinking:

> Even the *Theological Word Book of the Old Testament*, authored by conservative Evangelical Christian scholars R. Laird Harris, Gleason L. Archer, and Bruce K. Waltke, and published by Moody Press, agreed that the Bible verses in which the root of *raqia'* (i.e., *raqa'*)

appeared were all related to something solid being spread out, stamped, or pounded down... [Babinski 2010, loc. 1501]

The biblical authors were men of their time who lived, quite literally, in the world of their time. And that world had a solid ceiling. (Note that even the English word "ceiling" is based on the root word for heaven, as in "*cel*estial.") J. Edward Wright explains the Sumerian view, where

> the universe was a tripartite structure—heaven (the place of the high gods), earth (the realm of humans), and the netherworld (the realm of deceased humans and the mortuary gods). According to S.N. Kramer, since the Sumerian word for tin is "metal of heaven," it may be that the Sumerians thought that the floor of heaven was made of tin or some comparable metal. It also appears that the Sumerians considered the sky to be a vault or dome because we read of heaven having a zenith. [Wright 2000, 29]

One of the most ancient representations of a solid firmament is that of the ancient Egyptians, who depicted the firmament in the form of the Goddess Nut arched over the flat earth (Babinski 2010, loc. 1334), as shown in Figure 5.

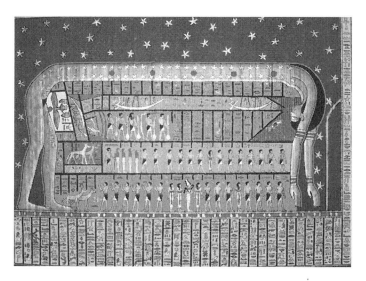

FIGURE 5: The Sky Goddess Nut.

The Egyptians also employed the less mythologized form of a "wall-ring" to represent the firmament above the earth. The Babylonians in their creation epic, *Enuma Elish*, depicted the firmament being constructed out of the body of a dead goddess/dragon named "Tiamat," who was cut in half to form the firmament and heavens above, and the earth below. One Babylonian tablet fragment even mentions a *"Tiamat eliti"* and a *"Tiamat sapliti,"* that is an Upper Tiamat (or Ocean) and a Lower Tiamat (or Ocean) that corresponds apparently to the Hebrew belief in "waters above and below the firmament" in Genesis 1:7 (Babinski 2012).

Tiamat or *Tiamu* is the sky in ancient Babylon, the goddess of watery chaos. Marduk "split her into two like a dried fish: One half of her he set up and *stretched out as the heavens*. He stretched the skin and appointed a watch with the instruction not to let her waters escape" (*Enuma Elish*, Tablet 4, lines 136-40, from Mark 2011).

Do *Tehom* (used in the Hebrew Creation account) and the Babylonian *Tiamat* share a common ancient root that refers to deep and chaotic waters? Bernard Ramm protests the tendency of "radical critics to play up the similarity of anything Biblical with the Babylonian, and to omit the profound differences or gloss over them" (1954, 69). He dismisses the association between the two terms as "simply [proving] the Hebrew account to be the original" (pp. 68-69). It's about as convincing as his special pleading regarding the Genesis Flood account vs. the *Gilgamesh Epic*:

> The Babylonian account represents the tradition freely corrupted by human imagination; the Hebrew account is that which was kept chaste and pure through divine providence and then recorded through divine inspiration. [Ramm 1954, 168-69]

Features of the Firmament

The heaven was equipped with two great lights, the sun and the moon (Gen. 1:16). The sun, moon, and stars appear to be set into the heavenly vault like ceiling lights—small and not so far above the earth. In his Internet essay *Interpretations of Biblical Cosmology*, Babinski notes that "the Hebrew term translated as 'great lights' in Genesis, means literally, 'great lamps.'" There was "no recognition of the fact that every twinkle in the sky might be another great lamp like the sun, or perhaps be a planetary body larger than the earth with many moons (great lamps) of its own" (*Interpretations*, Introduction).

Genesis 1:16 treats the stars as afterthoughts: "he made the stars also." Referencing Rev. 6:12-14; 7:1; 21:1, Babinksi says the Bible "depicts stars as relatively small objects, created after the earth and 'set' in the firmament above it, which may even 'fall' to earth at its end when the 'earth is shaken' and they 'fall like ripe figs' from the sky, and the 'heavens are rolled up like a scroll' to be 'created' anew" (*Interpretations*, Introduction).

It was these heavenly features that moved in the firmament, not the fixed earth under the ancient observers' feet. Babinski cites a number of Bible verses illustrating belief in the movement of the sun. They include Job 9:7, "He [God] can command the sun not to rise." That God would direct his command at the sun rather than the earth implies a belief in a stationary earth. Likewise, Joshua directed his commands at both the sun and moon, even commanding the sun to stand still "over Gibeon," and the moon "over the valley of Aijalon" (Joshua 10:12). The Bible also teaches that stars "course" through the sky each night (Judges 5:20), and God "brings them out one by one" and "because of His great power not one of them is missing" (Isaiah 40:26). Compare *Enuma Elish* VII:130, which states, "He [Marduk] shall maintain the motions of the stars of heaven." In addition, Job 38:31-33 (NASB) states that constellations are "led forth" by God, as when God asks Job rhetorically, "Can you lead forth a constellation in its season, And guide the Bear with her satellites? Do you know the ordinances of the heavens, Or fix their rule over the earth?" [Babinski 2010, loc. 1577]

In *Interpretations*, Babinski also mentions Eccl. 1:5 ("The sun rises and the sun sets, and hastening to its place it rises there again," NASB) and Ps. 19:4-6 ("In [the heavens] He has placed a tent for the sun, which is as a bridegroom coming out of his chamber; it rejoices like a strong man to run its course, its [daily] rising from one end of the heavens, and its circuit to the other end of them").

The Joshua 10:12 incident is worth singling out for a moment. The description that "the sun stood still in the midst of heaven, and hasted not to go down about a whole day" certainly bespeaks the writer's belief that the earth is a disk of fairly limited size, with the sun going up and down around it. In view of what we now know about the conservation of momentum and energy and the earth being a rotating sphere illuminated by a distant sun, it is hard to know where to begin in accounting for the necessary miracles.

The entire spinning mass must have suddenly stopped in place, the thin crust of the earth staying exactly in place atop the liquid mantle as an unimaginable amount of kinetic energy just disappeared into nothing. Then, when the fighting had finished, all that kinetic energy was miraculously inserted back into everything in and on earth again, in perfect timing and quantity, and the earth resumed spinning. And all this happened without any of the other peoples dispersed across the entire earth making any note of it, including the neighboring Egyptians who made detailed astronomical records and worshiped the sun as a god.

The science-bending that would have been involved makes it apparent that this is really the most incredible miracle of the entire Bible, if one is determined to remain shackled to it while also pursuing scientific truth.[18] One might reply that all this and more is possible for an omnipotent deity, but that is not the point. Is it really likely that so much is intended in the telling of the story? The event gets only a few verses in the Bible; the ancient writer probably didn't think it that significant for the bright little dot of the sun to just stop moving across the sky for a while.

Yet even in the ancient world, people held fast the flat earth view in the face of counterarguments and evidence that were already adding up. Toward the end of the second century CE, Tatian poo-poohed as mere arguments about words, "not a sober exposition of truth," the idea, unbelievable to him, "that the sun is a red-hot mass and the moon an earth" (*Address to the Greeks*, Ch. 27). Writing around the same time, Theophilus of Antioch was taking all that firmament business quite literally, writing that the heaven was "like a dome-shaped covering" and citing Isaiah:

> "It is God who made the heavens as a vault, and stretched them as a tent to dwell in." The command, then, of God, that is, His Word, shining as a lamp in an enclosed chamber, lit up all that was under heaven, when He had made light apart from the world. And the light God called Day, and the darkness Night. [*To Autolycus*, Book 2, Ch. 13]

More than a thousand years after Tatian and Theophilus, Luther remained mired in this pious ignorance. In 1535, he praised it as "a work of the Divine Majesty that the sun follows its course so exactly and in a most precise manner without deviating a fingerbreadth from the straightest possible line in any part of the heaven." The firmament was no longer quite so firm, however, being "softer and more tenuous than water and yet . . . no part of it has deteriorated or become weak in the course of so many thousands of years" (*Lectures in Genesis*, Ch. 1, v. 6).[19]

---

18. This discussion of the Joshua story was adapted from Suominen 2012, §6.6. Though it does not bear on our larger point about the Bible's scientific inaccuracies, it's worth noting that there's no point in seeking *moral* truth in the story, either. God's whole point of staging this stupendous disruption of nature was to allow Joshua to continue a bloodthirsty campaign of conquest against some neighboring tribes in a tiny corner of the Mideast. The whole thing is just a testament to a scientifically ignorant, barbaric, and tribalistic past that deserves no more from us now than a curious backward glance.

19. *Lectures in Genesis* includes an intriguing comment where Luther seems to reveal some insight beyond slavish literalism and its dome-over-the-earth cosmology. He gives credit to contemporary astronomers as "the experts from whom it is most convenient to get what may be discussed about these subjects," and acknowledges their claim that the stars (oops!) and the moon are lit by the sun. That "the moon derives its light from the sun . . . is really well proved at an eclipse of the moon, when the earth, intervening in a direct line between the sun and the moon, does not

Not quite ten years later, Copernicus finally published his *On the Revolutions of the Heavenly Spheres* that displaced the earth from its fixed throne at the center of things. He had already made his revolutionary idea known, and both Luther and his colleague Philip Melancthon made disparaging remarks about it. But it was actually a *Lutheran* minister, Andreas Osiander, who arranged publication of Copernicus's book, albeit with a soft-pedaling preface that sought to reassure the faithful that this was all just a hypothesis. Nearly a century later, the Catholic Church made its infamous attempt to suppress heliocentricism, when Galileo's telescopic observations made it uncomfortably real (Kobe 2004).

FIGURE 6: Copernicus's illustration of heliocentrism.

Geocentrism Is Egocentrism Writ Large

The earth from which the ancients looked up at their vaulted heaven was not just fixed, but flat. Ed Babinski cites five examples of the Bible writers taking this view. With his kind permission, we present a numbered list that loosely quotes and condenses what Babinksi writes in *Interpretations*, "How to Explain Away Flat Earth Verses in the Bible":

> **1.** Daniel 4:10-11 says, "There was a tree in the midst (or center) of the earth, and its height was great. It reached to the sky, and was visible to the end of the whole earth." Such visibility (i.e., "a tree of great height at the center of the earth and seen to the end of the whole earth") implies a flat earth. The verse might be explained away as

---

permit the light of the sun to pass to the moon" (Ch. 1, v. 14). That is correct, of course, and it's hard to see how one could picture it happening with the moon slid underneath the disk of the earth while the sun hovers overhead, relatively nearby.

depicting a "mere dream" of Daniel's, viz., a "metaphorical image" of the extent of Nebuchadnezzer's kingdom. However, the fact that flat earth imagery surprised no one in the story of Daniel also implies that it was taken for granted.

2. Isaiah 42:5 and 44:24 state that at creation God "spread out the earth"–the Hebrew verb for "spread" being used elsewhere in Scripture to depict a "flattening" or "pounding." If the earth was not "spread out," but "rolled up tightly like a ball" at creation, the writer could have said so. We find the requisite Hebrew construction in Isaiah 22:18, where a man is "rolled up tightly like a ball." Hence the earth was spoken of as being "flattened or pounded flat" at creation.[20]

3. Isaiah 11:12 declares, "Gather (them) from the four corners of the earth," and Revelation 7:1 adds, "I saw four angels standing on the four corners of the earth."

4. Matthew 4:8 states that "The devil took him [Jesus] to a very high mountain, and showed him all the kingdoms of the world [Greek, *cosmos*], and their glory." One could see "all" the kingdoms of the world from a very high mountain if the world were flat. This verse has been explained away as a "vision" of all the world's kingdoms, a vision received on a very high mountain. However, such a "vision" could have been granted Jesus in his sleep, so why did the devil "take Jesus" to a "very high mountain" to show him such things?

5. Throughout Scripture, the shape and construction of the earth is assumed to resemble that of a building having a firm immovable foundation, and a roof (or canopy). "He established the earth upon its foundations, that it will not totter, forever and ever" (Ps. 104:5). "The world is firmly established, it will not be moved" (Psalm 93:1). "For the pillars of the earth are the Lord's, and he set the world on them" (1 Sam. 2:8). "It is I who have firmly set its pillars" (Ps. 75:3). "Who stretched out the heavens ... and established the world" (Jeremiah 10:12). "Who stretches out the heavens like a curtain and spreads them out like a tent to dwell in" (Isa. 40:22). "Stretching out heaven like a tent curtain" (Ps. 104:1-2). "In the heavens ... in the true tabernacle (tent), which the Lord pitched, not man" (Heb. 8:2-3). "Praise God in his sanctuary, praise him in his mighty firmament" (Ps. 150:1).

---

20. Babinksi notes, "Round-earth creationists at this point usually change the subject by concentrating their 'scientific' attentions on another verse in the book of Isaiah, 'He who sits above the circle of the earth' (Isaiah 40:22), that they say implies a spherical earth." It doesn't, Babinski counters, providing two paragraphs of reasons why not (*Interpretations*, How to Explain Away Flat Earth Verses in the Bible).

The biblical writers took their fixed, flat earth seriously. To claim otherwise is to diminish their voice, to "erase the line between text and interpretation" (Price 2006, 311). We must accept the Bible's scientific errors for what they are, just as the liberal theologian Thom Stark does with the biblical stories of divinely directed atrocities when he says "allegorizing a problematic text is little different from simply discarding it–because the text's original meaning is simply ignored" (2011, 140). Enns recognizes this, too: "The most faithful, Christian reading of sacred Scripture is one that recognizes Scripture as a product of the times in which it was written and/or the events took place–not merely so, but unalterably so" (2012, loc. 168).

Supporting the Sky, and Earth

The sky, being a tangible structure, had to be propped up there somehow. In ancient Egyptian cosmology, Wright says, the "celestial plane itself is either supported by pillars, staves, or scepters, or is set on top of the mountains at the extreme ends of the earth." Egyptian iconography typically depicted those supports in pairs, but "the pair in fact represent four supports, thus 'the four corners of the earth.' The tombs of pharaohs Tutankamon, Seti I, and Ramses II in addition to the figures of the celestial cow or woman have alternative depictions of the sky being supported by pillars or by people holding staves of some kind. These images also appear in many texts" (Wright 2000, 13).

Wright supports his point with a number of pertinent quotations. This is from the *Victory Hymn of Thutmose II*: "I see the glory of thee and the fear of thee in all lands, the terror of thee as far as the *four supports of heaven*" (from Wright 2000, 13, our emphasis). The *Book of the Dead* (2:102) addresses the sun god "who shinest from thy Disk and risest in thy horizon, and dost shine like gold above the sky, like unto whom there is none among the gods, who sailest over *the pillars* of Shu" (from Wright 2000, 14, our emphasis). Shu was the air god who supports the heavens. Finally, we note Wright's citation of an inscription of Ramses II at Luxor concerning the "Great Circle, the sea, the southern countries of the land of the Negro as far as the marsh lands, as far as the limits of darkness, even to the four pillars of heaven."[21]

It turns out that the earth, too, needed to be supported somehow. Genesis 1 seems to have the flat earth resting upon a cosmic ocean, a view reinforced in Psalm 24:1-2: "The earth is the LORD's, and the fulness thereof; the world, and they that dwell therein. For he hath founded it upon the seas, and

---

21. So much for the oft-heard claim that the Bible is only referring to the four compass directions when it speaks of the "four corners" of the land (Ezek. 7:2) or earth (Isa. 11:12, Rev. 7:1).

established it upon the floods." The solid dome of the sky rests upon great pillars thrust up from below. The second commandment makes reference to the subterranean ocean when it urges us: "You shall not make for yourself a graven image, or any likeness of anything that is in heaven above, or that is on the earth beneath, or that is in the water under the earth" (Deut. 5:8; Exod. 20:4).

Under the earth? Good luck making sense of that passage in light of modern science! The first two lines in Genesis point the way to the truth of what biblical people believed. Genesis 1:1 says, "In the beginning God created the heaven and the earth." Nothing is said of creating water, at least not so far. Then Genesis 1:2, "And the earth was without form, and void; and darkness was upon the face of the deep. And the Spirit of God moved upon the face of the waters." Here are the waters, but their presence seems taken for granted. Nothing is said even here about their creation. Ramm tries to modernize this aspect of biblical cosmology, too: "As for the word under in the phrase 'under the earth' the Hebrew word tachath means not only under but lower. In our own day we speak of lowlands. Water in the form of seas is always in lower places" (Ramm 1954, 67). But who would ever bother to refer to "the oceans at sea level"?

Psalm 148:4 urges, "Praise him, you highest of heavens, and you waters above the heavens." Once again *shamayim* is used here in all instances of "heaven." It is interesting to note that water is *mayim* in Hebrew. In Aramaic: Heaven is *dabashmaya*; water is *maya*. As you can see, the *ya* (Aramaic) and *im* (Hebrew) allude to the plurality of what is being mentioned. As mentioned earlier, *shamayim* literally means "from the waters."

In the Book of Job, God harassed Job with questions not only about how the earth was "supported," but concerning how it was shaped, and those questions likewise implied a flat earth: "Where were you when I laid the foundation of the earth?" (Job 38:4). "On what were its bases sunk?" (Job 38:6). One could also compare Jeremiah 31:37: "If the foundations of the earth (can be) searched out below, then I will cast off Israel."

God Most High

Where is God in all this? On top of the vault of heaven, of course! As "God *Most High*," the deity is naturally to be depicted as sitting on a throne at the tip-top of the heavenly dome: where else? Thus it is common to find both literary and graphic representations of God enthroned atop the firmament, both in the Bible and in adjacent cultures.

> One such depiction is the ninth-century tablet of king Nabuapaliddina (885-882 BCE) from the temple of the sun god Shamash in the city of Sippar ... The wavy lines at the bottom of this scene indicate water, and beneath the waters is a solid base in which four stars are inscribed. These waters, then, are the celestial waters above the sky. This tablet depicts the god Shamash enthroned as king in the heavenly realm above the stars and the celestial ocean. [Wright 2000, 36-37]

In just the same manner, Yahweh "sits enthroned over the flood" (Ps. 29:10a, RSV). Isaiah 40:22 says the same: "It is he who sits above the circle of the earth, and its inhabitants are like grasshoppers; who stretches out the heavens like a curtain, and spreads them like a tent to dwell in" (RSV). This verse has occasioned considerable debate, as fundamentalist apologists have seized upon it as a foothold for their attempts to read modern cosmology anachronistically back into the text. "According to [Henry] Morris this verse describes a spherical earth. The Hebrew word is *hwg*. I believe that this refers to the circular horizon that vaults itself over the earth to form a dome" (Meyers 1989, 63-69).

Similar opportunistic use is made of Job 22:14: "Thick clouds enwrap him, so that he does not see; and he walks on the vault of heaven" (RSV), translating *hwg* as "vault." The KJV uses the phrase "on the *circuit* of heaven." But it is vain to use *hwg* in this way, because, as *Strong's Concordance* tells us, it is "a primitive root to describe a circle/compass." And *hwg* is most definitely not a sphere; there are perfectly adequate Hebrew words for sphere or spheroid if that is what one wanted to mention.[22]

The god of which the Bible speaks is a deity with a posterior who uses it to sit upon a throne, *up there*. Deuteronomy 26:15 beseeches God, "Look down

---

22. Hebrew alternatives to *hwg* include: the word meaning "ball," *rwd duwr*; the word for "pot," *dwd, duwd*, a pot for boiling, or, by resemblance of shape, a basket; and the word meaning "round," *tlglg gulgoleth*, a skull (as round) or a head. Then there is the Babylonian loan word *llg galal*, the verb "to roll," which is based on the description of a type of water pot shaped like a human skull. The Bible never once uses any of these fine words to describe the earth. The usage of the word *hwg* simply reflects the form that the Bible writers thought the earth had. It is entirely in accord with the Babylonian map of the world noted by Meyers (1989), one that "clearly shows a circular earth surrounded by a circular sea (*Cuneiform Texts from Babylonian Tablets in the British Museum* 1960, part xxii, pl. 48." Meyers also cites the "*äamaö Hymn*" which is written to the Sun-god [and] says, 'You climb to the mountains surveying the earth, you suspend from the heavens the circle of the lands.' The phrase 'the four corners of the earth' which in Akkadian is *kip-pát tu-bu-qa-at eerbitti*, can be literally translated 'the circle of the four corners' (Grayson, Albert, *Assyrian Royal Inscriptions*. Vol. 1. Wiesbaden: Otto Harrassowitz, 1972, p. 105)."

from Your holy habitation, from heaven, and bless Your people Israel, and the ground which You have given us, a land flowing with milk and honey, as You swore to our fathers" (NASB). And the Bible tells us that, sometimes when he looked down, he did not like what he saw. And then, we must assume, he took a climb down that ladder mentioned in Genesis 28:12-13. The divine reaction to the Tower of Babel (Figure 7) was a response to a threat, not offense at human ambition or pride.

FIGURE 7: *The Confusion of Tongues.*

"And the LORD [Yahweh] *came down* to see the city and the tower, which the children of men builded" (Gen. 11:5). Just before Yahweh stops work on the tower, he complains that now nothing will be impossible for them, nothing they can imagine doing (Gen. 11:6). The story demands a physical location of the deity in a realm vertically above the earth, and one who must engage in reconnaissance. There is not a theology of omniscience here. Nor even of monotheism, as Yahweh is depicted alerting his fellows to an unsuspected danger and urging action to nip it in the bud.[23]

Moving On

Since the invention of the telescope, our feelings of "coziness or peculiar dread" in the universe have diminished greatly from the time when the "god(s) were perched on a celestial balcony, so to speak, gazing at the drama below, handing out blessings and curses to individuals and nations alike" (Babinski 2010, loc. 1611). Christian or not, we view the cosmos very differently from the people of the ancient Near East, "who built the temples, founded the priesthoods, invented holy rituals, and performed burnt offerings" to have smoke "ascend to heaven as a 'soothing aroma'" (loc. 1611).

Now, with communications satellites whizzing around an oblate-spheroid shaped earth whose image has become iconic from NASA photographs, the Bible's cosmology has had to yield to the plainly evident reality of the earth's shape and orbit around the sun. Only the most ridiculous lunatic fringe of young-earth creationists still worships at this particular altar of literalism, an embarrassment of the past that most of Christianity has conveniently forgotten.

But as we chuckle about our previous ignorance, let's not forget the parallel between heliocentrism then and evolutionary theory now. There was plenty of denial about astronomy, too, as Andrew Dickson White recounts in a scathing historical review published over a century ago: "In vain did Galileo try to prove the existence of satellites by showing them to the doubters through his telescope: they either declared it impious to look, or, if they did look, denounced the satellites as illusions from the devil." The "hated telescope" agitated the churchmen by revealing that the moon had mountains and valleys (and thus was not a "great light," Gen. 1:16) and that the sun had spots. "Such are the consequences," White dryly observes, "of placing the instruction of men's minds in the hands of those mainly absorbed in saving men's souls" (1895, Ch. 3, §3).

The evidence for evolution is every bit as abundant and compelling as what the telescope was revealing four hundred years ago. The insights of

---

23. If Yahweh was troubled by man's accomplishments, why are the Egyptian pyramids still standing? At around 450 feet, the largest pyramid is much taller than any recorded ziggurat, and it's still there! Let's not forget our modern day architectural buildings, which are well in excess of a thousand feet tall. We have already climbed into Yahweh's heaven, and on the wings of modern aerospace we have flown far beyond it.

evolutionary biology are being put into practice with antibiotics, vaccines, genetic testing, and crop management. Yet the majority of people in the United States—among evangelical Christians, the *vast* majority—stubbornly disregard all that evidence in favor of scriptures with no more factual basis than Joshua's stopping the sun for an extended afternoon of battle.

Eden Disorder

> *Is there an intelligent man or woman now in the world who believes in the Garden of Eden story? If there is, strike here [taps forehead] and you will hear an echo. Something is for rent.*
>
> —Robert G. Ingersoll, *Lecture on Orthodoxy* (1884)

An important aspect of creationism is the belief that there were two actual people, Adam and Eve, who were formed full-grown either after vegetation (Gen. 1) or before (Gen. 2), wandered around some of it in the Middle East thousands of years ago, listened to some unfortunate (though ultimately accurate) advice from a talking snake, and ate a forbidden piece of fruit. Consequently, they noticed their nakedness and became the originators not just of the "R" movie rating, but the entire human race. Now, of course, modern science has cast them into a pit of mythology from which no rational argument provides any foothold.

Like Jesus in the Harrowing of Hell, evangelical evolutionists have gone to hell and back to rescue Adam. The legend (from the apocryphal fourth-century *Gospel of Nicodemus*) has Jesus spending his time between crucifixion and resurrection liberating all of the Old Testament righteous. But our theologians can't even seem to extricate just the first one from the abyss, despite all their reinterpretations and reconstructions.

Adams by the Sackful

In *Jesus Christ Superstar*, Pontius Pilate complains, "You Jews produce messiahs by the sackful!" Just as there were many prophet-messiah figures wandering around Roman-occupied Palestine, there were untold numbers of candidates for the "first" humans during the gradual, evolutionary transition to our current species. Every early human, no matter how many thousands and millions of years back you go in prehistory, had parents who were pretty much as human as he or she was. "In a series of forms graduating insensibly from some ape-like creature to man as he now exists, it would be impossible to fix on any definite point where the term 'man' ought to be used" (Darwin 1888, 279-80).

This makes things difficult for creationists, as Jim Foley illustrates with a table (Figure 8) that "summarizes the diversity of creationist opinions about some of the more prominent items in the human fossil record." Although "creationists are adamant that none of these are transitional and all are either apes or humans, *they are not able to agree on which are which*" (Foley

2008). But Christians who accept evolution must wrestle with this, too, unless they are willing to abandon Adam completely to myth or metaphor.

None of these fossils is of *Homo sapiens*. All but one of the creationists in Foley's table consider the *Homo erectus* fossils ER 3733 (Peking Man) and WT 15000 (Turkana Boy) as human. Turkana Boy is an impressive fossil, an 80% complete skeleton. But that prehistoric young *Homo erectus* who walked the shores of a lake in the Kenyan Rift Valley 1.6 million years ago (Gibbons 2007, loc. 1972) was no descendant of any biblical Adam. His location, date, and appearance all preclude that.

Specimen	Cuozzo (1998)	Gish (1985)	Mehlert (1996)	Bowden (1981) Menton (1988) Taylor (1992) Gish (1979)	Baker (1976) Taylor and Van Bebber (1995)	Taylor (1996) Lubenow (1992)	Line (2005)
 ER 1813 (510 cc)	Ape	Ape	Ape	Ape	Ape	Ape	Human?
 Java (940 cc)	Ape	Ape	Human	Ape	Ape	Human	Human
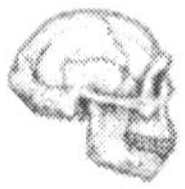 Peking (915-1225 cc)	Ape	Ape	Human	Ape	Human	Human	Human
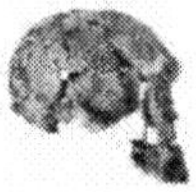 ER 1470 (750 cc)	Ape	Ape	Ape	Human	Human	Human	Human?
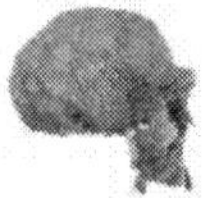 ER 3733 (850 cc)	Ape	Human	Human	Human	Human	Human	Human
 WT 15000 (880 cc)	Ape	Human	Human	Human	Human	Human	Human

FIGURE 8: Creationist classifications of hominid fossils.

The problem does not end even if you look exclusively within *Homo sapiens* for a first human ancestor. First of all, our species originated a long way from the Tigris and Euphrates, and a lot earlier than any stretching of biblical dates would plausibly permit.

> The Genetic case for an African origin for *Homo sapiens* seems overwhelming. The archaeologists have also stepped forward with new fossil discoveries, including a robust 195,000-year-old modern human

from Omo Kibish, in Ethiopia, and three 160,000-year-old *Homo sapiens* skulls from Herto, also in Ethiopia. Few anthropologists now doubt that Africa was the cradle of *Homo sapiens* and home to the remotest ancestors of the first modern Europeans—the Cro-Magnons. The seemingly outrageous chronology of two decades ago is now accepted as historical reality. [Fagan 2010, loc. 327]

Second, no first human ever existed who was a member of *Homo sapiens* but had parents who were not. (Neither we nor our theistic evolutionists take seriously the nonsense about Adam's *ex nihilo* special creation.) There was just a gradual transition to the cluster of attributes we now associate with our own species: walking upright, a large brain with an ample prefrontal cortex, the presence of a hyoid bone for vocalization, to name a few. Some of our supposedly "human" traits actually go back a lot further than us, like the use of tools and fire. *Homo heidelbergensis* was making sophisticated stone hand axes and wooden spears hundreds of thousands of years before the first humans did (Fagan 2010, loc. 770-795). There were probably human-controlled fires burning in hearths by 500,000 years ago (Goren-Inbar et al. 2004).

Even if it were possible to isolate a single, final mutation as the emergence of humanity, it is just not possible for the entire human race to have descended from a single father and mother. Genetic evidence now makes clear that there have never been fewer than about a thousand members of *Homo sapiens* throughout the more than 100,000 years of its existence (Coyne 2011b).[24] Actually, that was well understood before scientists started looking into DNA: The "paleontological record thoroughly establishes that one population is always preceded by another, making the idea of a single pair of humans procreating an entire species unthinkable" (Ronald Youngblood, quoted in Loftus 2008, loc. 4838).[25]

In the face of this reality, the suggestion has arisen that perhaps Adam and Eve were merely the representatives or leaders of a wider human

---

24. No *Homo sapiens* fitting the time or location for the biblical Adam can be imagined, either. "The Genetic case for an African origin for *Homo sapiens* seems overwhelming. The archaeologists have also stepped forward with new fossil discoveries, including a robust 195,000-year-old modern human from Omo Kibish, in Ethiopia, and three 160,000-year-old *Homo sapiens* skulls from Herto, also in Ethiopia. Few anthropologists now doubt that Africa was the cradle of *Homo sapiens* and home to the remotest ancestors of the first modern Europeans—the Cro-Magnons. The seemingly outrageous chronology of two decades ago is now accepted as historical reality" (Fagan 2010, loc. 327).

25. This paragraph and the first one of this chapter are adapted from Suominen 2012, §4.3.1.

community. Peter Enns seems to feel some guilt on behalf of his fellow Christians for raising such ideas. Yes, "it is *possible* that, tens of thousands of years ago, God took two hominid representatives (or a group of hominids) and with them began the human story where creatures could have a consciousness of God, learn to be moral, and so forth."[26] But, Enns calls that what it is: "an alternative and wholly *ad hoc* account of the first humans, not the biblical one" (2012, 139).

Enns knows it's just not a viable solution: "One cannot pose such a scenario and say, 'Here is your Adam and Eve; the Bible and science are thus reconciled." Whatever one might call them, such designated humans would not "satisfy the requirements of being 'Adam and Eve'" (p. 139). Anthropologist James P. Hurd shows us the contrast:

> The fossil and genetic records suggest that *Homo sapiens* originated about 100,000 years ago with a modern skeleton. In contrast, Genesis sets Adam and his immediate descendants in a much more recent farming culture. This farming family already possessed a relatively advanced culture including animal and plant domestication. Cain worked the soil and offered to God his fruits of the field. Abel offered the fat parts of the firstborn of his flock of sheep. Archaeologists trace the beginnings of plant cultivation to 7,500-7,000 B.C., with domestication of cattle and sheep appearing soon after. When Cain left home, he built a "city." However, no cities or even villages appear in the archaeological record until about 9,000 B.C. [Hurd 2003, 223]

C. John Collins notes a tentative speculation of Derek Kidner in which "we have the special creation of Adam and Eve, perhaps refurbishing an existing hominid." Then, according to Kidner,

> It is at least conceivable that after the special creation of Eve, which established the first human pair as God's vice regents (Gen 1:27, 28) and clinched the fact that there is no natural bridge from animal to man, God may now have conferred his image on Adam's collaterals, to bring them into the same realm of being. Adam's 'federal' headship of humanity extended, if that was the case, outwards to his contemporaries as well as onwards to his offspring, and his disobedience disinherited both alike. [Collins 2011, 124-25]

---

26. Traditionally, writers have referred to members of the human ancestral line as *hominids* as Enns does. Now the term covers all great apes, including those on both sides of the human-chimpanzee split some five million years ago. Direct human ancestors are technically called *hominims*. But Enns and most of the other writers we cite continue to refer specifically to "our walking ancestors" as hominids, and we will do the same (Winston and Wilson 2006, 22).

But, warns Collins, this "suggestion is moving us away from the simplicity of the Biblical picture." And "one would need to imagine Adam as chieftain, or 'king,' whose task it is not simply to rule a people but more importantly to represent them (the basic idea of a king in the Bible)" (p. 125).

Denis Alexander admits that humans evolved from simian ancestors. A friendly but skeptical Collins says that Alexander

> finds in Genesis 4 "the clear implication that 'the man' [i.e., Adam] and 'the woman' [Eve] were not the only people around at that time"; indeed, he reads the agriculture and gardening of Genesis 2-4 as implying a Neolithic setting (some time since 10,000 B.C.): "God in his grace chose a couple of Neolithic farmers in the Near East, or maybe a community of farmers, to whom he chose to reveal himself in a special way." That is to say, these were the first people to have a personal relationship with God, real spiritual life; presumably they were to spread the knowledge of God through the rest of mankind [Collins 2011, 125-26]

Collins says there were "between one and ten million people, spread over the globe" (p. 126) during the Neolithic, which makes it hard to take seriously an assignment of Adam as "the federal head of the whole of humanity alive at that time" (p. 126, quoting Alexander). He recognizes what is behind Alexander's intention "to take Adam as an actual person": the Christian obligation "to come to grips with all of the Bible; since Paul, for example took Adam to be as historical as Jesus" (p. 126).

Collins notes that the "consequence of Alexander's view is that death is a 'natural' end of the life of every human being," rather than a penalty imposed subsequent to the Fall. That fall of Adam imparted sin to the rest of mankind, in Alexander's words "spreading the spiritual contamination of sin around the world" (p. 126).

Note the cognitive dissonance-reducing *ad hoc* character of these speculations, jamming a square peg into a round hole to save face for faith. They don't even manage that; there are so many splinters that the peg itself has become round! The resultant story is a different one from what Genesis tells, as different as the version of the Kabbalist Isaac Luria. Sure, it involves the same names and a loosely parallel plot line, but it is a different tale. And Alexander's tale is reminiscent of the Book of Mormon, implying primitives building sea-worthy craft to cross the seas to evangelize other cavemen in Australia, North and South America, etc. We might as well envision an episode of *The Flintstones*: "Bedrock Mission Board!" Did they have little palm parchment *Four Spiritual Laws* booklets to lead converts into a "personal relationship with God?" Did Fred and Barney pray the sinner's prayer?

For Robin Collins (yes, another Collins is entering the fray),

> Adam and Eve play two ... representative roles, that of representing "everyperson"—that is each one of us—and that of representing the first hominids, or groups of hominids, who had the capacity for free choice and self-consciousness. ... These hominids also became aware of God and God's requirements, but more often than not rejected them ... They were subject to various temptations arising both from the desires and instincts they inherited from their evolutionary past and from various new possibilities for self-centeredness, self-idolization, self-denigration, and the like that came with their new self-consciousness. [Collins 2003, 470]

This notion "that Adam represents the very first self-aware, free-willed hominids instead of a single human being living in some state of spiritual and moral rectitude" Collins admits is "not an interpretation that one would come up with apart from modern science." But he finds "some basis in the text itself for saying that Adam is being used to represent the first acts of disobedience, not merely the first human being" (p. 481).

In one sense he is right: Hermann Gunkel would make it an ethnological myth, the character Adam representing a group, as in a political cartoon depicting America as Uncle Sam and Britain as John Bull. There is a sense in which both Robin Collins and Peter Enns—who views Adam as the symbolic progenitor not of all humanity, but of Israel—are correct. In fact, Enns's theory seems totally arbitrary until one realizes something unsuspected about the Adam character. He is not the only "First Man" figure in Genesis! In Genesis 4:26 and 5:6 we come across the name of *Enosh*, of whom precious little is said but whose name packs a wallop. For it means, simply, "man," "human being." We must suspect that Enosh was the original Hebrew First Man character, and that he was at some point supplanted by Adam.

Where did Adam come from? He must have been for some reason appropriated from Israel's cousins the Edomites. "Adam" and "Edom" appear to be variant versions of the same word or name. In Near Eastern languages like Hebrew, Ugaritic, Arabic, and Aramaic, words are formed from roots made of three consonants. New words are produced when the order of the consonants changes and new vowels are supplied. Think of "shalom," "salaam," and "Islam." Adam boils down to *Edom* just as the Arabic name *Umar* is also Omar, and *Uthman* becomes Othman or Osman.

Adam, then, was supposed to be the First Man *and thus* the eponymous ancestor of the Edomites who fancied themselves (as many groups do) to be the root stock of humanity from which all others descended. Even though the Israelites borrowed Adam as the First Man, they were unwilling to regard the Edomites as the first created race. They substituted the sun god Esau as the mythic progenitor of the Edomites, who were imagined to have appeared later in history. To these peoples, the first human being and the first ancestor of their tribe would have been the same thing. Naturally the chain of ancestors was eventually extended, with various patriarchs inserted between Adam and Abraham, Isaac, and Jacob. But at first Adam, like Enosh before him, was both the progenitor of Israel and of humanity in general.

Thus, the story of Adam works as an ethnological myth, Adam symbolizing all humanity. But then the whole genre implies *transmission* of characteristics from the eponymous ancestor to his descendants. "It's in the blood! What do you expect?" Thus, contrary to the intent of our evolutionary creationists, it all comes back to the Augustinian doctrine that Original Sin was transmitted from one man to his descendants. Or at least it would if the Adam story had anything to do with the origin of sin in the first place. But, as we will see in "Sinful Selection," it does not.

The Evolution of Genesis

It is important to grasp that the Pentateuch (the collection of five books that comprises the Torah proper) was not written by Moses. That ought to be obvious from the fact of several references to events long after the time of Moses, who is already regarded as an ancient figure even in Exodus. Rather, the Pentateuch is a compilation (probably made in the fifth century BCE or later) of earlier narrative and legal sources, collections of legends and laws. It is fairly simple to tell the sources apart even in English translation, once you know what to look for.

The earliest of these, as well as the most stylistically beautiful, is the **J** source, so called because it uses the name **Yahweh** (or **Jehovah**, represented in most English Bibles as "LORD" or "GOD") for God the whole way through and it is a collection of material from the southern Hebrew kingdom, **Judah** (**Yehudah**). (German scholars discovered all this, and Germans pronounce "J"s like we do "Y"s.) The J source gives us raw mythology (God often appears as a character on stage) and unretouched, entertaining portraits of biblical characters, not yet "sanitized" as holy saints. J is anachronistic, paying little attention to when various laws actually began to be observed in Israel, sometimes describing the ancients as if they followed the customs of his own day. J's story begins with the creation (the Garden of Eden story, Gen. 2:4b through ch. 3) and continues on through Moses. The J source was compiled, at the earliest, in Rehoboam's reign (just after Solomon), but may even be as late as Post-Exilic times.

The second source is a collection of similar materials, the **E** source, so called because it calls God Elohim ("God") until the burning bush story (Exod. 3:15), when it adds "Yahweh." It was a collection of stories from the Northern Hebrew kingdom, Israel or Ephraim. E is more conservative in its depiction of God, having him speak from offstage in dreams or visions. It also tries to clean up the faults of the characters, as we can see by comparing the J and E versions of the same story. E begins with Abraham and extends through Moses. E is later than J and may come from the ninth century, the dawn of the Prophets, since it calls Abraham a prophet. But it may be much later.

The third source is **D**, the basis of the Book of **Deuteronomy**, consisting of the sermons fictively ascribed to Moses from chapter 4 through 34. Though it seems to incorporate sermonic material from the old Shiloh shrine in the North, this book was put together by prophets and priests (Hilkiah, Huldah) in the seventh century BCE. Someone later added a historical preface (Deut. chapters 1-3), a summary of the Moses/Exodus sections of J and E.

Like D, the fourth source is mainly a vast law code, or set of them, called **P**, or the **Priestly Code**, compiled by the exiled Jewish priests while in Babylon in the sixth century BCE. It contains the sketchiest summary of Israelite history, beginning with the Six Day Creation (Gen. 1:1-2:4a) and going on through Moses. The Creation, Flood, and Moses stories are told at some length. But by far most of P is legal materials. We do not know to what extent these were actual laws that had governed Israel and Judah, or to what extent they may have been an ideal blueprint like Plato's *Republic* or Thomas More's *Utopia*. Like E, P calls God Elohim or El Shaddai until Moses is told to use Yahweh (Exod. 6:2-3)

This source-critical analysis is as incontrovertible in biblical studies as evolution is in modern science. But how is all this relevant to our subject? Simply that Genesis 1:1-4a constitutes the creation story of the P source, while the portion of Genesis from 2:4b through the end of chapter 3 is the J creation account. They profoundly contradict one another and cannot be harmonized.

The Priestly creation in Genesis 1 has everything emerging from water, whereas in the J version it is from the dry ground, reflecting a Babylonian (Ps. 137:1; Ezek. 1:1) and a Palestinian origin respectively. The Priestly version of creation covers one week (whether of literal days or longer periods), while the events of the Eden story would seem to have occupied a much shorter time. In Genesis 1, the basic "kinds" of creatures are spotlighted day by day, but not so in the second chapter, where all of them appear to have been conjured up, rapid-fire, to audition as possible companions (*not* sexual partners!) for the man. Of course this also means that the order of creation is irreconcilably different between the Priestly and the Yahwist accounts, since P has an unspecified number of men and women (collectively called "Adam," i.e., humanity, exactly like the Greek *Anthropos*) created in a single stroke, while J (the Yahwist) depicts the creation of the man and the woman being separated by the circus parade of newly-created beasts. (One strives in vain to picture the unwittingly implied scene of microscopic amoeba and paramecia passing before Adam, trying to gain his approval.)

It is seldom noticed (because who wants to notice it?) that the two accounts do agree—in their polytheism. The seven-day creation has "Elohim" ("gods," plural ending) create humans "in our image," while the Eden story has Yahweh Elohim compare the newly enlightened pair with "one of us." And there are goddesses as well as gods in both versions. That is why Genesis 1 parallels creation in the divine image with "creating them male and female." The Eden story implies that, like their counterparts in all mythologies and pantheons, Yahweh's fellow deities include females with whom the male deities share the secret (the "knowledge") of procreation. Remember, the tree of knowledge is for them, not for the help.

This, however, points up another difference between the two stories: The Priestly version has God command his new humans to breed, while such is the furthest thing from the mind of the Yahwist's God character, since he hides such knowledge from them and punishes them when they find out about it anyway. This knowledge makes them "like God," which Yahweh Elohim thus cannot have wanted, the very thing, however, God freely conferred upon his humans in chapter one, where he made them in his own image and likeness.

Now, some evangelical apologists for evolution accept the Documentary Hypothesis, or JEDP Theory, reasoning that, if they are going to wake up and smell the scientific coffee, they might as well accept the source-critical creamer, too. But the more conservative evangelicals, while willing to engage in some risky reinterpretations in order, they think, to make Genesis more plausible (while actually rewriting it), still cannot brook multiple authorship. Sure, admits C. John Collins, we often hear that the two accounts "may even be difficult to reconcile with each other." But "we have no reason to expect that whoever did put these passages together was a blockhead (or a committee of blockheads), who could not recognize contradictions every bit as well as we can" (Collins 2011, 52).

This is just passing the buck. We can trust the ancient harmonizer to have made the right judgment rather than making it ourselves! It is sadly reminiscent of the "implicit faith" criticized by Martin Luther, whereby the medieval clergy told the illiterate laity not to trouble their heads over complex doctrines but just to "leave the believing to us." Collins's alternative to distinguishing J from P is to "argue for a version of the traditional Rabbinic opinion." Rather than "two discordant accounts," we "should see Genesis 1:1-2:3 as the overall account of the creation and preparation of the earth as a suitable place for humans to live, and Genesis 2:4-25 as an elaboration of the events of the sixth day of Genesis 1" (p. 53). It is a viewpoint rooted in a denial of post-Mosaic sources, the best the Rabbis could do while believing themselves stuck with the traditional (but groundless) ascription of Genesis to Moses.

As different as they are, how and why did the two creations wind up cheek by jowl in the same scripture? Of course, each was the traditional lore of a different tribe, clan, sect, or faction among the Hebrews. And our Pentateuch represents an attempt to provide a national charter that everyone could get behind. In order to do this, all the rival, disparate sacred documents had to be included. It would not do to exclude this group's cherished version of the creation or that tribe's beloved version of the flood, or of Abraham and Hagar. It was not that the biblical compilers were or were not "blockheads." Surely they noticed the glaring differences, but what could they do? It is what happens with compromise documents. In the same way, David F. Wells once showed how the same factors led to the documents of Vatican II containing contradictory elements, the deposits of competing factions, to which rival Catholic factions still appeal today.[27]

Another conflict within Genesis is found in the Cain episodes of chapter 4, which appear to presuppose a world already populated. If you feel at liberty to rewrite the Eden story in order to place it plausibly within a context of scientific paleontology, that may seem like an alluring escape route: Were these shadowy people not descendants but contemporaries of Adam and Eve?

Alas, that is all based on a redactional phantom. The Cain material does not constitute a single, unitary narrative. What happened is that the Genesis redactor merely collected snippets from here and there in the larger cycle of Cain legends. They are not compatible. In 4:11-12 Cain is barred from any further settled existence, yet in verse 17b, he is building a city. In 4:1, he is the first human being born from a womb, yet, famously, in verse 17a there is a woman available for him to marry, presumably not his own sister. After he murders his brother (verse 8), reducing the fledgling human race by 25%, he is fretting that any of an implicitly large number of contemporaries will kill him if they chance upon him (verse 14).

None of this fits together at all. Cain's bouncing all over the timeline is exactly like what happens in the TV show *Xena: Warrior Princess*. From

---

27. David F. Wells, *Revolution in Rome* (Downers Grove: InterVarsity Press, 1972).

one week to the next, we find her hob-nobbing with Goliath in the eleventh century, and with Julius Caesar in the first century BCE. Then again, she fights in the Trojan War in archaic Greece, but soon pops up in the second century CE with Galen the physician! Of course, Xena lived in none of these centuries; she is a fictional character. And so is Cain.

Creation by Combat

All but the most fundamentalist of Bible scholars recognize that Genesis contains two disparate accounts of Creation. As a former conservative Presbyterian recalling how he finally confronted this fact, Arch Taylor says he "had conveniently ignored the details of the two accounts that were mutually inconsistent: the difference in the numbering of the days and the order in which God was said to have created various things. In keeping with the children's story Bible on which he had been brought up, he "had unquestioningly assumed that the story about Adam and Eve was just a more detailed description of what had taken place when, as chapter 1 had said, God created them in the divine image on the sixth day." But, as anyone reading it with a clear mind must, he was forced to acknowledge that the "Bible doesn't say that; it obviously says something different" (Taylor 2003, 160).

Far less recognized is that the Bible includes a *third* Creation account, a "creation by combat" version that is actually the oldest. It is part and parcel of the royal ideology of the kingdom of Judah (and presumably of northern Israel, too, though we know less about that: the compilers of scripture were southerners, Judeans). Ancient Israel borrowed the whole institution of the monarchy from the surrounding nations, replacing an earlier, much looser tribal confederation (1 Sam. 8:4-5). With it came the accoutrements of the institution. Among these were an ideology exalting the king's authority to that of a god on earth.[28]

The king was Yahweh's vicar; hence, he could be addressed simply as "God," as in Psalm 45:6 (a royal wedding song). Accordingly, one of his royal titles was "Mighty God" (Isa. 9:6). Any Judean king, once inaugurated, was considered the deity's son (Ps. 2:7, a coronation hymn, and 89:26-27) and his anointed, i.e., Messiah, as in Psalms 2:2 and 89:20.[29] The king annually

---

28. The remaining paragraphs of this section are adapted from Price 2011.

29. Several passages reinterpreted by New Testament writers as predictions of a messiah were first intended as birth or enthronement oracles, or as coronation anthems. The "messiah" and "son" of Yahweh in Psalm 2 is every new king of Judah. Psalm 110 makes *pro forma* predictions for military victories by the new sovereign and secures for him the hereditary prerogatives of the old Melchizedek priesthood (taken over by David when he annexed Jebusite [Jeru-]Salem). Psalm 110:3 also makes him, like the king of Babylon (Isa. 14:12), the son of the Semitic dawn goddess

renewed his divine mandate to rule, and with it the very vigor of the cosmos and fertility of the land, by ritually re-enacting the myth of how Yahweh became king of the gods. As in Babylonian myth, the gods were frightened (Job 41:9) by the menace of the chaos dragon, seven-headed Leviathan, mentioned in Job 26:12-13, as well as in Psalm 74:14 and Isaiah 27:1.

Leviathan appears again as late as Revelation 12:3, 7 where he has become the apocalyptic Beast from the sea. He was, under the synonymous name *Nehushtan* (both *levi* and *naas* meant "serpent," plus the honorific suffix "-than" or "-tan"), worshiped right in the Jerusalem temple (2 Kings 18:4). Called *Rahab* in Psalm 89:10 and Isaiah 51:9, the dragon was apparently attended by sacred prostitutes like the one in Joshua 2:1, who was named for her deity.[30] (Similarly Job 3:8 mentions priests who knew the rituals to invoke Leviathan in worship.)

Behemoth (Job 40:15-24) was the land-going counterpart of Leviathan-Rahab, probably to be identified with the Second Beast of Revelation 13:11. *Yamm* (Pss. 74:13a, 89:9) and *Tiamat* were the sea personified, while Leviathan, or Lotan (as he was known in Canaan) was the winding Litani River personified. Then Yahweh, the young war-god (Exod. 15:3) and storm-god (Exod. 19:16; Pss. 18:7-15, 29:3-9; Gen. 9:11), stepped forth. El Elyon was the "Most High God," the name implying the head of a pantheon. Yahweh was one of his sons among others (Pss. 29:1, 89:6-7; Gen. 6:1-4; Job 1:6, 2:1), and had originally been placed in charge of a single nation, Israel (Deut. 32:8). But now he made his bid to take the throne over them all.

Yahweh volunteered to destroy the dragons if the gods would make him king. They agreed, and he did destroy (as in Pss. 74:13-14, 89:10; Job 26:10-13; Isa. 27:1, 51:9) or subdue and tame (Job 41:1-5) the monster(s). Then he took the throne (Pss. 74:12, 89:13-14, 93:2a, 95:3, 97:1-2) alongside El Elyon, perhaps as co-regent (as in Dan. 7:2-7, 9-10, 12a, 13-14). From the remains of

---

Shahar (usually mistranslated as a common noun, "dawn").

Isaiah 9:2-7 is either a coronation oracle or a birth oracle in honor of a newborn heir to the throne, depending on whether "unto us a child is born, unto us a son is given" (verse 6) refers to the literal birth or the adoption as Yahweh's "son" on the day of coronation ("this day I have begotten you," Ps 2:7). The epithets bestowed on the king in Isaiah 9:6, "Wonderful Counselor, Mighty God, Everlasting Father [cf., 1 Kings 1:31: "May my lord King David live forever!"], Prince of Peace," echo the divine titles of Pharaoh. Isaiah 61:1-4 is an inaugural oath, pledging universal justice and amnesty to prisoners. Isaiah 7:14 began as a similar birth oracle, casting the newly conceived or newborn royal heir in the role of the son of the virgin goddess *Anath* (equivalent to *Shahar* as in Ps. 110).

30. Thanks to Geoffrey Tolle for this observation.

the dragon(s) he created the world, a feat for which he is praised in Psalms 74:15-17; 89:11-12; and 93:1b.

Eventually, the two gods Yahweh and El Elyon were merged into one (Gen. 14:22). The sons of El became a council with whom Yahweh consults (as we glimpse in Ps. 89:5; 1 Kings 22:19-22; Gen. 3:22, 11:6-7) in their meeting place on Mt. Zaphon in Lebanon (Isa. 14:13). That is, until Yahweh finally condemns them for their misrule of their nations and sends them to the netherworld of *Sheol*, where their blind stumblings cause earthquakes (Ps. 82). They become both the fallen angels imprisoned underground (2 Pet. 2:4; Jude 6) and the Principalities and Powers ruling this age (Rom. 8:38-39; 1 Cor. 2:6-8; Col. 2:15; Eph. 6:12).

When evangelical scholars give any notice at all to this myth, attested here and there throughout the Bible, they usually seek to escape the implications by dismissing all the references as mere literary allusions to already obsolete myths, perhaps borrowed from foreigners. But this seems disingenuous. The various Psalms that refer to the defeat of Leviathan and Rahab glorify Yahweh for his primordial victory just as the *Rig Veda* praises Indra for slaying the dragon Vritra and saving the world from perishing.

It's not like making an allusion to Paul Bunyan and Babe the blue ox, which are admitted folklore. It's more like comparing Christ to the savior god Mithras who slew the cosmic bull to save humanity. The only reason to take the biblical references to divine dragon-slaying as figurative and the numerous Babylonian, Iranian, Canaanite, Norse, and Vedic parallels as intended to be real is to give the Bible special treatment. One does not like to reckon with a Bible that assumes belief in dragons. That would be just too much to swallow, even for a creationist who would like to think he takes everything in the Bible literally. Miracles, yes; dragons, no.

And so the Bible defenders pretend Leviathan is a crocodile, even though Job 40 describes him in great detail, including his fiery breath—just like Godzilla. And if you face up to the fact of the Combat myth honestly and admit the Bible, too, contained belief in mythical monsters, on what basis could you go on to defend what it says elsewhere about a creation done in seven days, or a talking snake in a garden? We can point to no "Scientific Dragon-Slaying Creationism Council" or "Creation from Leviathan's Carcass Institute." Yet that is as much part of the Bible as traditional Adam-and-Eve Creationism.

Yahweh versus Nehushtan

The ancient Gnostics probably had it right when they suggested that Yahweh (and his fellow gods whom he addresses in tones of panic at the end of Genesis 3) meant to prevent the man and the woman from learning about procreation. The Tree of Knowledge stands for *carnal* knowledge (as "knowledge/to know" does ten other times in the Old Testament). Yahweh tells them they will die as soon as they eat it: It is poison. But as the Serpent tells them, it is not. Instead, it will impart divine wisdom. They eat and gain wisdom, especially the ability to reproduce ("even *educated* fleas do it"!). Since their species can thus continue in perpetuity, Yahweh now forbids individual immortality, so the man and the woman are exiled from the tree of eternal life, from which Yahweh himself eats, as the Greek gods ate ambrosia, to maintain immortality.

Because they automatically import later God concepts as they read the story, traditional Christians, even our Christian evolutionists, imagine that God cannot need nourishment. So they speculate that he placed the tree in the Garden to test Adam and Eve, like a stunt from the long-running TV show *Candid Camera*. But the story never implies that the garden was designed as a habitat for man, like a creature in a zoo. The man is created to serve as caretaker of Yahweh's garden—from which *he* eats and in which *he* takes a stroll in the cool of the evening. Thus Denis Lamoureux, speaking for many, is just plain wrong when he assures his readers:

> The gods in many of these pagan accounts create humanity in order to free themselves from work. In effect, men and women are slaves. But in sharp contrast, Gen 2 features a faith message that the Lord cares for humanity and meets their physical (food) and psychological (companionship) needs. [Lamoureux 2008, 200]

Unlike the Priestly creation story of Genesis 1, Yahweh here does not want a human *race*. One man was all he needed, and he added a companion for him only as an afterthought. There is nothing about him wanting the caretaker to procreate with her. Genesis 2:24 is just a parenthetical pause to note, in the manner of an etiological myth (see below, under the subheading "Ancient Nomads Intents") that this is where sex came from, not that it was supposed to happen. There is no command here, as there is in the Genesis 1 creation story, to "be fruitful and multiply." That is someone else's theology. God has no need for a whole slave race, since he needs only two people, one really, to tend the garden, and the gardener and his companion can retain eternal youth and vitality by eating of the tree of immortality. They will not need to be replaced by a subsequent generation. It is just like the *Atrahasis Epic*, where the gods flood the world because they are sick of the incessant noise made by the teeming multitudes of pesky humans (Tablets I, II).

Sex equals "knowledge," as it does for the Adam-like Enkidu, made from the clay by the goddess Aruru, in the *Gilgamesh Epic*. Enkidu has sex with Aruru's priestess Shamhat, then returns to his old pals the animals. When they see him, however, they flee from his presence (Tablet I, lines 195-98). He is a changed man now, and can't chase them:

> Enkidu had defiled his body so pure,
> his legs stood still, though his herd was in motion.
> Enkidu was weakened, could not run as before,
> but now he had *reason*, and wide understanding. [lines 199-202]

Shamhat consoles the baffled Enkidu, "You are become wise Enkidu, you are just like a god!"[31] He had reason and understanding, and was not to wander the wild with the beasts anymore. Instead, he was to go "to the sacred temple, home of Anu and Ishtar" (lines 207-10). Now you can see why Yahweh's punishment for the woman was pain in *childbirth*, of all things. He had proven unable to prevent the transformation associated with "knowledge," so now he would make her pay dearly for the privilege!

Please note that the Eden story nowhere mentions Satan. Does that matter? Not to C. John Collins, who refers us to "the serpent's knowledge of what God said in Genesis 2:17" and "the evil that the serpent speaks." This nasty reptile "urges disobedience to God's solemn command, calls God a liar, and insinuates that God's motives cannot be trusted." In all of this, we find a "firm footing" for "the Jewish and New Testament interpretive tradition (e.g., Wisdom 1:13; 2:24; John 8:44; Rev. 12:9; 20:2) that sees the Evil One ('Satan' or 'the devil') as using this serpent as its mouthpiece" (Collins 2011, 63).

But Collins needs a bite of the Fruit of Knowledge to open his eyes. He is too myopic, seeing his familiar dogma as the only alternative when in fact it is merely his theological default mode. In fact, the Serpent is probably to be identified with the serpent god *Leviathan (Nehushtan)*. In the Genesis story, it plays the role of Prometheus, of whose myth this one is obviously a cognate, a parallel.

Let us refer now to Robert Graves's summation of the relevant portion of the Prometheus story, from his *Greek Myths*. The Titan Prometheus ("forethought" or "foresight") was the son of Atlas and the brother of Epimetheus ("afterthought" or "hindsight"). Prometheus was the "wisest of his race." Like God in the Flood story of Genesis 6, Zeus "had decided to extirpate the whole race of man, and spared them only at Prometheus' urgent plea," recalling both the unsuccessful intercession of Abraham on Sodom's behalf in Genesis 18:23-25 and Moses' success on Israel's behalf in

---

31. "Handsome" is a possible translation but misses the point, omitting the crucial parallel: Enkidu, like Adam and Eve, becomes wise once he partakes of carnal knowledge.

Exodus 32:11-14. Nonetheless, Zeus "grew angry at their increasing powers and talents" (Graves 1960, 144), just as Yahweh did in Genesis 11:6-7.

Prometheus soon earns Zeus's hostile suspicions, too. He was more sympathetic to the human cause than Zeus was, and so Prometheus decided to help them, even if it meant outwitting Zeus, when he was invited to arbitrate a dispute about a sacrificial bull. Which parts of it should be offered to the gods, and which parts should be kept for this weekend's barbeque? Prometheus flayed and butchered it, and made two bags from its hide. "One bag contained all the flesh, but these he concealed beneath the stomach, which is the least tempting part of any animal; and the other contained the bones, hidden beneath a rich layer of fat" (p. 144).

He then gave Zeus the choice of one of the bags. Zeus made the wrong selection, picking "the bag containing the bones and fat," though they "are still the divine portion" (p. 144). He "punished Prometheus, who was laughing at him behind his back, by withholding fire from mankind. 'Let them eat their flesh raw!' he cried" (p. 145).

We think here of the sad altercation in Genesis 3 where the Serpent lets the man and the woman in on the secret of the Tree of Knowledge, which Yahweh seeks to safeguard for himself by promising death to the humans should they eat of its fruit.[32] He is furious when he learns that the Serpent (like Prometheus, wiser than God) has divulged his secret. Unable to undo what has been done, Yahweh decrees onerous repercussions henceforth, just as Zeus did. If Zeus condemns mankind to eating meat raw, Yahweh imposes painful childbirth and domestic subjection for the woman and a life of hard labor for the man, ending in futile death. The element of knowledge, explicit in the Eden story, is implicit in the Prometheus myth.

Our hero hurried off to see the goddess Athena, asking her to let him inside Olympus. She did, and there he lit "a torch at the fiery chariot of the Sun and presently broke from it a fragment of glowing charcoal, which he thrust into the pithy hollow of a giant fennel-stalk. Then, extinguishing his torch, he stole away undiscovered, and gave fire to mankind" (p. 145).

This is the fire of knowledge, the beginning of technology, as when we later read of Tubal-Cain's discovery of metallurgy (Gen. 4:22), which required fire. Zeus was none too happy about this, swearing revenge and ordering the creation of Pandora, "the most beautiful ever created," from clay (p. 145). He sent Pandora

---

32. Actually, "safeguard for *themselves*" might be more accurate, given the vestigial polytheism that is evident, e.g., in Genesis 3:22.

> as a gift to Epimetheus, under Hermes' escort. But Epimetheus, having been warned by his brother to accept no gift from Zeus, respectfully excused himself. Now even angrier than before, Zeus had Prometheus chained naked to a pillar in the Caucasian mountains, where a greedy vulture tore at his liver all day, year in, year out. There was no end to the pain, because every night (during which Prometheus was exposed to cruel frost and cold) his liver grew whole again. [p. 145]

Can anyone miss the parallel between Pandora's creation and that of Adam? Both are shaped from clay (Gen. 2:7), and God (or the divine Winds) breathes life into them, as by artificial respiration. "Pandora" appears to be a divine name, originally denoting "all gifts." Pandora was at first a creator goddess. But her name has been reinterpreted in the Prometheus myth as denoting that she embodies divine attributes contributed by the gods. Eve, too, is called "the mother of all living" (Gen. 3:20), a title of the goddess Aruru.

Like Pandora, Eve has been demoted to become the first housewife, and a misbehaving one at that. We finish up our quotations of Graves with his description of Pandora's version of the Fall:

> Epimetheus, alarmed by his brother's fate, hastened to marry Pandora, whom Zeus had made as foolish, mischievous, and idle as she was beautiful—the first of a long line of such women. Presently she opened a jar, which Prometheus had warned Epimetheus to keep closed, and in which he had been at pains to imprison all the Spites that might plague mankind: such as Old Age, Labour, Sickness, Insanity, Vice, and Passion. Out these flew in a cloud, stung Epimetheus and Pandora in every part of their bodies, and then attacked the race of mortals. Delusive Hope, however, whom Prometheus had also shut in the jar, discouraged them by her lies from a general suicide. [pp. 145-46]

The extensive parallels indicate either that the Prometheus story is another version of the Eden myth or at least that the two myths are aiming to make the same point. It is interesting that Prometheus, Zeus' cousin from the previous dynasty of immortals, the Titans, is the true friend of man. Not Zeus, who created them but is wary of their abilities he had not counted on. Like Abraham and Moses, Prometheus intervenes to dissuade Zeus (who is a war-like storm deity corresponding to Yahweh, Marduk, Indra, Thor, etc.) from his plan to eradicate his people. As the Edenic Serpent is called the wisest of the creatures the Lord God (*Yahweh Elohim*) had made, Prometheus is said to be the wisest of his race. Prometheus provokes Zeus' wrath by managing to give to humans the food to which Zeus was properly entitled, just as the Serpent gets in big trouble for defying Yahweh's will by giving the man and the woman access to one of the trees in God's garden, intended for immortals alone.

Divine dining prerogatives are at stake again when Prometheus defies Zeus' ban on the human use of fire for cooking, and again Zeus fails to keep it from them, thanks to Prometheus' sneaky intervention. In the Eden story the eating element, though clearly present, shifts over to sexuality and procreation, connecting with the element of the *Atrahasis Epic* in which the gods are annoyed at the teeming, noisy human race. And the awakening of sexuality brings with it the knowledge, via carnal knowledge, that separates adults from children who have hitherto shared the oblivious innocence of animals, e.g., no qualms about nudity. Think here of the Tarzan-like Enkidu in the *Gilgamesh Epic*.

Prometheus becomes the victim of terrible torment, while his counterpart, the Serpent, loses his limbs and wings. (He must first have been one of the winged serpents, *seraphim*, depicted in Isaiah 6 and in Egyptian iconography.) He and his descendants are then subject to being trod underfoot by humans forever after. The punishment of humans follows a step or two down the line as Zeus schemes to unleash all "the thousand natural shocks that flesh is heir to," as Shakespeare put it in *Hamlet*. He provides Pandora to Epimetheus just as Yahweh provided the man with the first woman in Eden, though she proved a bane, not a blessing. Zeus certainly had something up his sleeve, as Prometheus had foreseen any gift from Zeus was by no means to be unwrapped.

And it turned out Zeus had sent Pandora to sabotage the welfare of humanity forever after. In Eden, of course, a similar trove of vicissitudes is emptied out on the spot. A furious Yahweh (who, even as a god, cannot put the toothpaste back in the tube) makes sure the man and the woman will rue the very blessings the Serpent had provided them by letting them in on the secret of the Fruit of Knowledge. His curse will be passed onto the whole species that he has been unable to prevent them from producing. Welcome to labor pains, male domination of women, back-breaking labor, and finally the despair of death.

The Eden story ends as pessimistically as that of Pandora and Epimetheus: no prospect of deliverance. Christian readers, even our sophisticated "evolutionary creationists," cannot resist misreading the Eden story as pointing to the light at the end of the tunnel, namely a future redemption in Christ. Alas, they are reading the wrong testament! There is nothing about Jesus Christ or even about life after death in this story. It is very much in the same vein as Psalm 90, which actually seems to be intended as a comment upon the Eden story. It is ironic indeed that some will object to our invoking the Prometheus myth from outside the Old Testament in order to interpret the Old Testament (though in fact the ancient Greeks were just up the block in the same neighborhood), while they happily haul in the gospels as if they were the direct continuation of Genesis 3.

It might be asked: Who would have penned such a tale, a story in which the acknowledged God, or chief god, is the peevish, grudging foe of his own creatures, who are aided by a seditious rival entity? The answer is simple. We only need to seek an author who gave required lip service to a deity whose worship had supplanted his own god.

Keep in mind that Prometheus was one of the Titans, the dynasty of gods who ruled before Zeus overthrew them. This situation no doubt reflects a religious revolution in archaic Greece in which the Titan pantheon was cast aside, presumably by invaders, who replaced their worship with the cult of their own gods, the Olympians, just as the Aryan pantheon superseded the native deities of the Dravidians whom they subjugated in India. And as certain of the Dravidian gods maintained a foothold in Aryan (Vedic) mythology/theology (the dark-skinned ones, Siva, Kali, Krishna, etc.), so did certain of the Titans retain a place in Greek mythology, though no longer at center stage.

The Prometheus story is surely the product of a Titans loyalist who nonetheless could not deny that his favorite god was on the outs, as symbolized by Zeus' punishment of Prometheus. In just the same manner, the author (or first teller) of the Eden story must have been one of the priests of Nehushtan displaced during the housecleaning and theological streamlining decreed by King Hezekiah in 2 Kings 18:4. (*Nehushtan* is equivalent to *Leviathan*, both terms being the combination of "serpent" with the same honorific suffix.) The Levites had functioned as the Serpent deity's priests until this happened ("those skilled to rouse up [i.e., invoke] Leviathan," Job 3:8). After that, they were demoted to a subordinate role as gatekeepers, singers, night watchmen, etc., with no more right to a share of the sacrificial meat. They had to make the best of it.

The Eden story must have been the product of a Levite priest who could not deny that Yahweh ruled the roost; Nehushtan was out. His old effigy, the brazen serpent, had to be re-explained as an apotropaic device once fashioned by Moses to magically cure snake-bite (Num. 21:6-9). Nonetheless this die-hard partisan managed to get his licks in. He made the Serpent the true champion of the fledgling human race against a jealous God who had created them but would not share his blessings with them, lest they become gods themselves and (who knows?) threaten to one day displace him.

This is not to say, however, that the Edenic reptile's identity as the interdicted god Nehushtan-Leviathan is explicit in the story. He has been, in a sense, demytholgized as the prototypical snake, the progenitor of all snakes we see today (which Nehushtan was no doubt supposed to be anyhow). He plays the role of a classic Trickster familiar from mythology the world over.

Deity Deception

It is no surprise that orthodox interpreters just cannot see what is on the page before them when it comes to God himself deceiving the humans (though perhaps 1 Kings 22:22 should make that sound a bit less strange). And this despite the insight of C. John Collins: "The narrator is *reliable* and *omniscient*: that is, he serves as the voice and perspective of God" (2011, 24). That's right, and note that it is the narrator himself who confirms the words of the Serpent in Genesis 3:7, and Yahweh himself who corroborates the avowal of the Serpent in 3:22. Both of those recall Genesis 3:5.

According to both the narrator and Yahweh as a character in the story, the Serpent was telling the truth and Yahweh must have been lying to them in order to keep his "private stash" of immortality to himself. As the clear-eyed author of the Nag Hammadi text *The Testimony of Truth* put it:

> What kind of a god is this? First, he begrudged Adam's eating from the tree of knowledge. Second, he said, "Adam, where are you?" God does not have foreknowledge; otherwise wouldn't he have known from the beginning? He has certainly shown himself to be a malicious grudger. [codex pp. 47-48, from Meyer 2007, 623]

You just need to see that the narrator is on the side of the man and the woman, and especially of the Serpent, the forbidden god Nehushtan. He's not on the side of Yahweh, whom he means to cast in the same derogatory light as Zeus in the parallel Prometheus myth. But what believer wants to see *that*? What a canonical headache! Admitting that the Bible employs ancient mythic genres is dangerous enough without acknowledging that scripture includes even an anti-Yahweh voice. But it does; just look at the text.

Having a Yahweh who lies also explains what is otherwise a great and notorious problem: Why did the first couple *not* in fact die the very same day they ate from the Tree? Perhaps they called a priestly poison control center? The famous though wholly implausible harmonization traditionally offered is that Adam and Eve would and did "die spiritually" that day. "And what shall we make of the 'death' that God threatens in Genesis 2:17? I have argued that the primary reference is 'spiritual death,' as exhibited in

Genesis 3:8-13" (Collins 2011, 62).[33] That won't work if you're a Baptist and Calvinist, however, because it means you can slip badly enough to lose a salvation once possessed.

At any rate, this strained explanation is of a piece with Jehovah's Witnesses trying to get us to believe that their prediction of Christ's return to rule the earth in 1914 did not fail; no, it happened *spiritually*, invisibly in outer space above the earth whence Jesus now reigns. Evangelical evolutionist Peter Enns sees the problem with this one: "spiritual death does not do justice ... to the physicality of death pronounced in Genesis 3:19–a return to dust" (2012, 67). But he can do little better.

> More likely the pronouncement of death should be understood in the narrative's logic as the physical death that becomes Adam's inevitable end once he eats of the forbidden fruit. Access to the tree of life, available to them before, is now denied (3:22), and so mortality is introduced. [Enns 2012, 67]
>
> The serpent's words to Eve are a half-truth: "When you eat of it, ... you will be like God" (3:5). Often this is read as an indication that the eating of the fruit represents an illicit attempt to become like God, in other words, to be proud. But this is not the case. Becoming like God *in knowing good and evil* is precisely what God wants for Adam and Eve .... The problem is the illicit way in which Eve tries to attain wisdom—quickly, prematurely, impatiently. As in Proverbs, God wants his people to be like him *eventually*, to grow *through training* to have a godlike knowledge of things—to have wisdom. [p. 89]

This is pathetic, comical even. This need to harmonize "in the day you eat from it you will die" only results from the refusal to see, as Gnostics did, that God was lying. The woman, after the Serpent sets her straight, realizes that the Knowledge fruit is in fact "good for food" (Gen. 3:6), which means that God had told her that it *wasn't* good for food. She would supposedly die the same day she ate it, hence he had told the man and the woman that the

---

33. Writing around 18 centuries before Collins, Ireneaus offered two other explanations that don't resort to a spiritual death. Though they are still interpretive word games, at least they have Adam and Eve physically dying on "that day." The first approach is dispensationalism, with Jesus undergoing "His sufferings upon the day preceding the Sabbath, that is, the sixth day of the creation, on which day man was created; thus granting him a second creation by means of His passion, which is that [creation] out of death." The other uses the day-age scaling factor of 2 Peter 3:8 to "relegate the death of Adam to the thousandth year; for since 'a day of the Lord is as a thousand years,' he did not overstep the thousand years, but died within them, thus bearing out the sentence of his sin" (*Against Heresies*, Book 5, Ch. 23). There is seemingly no problem that a little creative hermeneutics can't fix!

fruit was poisoned. And it wasn't. The awkward reality for the inerrantist is that *Yahweh was the one doing the lying*, not the Serpent.

In the Image of the Gods

Many a theological debate has raged over the meaning of the divine image in Genesis 1. We will be discussing evolution's implications for that in "Apex or Ex-Ape?" when we turn to the second of our three branches of theological conception, the Creature. There's plenty to cover on that topic right now, though, as it relates to the first branch, the Word.

Traditionally, theologians have interpreted the divine image as the spiritual or rational aspect of human beings. One may be forgiven for wondering if this is not really an attempt to evade the meaning of Genesis 1 that seems obvious: God himself has a humanoid form (Muslims still believe this) and our physical bodies are just small, frail copies of his. The God of Genesis strolls around Eden in the cool of the day (3:8), worries about man's godlike status and potential immortality after the Fall (3:22), chats with Cain (3:6-15), and has sons who intermarry with human females (6:2). He has extensive discussions with Abram (13-15). In Genesis 18, he appears to Abram personally as one of three men and stands beside him to haggle about how many righteous there must be in Sodom in order for it to be spared.[34]

Peter Enns rejects the classical interpretation in favor of another he believes to be more culturally authentic:

> [A]lthough what "image of God" means in its fullest biblical witness may be open for discussion, in Genesis it does not refer to a soul or a psychological or spiritual quality that separates humans from animals. It refers to humanity's role of ruling God's creation as God's representative. We see this played out in the ancient Near Eastern world, where kings were divine image-bearers, appointed representatives of God on earth. This concept is further reflected in kings' placing statures of themselves (images) in distant parts of their kingdom so they could remind their subjects of their "presence." ... Genesis 1:26 clearly operates within the same thought world: "Let us make humankind in our image, according to our likeness, and let them have dominion over the fish, ... birds, ... cattle, ... all the wild animals, ... every creeping thing." ... Humankind, created on day 6, is given authority to rule over what God had made on days 4 and 5. [Enns 2012, xv]

---

34. Adapted from Suominen 2012, §4.4.1.

Nice try, but no cigar. He is right about the anachronistic attempt to read reason and spirit into the ancient text, but he is no more successful, in our judgment, in escaping the implication of an anthropomorphic deity. Though the linking of dominion with being the image, i.e., vicar, of the Creator on earth is not implausible, there is another element in the immediate context that renders it questionable as the meaning intended here. As soon as it says, "in the image of God he created them," it follows with "male and female he created them" (Gen. 1:27), a bit of parallelism indicating that the one phrase is tantamount to the other or follows directly from the other. What does the creation of the two sexes have to do with dominating the creation, if that is what "image of God" denotes? Not much. The parallel must indicate that the physicality, especially the gender, of human beings reflects that of *the gods and goddesses*.

That God has a body is clearly implied next door in the Eden story, where God takes a daily walk in the evening shadows with the man and the woman. Is he there for their sakes? No, remember, it is *his* garden or oasis. The trees and their fruit are for *him*. He allows his servants access to most of them, but not all, since some remain the prerogative of the gods. Exodus 33:20-23 says God can walk past Moses, shelter him with his (gigantic) hand, and let his broad back be seen. That would be a pretty neat trick for a metaphor. In Exodus 24:9-10, Moses and his associates "saw the God of Israel," and they saw his feet standing upon the sapphire pavement of the firmament. In Ezekiel 1:26-28, God, despite all the hedging terminology, is shown sitting on a throne in his chariot drawn by the cherubs, winged creatures personifying the storm clouds. This is especially significant for our purposes, since Ezekiel is a Jewish priest in exile in Babylon—just like the Priestly writer of Genesis.

It is obvious that someone catechized on John 4:24 ("God is spirit," or "a spirit") and 1 Kings 8:27 ("the highest heaven cannot contain thee"), much less on Anselm and Aquinas, would find the notion of a man-shaped God a dire affront. But that doesn't mean the texts don't say it. And it gets worse.

The text is polytheistic. Granted, someone has tried to adjust the text to make it compatible with monotheism, but he has not managed to erase all vestiges of the polytheism of the original. First and most obviously, "Elohim" has a plural ending, meaning "gods," though scholars have tried all sorts of linguistic gymnastics to make it mean something else. Nor is Genesis the only place from which the polytheism was scrubbed. Nevertheless, there remain a few passages where scribes neglected to change plurals to singulars: Genesis 20:13; 35:7; Deuteronomy 5:26; 2 Samuel 7:23; 17:26, 36. Joseph Wheless comments: "the editorial blue-pencil overlooked the little 'u' plural-sign of the Hebrew verbs and the unobtrusive 'im' of the adjective" (1926, 203). It is a case of what Mark Goodacre calls "editorial fatigue."

Thus when Elohim says, "Let *us* make man in *our* image" (Gen. 1:26), he is making a proposal to his fellows in the pantheon. Likewise, over in the J story of the Garden of Eden, Yahweh Elohim speaks in a panic to his fellow deities, urging immediate action, the expulsion of the man and the woman from Eden, now that they have mastered the knowledge proper to gods alone, and before they can gain divine immortality, too: "Behold! The man has become like *one of us!*" (Gen. 3:22-23). Good luck making that into the royal or editorial "we"! The same divine dismay occurs in chapter 11 when Yahweh examines the Babel construction site and advises the pantheon to nip human potential in the bud: "Come, let *us* go down and there confuse their language" (Gen. 11:7).

How do we know there were goddesses among the Israelite deities? One of them is named more than once in the Hebrew Bible, though for obvious reasons her name is rendered as a common noun: "the morning" or "the dawn." Her name is Shahar, the winged dawn goddess (Ps. 139:9), who was the mother (Ps. 110:3) of every king of Judah, Yahweh having begotten him (Ps. 2:7). Isaiah 14:12 makes her the mother of the godling Helal, the morning star, or Venus. She would have been sister to the sunset god Shalman, for whom David named two of his sons: Solomon and Absalom (the latter meaning "Shalman is my father"). There was also Ishtar Shalmith, still visible despite the editing as "the Shulamite" in Song of Solomon 6:13.

One more thing. You can see what sense it makes for the "knowledge" imparted by the forbidden tree in Eden to be *carnal* knowledge: The gods procreated to make new gods, just as in all mythologies.

Ancient Nomads' Intents

The great Old Testament scholar Hermann Gunkel delineated several major types of stories and legends in Genesis, though we find many of them elsewhere in the Bible as well. They are differentiated according to what question the ancients were seeking to answer in a pre-scientific, yet ingenious and inventive, way.

Etiological stories are narrative attempts to explain why remarkable things are the way they are, how they came about. Why is there death? Why do we wear clothes? Why must we work for a living? Why do people hate snakes? What is the rainbow? *Etymological stories* explain, by means of puns, the supposed origins of names no one understood anymore or that had unsavory pagan origins and needed a more wholesome Hebrew explanation. *Ethnological stories* depict the relations of nations, tribes, groups in the narrator's day, symbolizing each group as a fictive ancestor who has the stereotype traits of the group. Why do the Israelites and the Edomites always fight, despite being neighbors and even kin? Because Esau and Jacob got off to a bad start, and now it's in the blood.

To these, anthropologist Branslaw Malinowski added *legitimization stories*. In order to reinforce a society's laws and customs, such a story tells how at the dawn of creation God or the gods decreed that it be this way. Who are we to change it now? Of course, the story is told after human beings have created the custom, not before.[35]

The Garden of Eden story is replete with such mini-stories. Genesis 3:14 accounts for why snakes crawl, as well as why we and snakes hate each other. Verse 16a explains why childbirth is painful. Verse 7 tells why we wear clothes. Verses 17-19a tell why we must work for a living. Why is there sex? See 2:21-24. Why must we die? See 3:19b. Why should women submit to male domination? Genesis 3:16b tells us: They asked for it.

This will be important to remember when we get to evangelical evolutionist attempts to demythologize and to rewrite the Eden story. They will piously tell us that we are free to disregard the quaint ancient "science" of the stories in favor of their supposed "intent" to teach us the basics of Christian theology and redemption. According to C. John Collins, for example, the Old Testament "is the story of the one true Creator God, who called the family of Abraham to be his remedy for the defilement that came into the world through the sin of Adam and Eve." God's rescuing Israel from slavery in Egypt was part of God's plan (Collins 2011, 41). So was the call of Abraham. The way Genesis presents the call supposedly indicates that Abraham is "a kind of new Adam, whose task is to undo what Adam did" (p. 43), to start "the family through which all mankind, which is now estranged from God, will come to know the true God" (p. 59).

Collins is reading both the Garden of Eden story and Genesis 12 through the lens of 1 Corinthians 15. But does the Eden tale have any conceivable relevance or reference to those of the Patriarchs, the Davidic dynasty, or the Moses story? Critical scholarship—which dares to study each text on its own, and not as proof texts for a system of theology—says no. It makes clear that the Adam stories have virtually no further role in the Old Testament and that the Old Testament compilers had their work cut out for them trying to tie the Patriarchal tales with the originally independent Moses stories.

As an example, consider how Moses is depicted as leading his people back into the nomadic lifestyle of their ancestors. They were not wandering

---

35. The preceding two paragraphs are adapted from Price 2007, 126.

aimlessly in the Sinai desert because God stole their compasses for forty years. The desert is where they wanted to be. They were sheep-herding Bedouins, for Pete's sake! But once the Moses tales were stapled onto the stories of Abraham, Isaac, and Jacob, Moses was re-imagined as having been trying to lead the Israelites into the territory which Yahweh had anciently promised to the Patriarchs. This is why Moses does not himself enter Canaan: The originally independent Moses cycle had him die where he lived and wanted to live, in the desert. Similarly, neither the Patriarchal cycle nor the Moses traditions have anything to do with the stories of the Davidic dynasty, which is why Jerusalem is said to have been conquered by both Joshua (Josh. 10-12) and David (2 Sam. 5).

It's clear that the etiological myths of Genesis exist simply to satisfy the curiosity of an intelligent and imaginative ancient people who lacked the observational technology to learn the truth about the origins of natural phenomena. By contrast, as we will discuss in "Everybody's Working for the Weekend," the Priestly creation account represents ancient science in the true sense: "natural philosophy" such as we find among the Pre-Socratics in Greece and Ionia. But our evangelical reinterpreters of Genesis who aim to remove the text as an impediment to embracing evolution assure us the stories are not intending to teach us about nature, but only theological generalities. Thanks to the Bible, we know that it was God who invented humanity and not Dr. Frankenstein! This tendency to blur the vivid stories into a Christian catechism is virtually to embrace Marcion's rejection of the whole Old Testament in favor of the New, only without being forthright about it as Marcion was (Bright 1971).

In-A-Gadda-Da-Vida

Today's evangelical evolutionists (and their Roman Catholic brethren) are fighting a two-front war regarding the Garden of Eden. On the one hand, they want to make the Bible look good to secular scientists by lowering its absurdity factor. On the other, they want to try to sell to more conservative Christians the notion that to accept evolution and modern science does not require dismissal of scripture but only a new and "proper" interpretation of it. It seems, however, that these apologists just wind up rewriting the Eden story *as they wish it had been.*

C. John Collins, not an evolutionist, dismisses revisionist theories of Original Sin that reject a historical Adamic fall and view sin as inherent in human freedom, saying "they end up telling a very different story from the one we find in the Bible" (2011, 48). But this appears to be equally true of rewrites that do retain a historical fall—of *someone.* Denis Lamoureux, looking very much like a revisionist, admits that he and his fellow "evolutionary creationists contend that Gen 1-11 is to be read in a very unnatural and utterly counter-intuitive way." They, and he, "suggest that Christians must move beyond the scientific and historical concordism that has marked the interpretation of these chapters for generations" (2008, 34).

Collins acknowledges that "the Mesopotamian origin and flood stories provide the context against which Genesis 1-11 are to be set." He seems to seek some cover for the obviously non-scientific elements of Genesis, divine inspiration aside, by asserting that the Mesopotamian stories

> also provide us with clues on how to read this kind of literature. These stories include divine action, symbolism, and imaginative elements; the purpose of the stories is to lay the foundation for a worldview, without being taken in a "literalistic" fashion. We should nevertheless see the story as having what we might call an "historical core," though we must be careful in discerning what that is. Genesis aims to tell the story of beginnings the right way. [p. 35]

He finds helpful the observation by Henri Blocher that the

> real issue when we try to interpret Genesis 2-3 is not whether we have a historical account of the fall, but whether or not *we may read it* as the account of a historical fall. The problem is not historiography as a genre narrowly defined—in annals, chronicles, or even sagas—but correspondence with discrete realities in our ordinary space and sequential time. [from Collins 2011, 37, our emphasis]

If all that Genesis wanted us to do is believe there was some kind of historical Fall into sin, we might feel ourselves entitled to disregard the details of the Eden story and replace them with a more believable version of our own. It's the thought that counts, apparently: We shouldn't take it "in a 'literalistic' fashion," yet we should "see the story as having what we might call an 'historical core,' though we must be careful in discerning what that is. Genesis aims to tell the story of beginnings the right way" (p. 35), i.e., the way that C. John Collins apparently learned in Sunday School.

C.S. Lewis was in the vanguard when he sought to replace the embarrassments of the story about day-old, reptile-conversing vegetarian nudists with a more vague and abstract version. It is blurry enough that one need not be disturbed at details seen too closely, like those of Genesis 2-3. "For long centuries," Lewis says in *The Problem of Pain*, "God perfected the animal form which was to become the vehicle of humanity and the image of Himself" (from Collins 2006, 208).

> Then, in the fullness of time, God caused to descend upon this organism, both on its psychology and physiology, a new kind of consciousness which could say "I" and "me", which could look upon

> itself as an object, which knew God, which could make judgments of truth, beauty and goodness, and which was so far above time that it could perceive time flowing past. [from pp. 208-209]

This is Lewis's version of Adam and Eve. He won't have us pretend to "know how many of these creatures God made, nor how long they continued in the Paradisal state." But he is confident that

> sooner or later they fell. Someone or something whispered that they could become as gods . . . They wanted some corner of the universe of which they could say to God, "This is our business, not yours." But there is no such corner . . . We have no idea in what particular act, or series of acts, the self-contradictory, impossible wish found expression. For all I can see, it might have concerned the literal eating of a fruit, but the question is of no consequence. [from p. 209]

Lewis seeks to strip away the offensive, mythical details of the actual story, urbanely pooh-poohing the implications. "Oh, it might be literal, old bean, but it doesn't matter, eh what?" The strategy is akin to that of apologists who admit the *Testimonium Flavianum* cannot be by Josephus—unless we trim some of it away! That's the ticket! It would be better vaguer, so let's cloud it up! It is one thing to find the Bible to be ambiguous when you seek clear teaching from it, but quite another to find it too darn clear and to try to make it more vague.

Everybody's Working for the Weekend

> *We assert that Moses spoke in the literal sense, not allegorically or figuratively, i.e., that the world, with all its creatures, was created within six days, as the words read. If we do not comprehend the reason for this, let us remain pupils and leave the job of teacher to the Holy Spirit.*
>
> —Martin Luther, *Lectures on Genesis*

One of the greatest of the many ironies entailed in creationism and Christian anti-evolutionism is the rejection of modern science in favor of ancient narratives that embodied the natural science of the biblical writers' era, rather than our own. The author of the seven-day creation story never says, as Jeremiah does, "The Word of Yahweh came to me, and he said . . ." The writer does not say, as John did in the Book of Revelation, that the heavens opened and he saw what he describes in a spectacular vision. No, the whole account can be shown to be cut from the same cloth of ancient science as practiced in the surrounding cultures.

The Eden story is a piece of mythology, but the seven-day creation is ancient "natural philosophy," the speculations of the wise based on the pre-technological observation of the world around them. If you could hop in a time machine and go back to explain to the Genesis 1 writer what we know of earth's and mankind's beginnings, he would have crumpled up his papyrus and bidden us to write down the facts as we know them. That is the attitude of proper scientists, ancient or modern. They do not defend a sentimental favorite come hell or high water, as modern Christian creationists are forced to do.

Days and Confused

Playing in the Creationist Sandbox

Let's get something out of the way right up front: No sensible person actually takes the six-day creation account literally anymore. Anyone with the faintest knowledge of geology or astronomy knows that it is ludicrous to posit a one-day interval between the formation of dry land and the appearance of vegetation (Gen. 1:9-13), to cite just one entry in a catalog of absurdities.

A faith that tries to adhere to the traditional reading of Genesis 1 as saying that creation took God a week is going to trip over *any* awareness of natural science, even without getting into the evolution part. Pentecostal healer William Marrion Branham used to give out that he was the predicted Elijah to precede the Second Coming (uh, where was *that* predicted?), but whose followers believed him to be God incarnate. A follower once heard Branham toss off a reference to the speed of light being 186 *million* miles per second. Of course it was a casual slip of the tongue. But this Branham devotee, believing implicitly in his messiah, remarked, "How could the scientists be so wrong?" Imagine if you will some Institute for a Faster Speed of Light coming out of this little conflict between science and "revelation." It may sound far-fetched, but not really much more than the mischief stemming from the Genesis writer's errors.

Most people have long since given up pretending that all the geological data for a billions-of-years-old planet are bogus propaganda (for what conceivable reason?) or Satanic illusion. Even before Darwin, there was abundant evidence contradicting the idea of a young earth. In his thoughtful analysis, *Living with Darwin: Evolution, Design, and the Future of Faith*, Philip Kitcher summarizes the "obvious story" faced by the "devout geologists" of the early nineteenth century. They realized that sedimentation

> rates would require vast stretches of time (at least hundreds of thousands of years) to lay down rocks to the depth observed. These strata were deposited sequentially, and the oldest almost always lie at the bottom. Most of the organisms they contain belong to species that have now vanished from the earth. The deepest rocks ... contain residues of marine invertebrates, some of which, like mollusks, are very familiar, others of which, like the trilobites, are very different. Above them lie layers in which there are both marine invertebrates and some fish, with an increasing diversity of fish as you climb the rock column. Higher still are strata with marine invertebrates, fish, and amphibians. Ascending further, these kinds of organisms are joined by reptiles, including huge reptiles of kinds that no longer exist —some of the dinosaurs—and later, after the vanishing of the dinosaurs, by birds and mammals, both of which become increasingly diverse as you approach the top. Near the surface, in the shallows of the rocks, there are finally traces of apes, and, eventually, of human beings. [Kitcher 2007, 29-30]

Since then, of course, the evidence has continued to pile up. Numerous radiometric dating methods are now well established that all converge with remarkable precision on an age of the earth in excess of 4 billion years (Prothero 2007, 75-76). Astronomers are observing galaxies outside our own Milky Way, not just its own stars, and the light from them has taken millions and even billions of years to arrive at their telescopes (Sparrow 2006).

Scientists in Antarctica are extracting ice cores from snow that fell over a million years ago.[36]

Even the genome provides evidence of ancient age, based on the statistical limits for the rate at which mutations occur and the track record of changes from one species to the next. It's called the molecular clock, and is based on the idea that

> as species diverge from a shared ancestor, each separate lineage accumulates mutations in their DNA at a steady rate on average over long spans of time. Therefore, as more time passes, their DNA (and certain proteins they encode) becomes more different—and the number of differences is directly proportional to the time that has passed since two species separated from the common ancestor. [Gibbons 2007, loc. 1240]

The probability density function of the mutations is too widely dispersed for very precise estimates, but the molecular clock has been ticking for a long time. From it, we can confidently assign an age of the last common ancestor of all eukaryotes (organisms whose cells have complex structures enclosed by a membranes) of about a *billion* years (Koonin 2011, loc. 3088).

Young-earth creationists who deny all this have utterly surrendered their intellect to the worship of a holy text. They are living in the medieval fantasy world of Martin Luther's *Lectures on Genesis*, which began with an admonition to "confess our lack of understanding rather than distort the words, contrary to their context, into a foreign meaning" if "we do not understand the nature of the days or have no insight into why God wanted to make use of these intervals of time." For Luther, the author of Genesis (Moses, naturally) had this purpose:

> to teach us, not about allegorical creatures and an allegorical world but about real creatures and a visible world apprehended by the senses. Therefore, as the proverb has it, he calls "a spade a spade," i.e., he employs the terms "day" and "evening" without allegory, just as we customarily do.

Unfortunately, there are still slaves to this kind of thinking, and they are forced to blame *themselves* for noticing the undeniable conflicts between the ancient words and the obvious realities of our world. After sermonizing about the retrograde motion of the planets as "a work of God, created

---

36. See, e.g., "The oldest ice core: A 1.5 million year record of climate and greenhouse gases from Antarctica," a white paper published by the International Partnerships in Ice Core Sciences, pages-igbp.org/ipics/data/ipics_oldaa.pdf (accessed June 2012).

through His Word," about God being the "One who commanded the sun to run, but the firmament to stand," Luther urged Christians to

> be different from the philosophers in the way we think about the causes of these things. And if some are beyond our comprehension (like those before us concerning the waters above the heavens), *we must believe them and admit our lack of knowledge rather than either wickedly deny them or presumptuously interpret them in conformity with our understanding.* We must pay attention to the expression of Holy Scripture and abide by the words of the Holy Spirit, whom it pleased to distribute His creatures in this way: in the middle was the firmament, which was brought forth out of the unformed heaven and the unformed earth and spread out through the Word; furthermore, above and below the firmament there were waters which were also drawn from this unformed mass. [*Lectures on Genesis*, Ch. 1, v. 6, our emphasis]

For a person to maintain this level of pious delusion today simply disqualifies him from rational discourse. We share the view of the accomplished paleontologist Donald Prothero (a religious believer, by the way) about one notorious example, Kurt Wise, who managed to obtain a Harvard Ph.D. in paleontology and still be a young-earth creationist. Wise "continues to pretend that he is following the rules of science," wearing "the label of scientist" and promoting "his particular brand of 'science' to unsuspecting people who are impressed with his Harvard Ph.D." What Wise admits but his followers don't realize is that he stopped doing science long ago (Prothero 2007, 23).

The story of Wise's capitulation to his fundamentalism is "just plain pathetic—pathetic and contemptible" (Dawkins 2006, 322). Rather than pretending that all his secular education about the natural world really means anything at this point, he might as well study the danger of airplanes ripping a hole in the firmament, letting the waters above it pour out and cause another worldwide flood. Let the rest of us engage in adult conversation while those who refuse to acknowledge reality play in their sandbox of infantile faith. "When I was a child, I spoke as a child, I understood as a child, I thought as a child: but when I became a man, I put away childish things" (1 Cor. 13:11).

The Time Telescope

Thinking people realized that they had to reinterpret the words of Genesis long before Luther sputtered his disdain for human reasoning. He knew about that and warned against it, saying "the ideas of Origen and Jerome, together with those of other allegorists" about Eden not being a historical place were "silly," and urging the "unwary reader" of Genesis 2:9 not to "be led astray by the authority of the fathers, who give up the idea that this is history and look for allegories" (*Lectures on Genesis*, Ch. 2, vv. 8-9). A prime target for reinterpretation was the length of the Genesis writer's "day." It's not a new idea. Sometime in the second century, the writer of the *Epistle of Barnabas* sought to reassure readers that Genesis 2:2 ("By the seventh day God completed His work which He had done") was compatible with the fact that, all these thousands of years later, things were still humming along.

> Attend, my children, to the meaning of this expression, "He finished in six days." This implieth that the Lord will finish all things in six thousand years, for a day is with Him a thousand years. And He Himself testifieth, saying, "Behold, to-day will be as a thousand years." Therefore, my children, in six days, that is, in six thousand years, all things will be finished. "And He rested on the seventh day." [Ch. 15; see Paget 1994, 168-69]

Augustine famously avoided speculation about the days of creation: "What kind of days these were it is extremely difficult, or perhaps impossible for us to conceive, and how much more to say!" (*City of God*, Book 11, Ch. 6). Luther the literalist characterized Augustine's view as "resort[ing] to extraordinary trifling in his treatment of the six days, which he makes out to be mystical days of knowledge among the angels, not natural ones" (*Lectures on Genesis*, Ch. 1).

The attempted metaphorizing of the word "day" allows old-earth or "progressive" creationists to insert into Genesis 1 a good bit of wiggle room. Even the writer of *Barnabas* saw the problem with the end of the story speaking of "the day God created the heavens and the earth." Since God couldn't have made the whole thing in a single 24-hour day, the reasoning goes, the six "days" of creation might have been intended less than literally. Maybe it just means "seven periods," and in that case, maybe these periods lasted for untold zillions of years.

One example of "day-age" creationism comes from the fundamentalist Protestant sect from which one of us (Ed) recently departed, in a book (Reinikainen 1986) published by the Finnish organization of Conservative Laestadians. The book acknowledges research in the natural sciences to the extent that it shows "that the origin of the earth, just as the origin of the total universe, dates back billions of years." The biblical "account of creation states that the formation of the Earth and the origin of all forms of life happened in six days, although all the evidence found in nature shows the timeframe of the origin of creation to be very long." But the creation account does not have the sun, along with day and night, being created until the fourth day, so one is entitled "without arbitrarily interpreting the

Bible . . . to the conclusion that the days of creation were eras, God's days, in His creation work" (p. 14).

Conservative Laestadianism is a very Lutheran group—indeed, it claims to be the *only* true Lutheran church, outside of which no one will be saved. But the book does not overly trouble itself with Luther's statement at the beginning of his *Lectures on Genesis*, "We know from Moses that the world was not in existence before 6,000 years ago." That figure was just plain wrong, it says, being based on "written genealogies, which are clearly incomplete" (Reinikainen 1986, 16). Instead, we "see, as a natural phenomenon, that people have been children of their time, interpreting these things in the Bible from their own presumptions, without contradiction. At this time we know more about natural science and history, as well as other things" (p. 17). Now, in our newfound enlightenment, we can see that "the listing of ancestors from Adam to Abraham [in Gen. 5 and 11] is not complete, but many generations in between are missing." Otherwise, "no more than about 100 generations would have lived on the earth up to this time."

Rather than admit any biblical error, the book concludes that it "has clearly been quite common for biblical authors to shorten the genealogies, leaving out those generations which they thought were not important [Matt. 1:8 vs. 1 Chron. 6:7-9 cited]" (p. 16). Based on Deuteronomy 7:9, it proposes an alternate calculation, in which "the Bible verifies that from Adam until the second coming of Christ there are about 1,000 generations." Thus, able to acknowledge "about 10 times more than the amount verified earlier" of human existence, a "Christian believes that the Bible speaks the truth" (p. 16). Even with such clerically sanctioned day-stretching, the poor inerrantist is still left wondering how Adam's descendants, i.e., tool-making, anatomically modern *Homo sapiens*, could have been leaving their bones and artifacts strewn across the Mideast, Southern Europe, Asia, and Australia by 40,000 BCE., after having already occupied parts of Africa for over 100,000 years.[37]

---

37. See, e.g., Fagan 2010, loc. 327 & 1542; Hurd 2003, loc. 2492; Wade 2006, 8-9; Wells 2006, loc. 1319. These paragraphs about Reinikainen's book were adapted from Suominen 2012, §4.3.1. In the decades since publishing it, the Finnish Conservative Laestadian organization has been doing some backtracking. Rather than calling theistic evolution "an outrage to the word of God" (Reinikainen 1984, 17), it is now making statements that are quite accommodating of modern science, with one recent article in a monthly church newspaper going so far as to criticize creationism as something *foreign* to Christian faith (Suominen 2012, §4.3.1). Perhaps it shouldn't be a surprise, even coming from a sect with doctrinal views that are quite extreme otherwise. The current generation of Finns has received sound scientific education untainted by the hysteria about evolution so prevalent in the U.S.

To accomplish pseudoscientifically what Reinikainen did with creative biblical exegesis, Gerald Schroeder makes an equally imaginative use of the "immense stretching of space since the Big Bang" and its "strong implications for our cosmic clock" (1997, 54). The whole effort is still driven by an allegiance to the Bible (only the Hebrew part, for him), and indeed he refers to the biblical "1000 years = 1 day" scaling factor, which he gets from Psalm 90:4 rather than 2 Peter 3:8. Schroeder admits that skeptics will call his efforts "a rationalization for squeezing fifteen billion years into six days." But he protests that he isn't calling for "a trivial correlation between science and theology such as 'Let's call each day a very long epoch, lasting millions or even billions of years.'" (p. 43).

So how does he try to pull it off? By claiming that the "cosmic perception of time" has been stretched, and by a factor of 10 to the 12th power! The "division of fifteen billion years by one million million reduces those fifteen billion years to six days!" (p. 58). Schroeder has God hiding the truth about the vast age of his creation behind an accounting trick that couldn't be figured out until Einstein came along thousands of years later. It's an impressive display of creative ability—not by a creator deity, but by a clever apologist who makes his god do whatever is needed to fit the dogmatic agenda.

Not so Fast!

The "day-age" interpretation is not out of the question, but there are some difficulties besides what scaling factor to use. The writer of Genesis 1 seems to have meant to say that creation took six literal, twenty-four hour days. He frames each stage of creation with the accoutrements of the Hebrew calendar: "There was evening and there was morning, the first day." And why does this "day-age" business look good to anyone? Is it the most natural reading of the text? Or is it that, if true, it would get creationists out of a tight spot? Certainly it is the latter. This interpretation is an *ad hoc* hypothesis.

The young-earth creationists still continue in Luther's literalist footsteps, and their criticisms of "compromises on Genesis" certainly apply to the conciliatory efforts of their old-earth counterparts. Here is another salvo fired against modern science from one party in the unwitting alliance between creationists and atheists, who both deny any compatibility with Christianity. It is telling that the quote itself provides little indication of the camp from which it originated. The issues are often presented with equal clarity by believer and unbeliever alike, so long as they both disagree with the middle-of-the-road position being criticized:

> Despite attempts to minimize the differences between biblical creationists and old-earth creationists or theistic evolutionists, the

theological problems created by such attempts to harmonize the Bible with billions of years are indeed significant. The order of events in Genesis 1 often have to be rearranged to accommodate such views, and those who wish to mesh evolution with the Bible must accept that there was death, suffering, diseases like cancer, and even thorns before the Fall—but all those things are a result of sin! The Scripture is explicit that thorns came after the Curse. How could God call cancer "very good"?

It's a valid criticism, even if it comes from someone—Ken Ham (2012)—who we find profoundly and disturbingly wrong about most everything pertaining to evolution other than its theological difficulties.

Ingersoll saw through the "day-age" charade over a hundred years ago when he said creationists were making the six days into "a kind of telescope, which you can push in or draw out at pleasure. If the geologists find that more time was necessary they will stretch them out. Should it turn out that the world is not quite as old as some think, they will push them up. The 'six days' can now be made to suit any period of time" (interview with the *Detroit News*, 1884).

Finally, why would God need all the time that the geological record gives him, even though this interpretation allows it? Old-earth creationists aren't concerned about evolution needing eons to work its way through to today's species, as Darwin was. ("We see nothing of these slow changes in progress, until the hand of time has marked the long lapse of ages, and then so imperfect is our view into long past geological ages, that we only see that the forms of life are now different from what they formerly were." Darwin 1859, 84) They think God just snapped his fingers, and it was done—just at intervals of millions of years.

Ramm wrote, using his frequent and annoying lapse into the first-person plural, "We believe that the fundamental pattern of creation is progressive creation" (1954, 76, emphasis omitted). He added, "We believe in several acts of fiat creation in the history of the earth, and this clearly differentiates this view from theistic evolution" (p. 78). This view at first seems to fit naturally with the "age-day" theory, "which considers the days of Genesis as being periods of time" (p. 145). So what caused all the delay? Was God like some tropical fish fancier who sat there watching his tank full of trilobites for eons?

There is a problem, too, with the order in which Genesis 1 has things being created. In what possible sense could there be an "evening" or a "morning" without the sun? Augustine couldn't answer that and punted to claiming that "what kind of light that was, and by what periodic movement it made evening and morning, is beyond the reach of our senses." We can't understand that, he admitted, "yet must unhesitatingly believe it" (*City of God*, Book 11, Ch. 7).

Ramm admits that the age-day model runs into real trouble, showing how little this model actually squares with the text: "The order in Genesis might be part chronological and part logical" (Ramm 1954, 148). So "if the organization is logical or topical then ... any specific chronological objection is invalid" (p. 149). "Further, there is nothing to prevent the creation of each day from overlapping in its development with the other successive days" (p. 149).

However, Ramm is not convinced the word *yom* ("day") can be stretched so far, so instead of the "age-day" theory he prefers the "pictorial days" theory: "creation was *revealed* in six days, not *performed* in six days ... the six days are *pictorial-revelatory* days, not literal days or age-days" (p. 151). So Moses was basically sitting in a theatre for a week with nightly intermissions. This theory was first proposed by the unfortunately but aptly named Benjamin Silliman (1779-1864). Again, would this occur to any reader of Genesis who was not looking for a fire escape?

It reframes the problem once one realizes that the division of creation into six days and a seventh day of down time is a secondary layer of the story. Just bracket all the business about the evening and the morning and which day it was, and you have a better-flowing narrative sequence. "He made this, then that, then the other thing." The days add nothing to the description. What they do add, though, is a lesson about the sabbath. The writer, or a subsequent scribe, must have introduced the day-divisions in order to make God into a good Jew who worked a full week and then took the sabbath off.

The goal was to say to readers, "Look, if God Almighty himself keeps the sabbath, shouldn't you?" The danger was assimilation to the neighboring peoples, whether Canaanites in the Holy Land or Babylonian or Persian neighbors during the Exile. You couldn't keep many of the Jewish laws outside the Holy Land, where the temple had been and the only place sacrifices could be offered. So it would have been easy for Diaspora Jews to assimilate to their Gentile neighbors' ways. What could a Jew still do away from home? Observe dietary laws (a big issue in Daniel, no surprise), circumcise your baby boys, and observe the sabbath (which Jews did: Romans thought them lazy for it, so we know they actually did it; it didn't remain just on paper).

Without the "day" punctuation, there is no telling how long the original writer thought it took God to make the cosmos: a single day? A second? Six billion years? Could be anything. But it complicates things that we do find the version with creation divided into days in the canonical text. Fundamentalists will naturally feel they are stuck with that. Unless, of course, they pulled a Warfield and speculated that, since a literal six-day creation is an error, *it cannot have read that way in the lost original autographs!* Why not?

Denis Lamoureux has a peculiar way of "defending" Genesis. Indeed, when biblical inspiration can count friends like this, it hardly needs enemies:

> The redactor of the book of Genesis clearly intended to present a record of how the universe and life originated. . . . [H]e intended Gen 1 to be understood as a natural history. However, the science in Gen 1-11 is an ancient science . . . based on an ancient phenomenological perspective of nature and conceptualized through ancient notions of causality. [Lamoureux 2008, 265]

But then Lamoureux drives a wedge between the intent of the author and that of the inspiring Spirit:

> In order to understand these chapters fully, they must be read in a counterintuitive fashion. Christians today must look beyond the intended literal statements of the human authors and the intended accommodation of the Holy Spirit. [p. 272]

Thus he flirts with allegorical interpretation, secret levels of meaning. For a Protestant, that is quite a price to pay. If the true meaning of scripture is to be found between the lines, not on them, then the grammatico-historical method goes out the window, and the Bible can mean whatever we, or the Pope, want it to mean. And then the Bible has been reduced to a ventriloquist dummy.

So the Priestly writer was offering an attempt at a cosmogony, just like his Greek neighbors. That was his intent. He would have taken umbrage at modern attempts to wave his scientific theorizing away as mere window dressing for the truism that Elohim made the world and liked it. Why do even more liberal-minded Christian evolutionists try to reduce his science to religion? They do not want to have to admit that what the biblical writer was affirming, the point he wanted to get across, was *wrong*. Admitting this would mean that the Bible's value is not after all in providing us with infallible information about secret things. If it only presents us with ancient *speculations* about the beginning, why should we not conclude it's doing the same thing when it comes to the *end*? Maybe all the things said in the Bible about heaven and resurrection are just human theorizing, too? Indeed, they are. But a conclusion is not to be rejected merely because we don't like it. That is the most anti-scientific attitude of all, and it has too often marred religion.

The Gap Theory

Another attempt to lengthen the time span allowed by a literal seven-day reading of Genesis 1 is the notorious and fascinating Gap theory, first proposed by Simon Episcopius (1583-1643) and popularized by Thomas Chalmers (1780-1847), William Buckland (1784-1856), and others. What if you translated "was without form and void" in Genesis 1:2 as "*became* without form and void"? That might open up room between verses 1 and 2 for, um, *a whole previous creation not actually mentioned there!* Yeah, that's the ticket! And this one had dinosaurs! And it was ruled by Lucifer! It is a scenario worthy of Edgar Rice Burroughs and even sounds kind of like his Pellucidar novels. And that's what the Gap theory says: Lucifer was the vicar of God on earth until he got too big for his britches and declared himself a god. This was something Yahweh could not brook, so he flooded out this creation, reducing it to the swampy mess, "without form and void," that we find in verse 2.

Gap theorists appeal to Isaiah 45:18: "For thus says the LORD, who created the heavens (He is the God who formed the earth and made it, He established it and did not create it a waste place, but formed it to be inhabited), 'I am the LORD, and there is none else'" (NASB). They take this to mean that the chaos void in Genesis 1:2 cannot refer to the world as God first created it, in which case Genesis 1:2 must refer to the debris of the lost, first creation. The problem here is, of course, the sheer enormity of the untold story that is being inserted with no hint to suggest it. The Isaiah passage obviously speaks of God's intention in creating the world, not the initial state of his raw material. He wasn't going to leave it a mess; what would be the use of that? And as for translating "was" as "became," well, sure it's possible and occasionally happens in Old Testament Hebrew, but context determines that. And nothing suggests it here.

The Gap theory is a modern instance of what is called a scribal legend, the formulation of a brand new, hypothetical back story that would provide harmonization between two apparently contradictory texts. A famous ancient example is the story of Lilith. The rabbis puzzled over the contradiction between Genesis 1, where men and women are created at the same time, and Genesis 2, where the man is created first, then the various animal species, and finally the woman. What gives? They proposed that God had first made Adam and Lilith at the same time, but that Lilith had proved "uppity," refusing to submit to Adam, so she left and shacked up with Samael (Satan). Henceforth she was known as the child-snuffing night hag of Babylonian legend mentioned in Isaiah 34:14. Then God made Eve as a replacement, a Stepford wife, though that didn't work out too well, either!

It is a legend, but unlike most legends which grow up from the soil of folk-belief, the Lilith legend was a conscious invention by scribes who wanted to use it as glue to repair a fissure in the scriptural text. The Gap theory belongs to the same species. It has become a modern midrash, part of the lore of the Bible, and it originated as a theory to harmonize Genesis 1:1-2 with Isaiah 45:18 and with the geological record.

Last Thursdayism

Surely one of the most bizarre efforts to defend young-earth creationism was that of Philip Gosse in his nineteenth-century work *Omphalos*. The word is Greek for "navel," and the book addressed itself to the old biblical stumper, "Did Adam have a belly button?" Why should he, if he were created, *ex nihilo*, as an adult?[38] Gosse contended that Adam indeed had a navel and that he was not alone. For though God created the world in 4000 B.C., a la Genesis, he created it with simulated signs of age and development. A version of this idea was appropriately mocked as "Last Thursdayism," after an end-of-the-world prediction in 1992. Since the latter-day prophet had claimed the world should have ended by this point, "the entire universe must therefore have been recreated, complete with an apparent 'history,' last *Thursday*. QED."[39]

Gosse's "omphalos" argument allowed him simultaneously to admit and dismiss all the biological and geological data for the great age of the earth and the evolution of life. He reasoned that if God were to create a functioning planet, he must have created it already "rolling." Understood this way, the creation might be compared to a movie, the first frame of which depicts an action scene. No sooner does the film start than a holdup or an air battle is already in progress!

Now if the earth was born full-grown (like the legendary sage Lao-tzu, fully age 75 from the womb!), there must have been telltale signs of age, but the tale they told was false, at least fictitious. A flowing river (let us say, the Euphrates at the border of Eden, Gen. 2:14) from the first moment of its creation must have already possessed an alluvial deposit along its banks. But, strictly speaking, it was never deposited! So with Adam, who had the mark of an umbilical cord which never existed save in the mind of God. And so with the earth's crust, pregnant with fossils of strange life-forms which never walked the earth. All were created *as if*. Therefore, all those

---

38. Irenaeus thought Adam and Eve had yet to "come to adult age" when they were created (*Against Heresies*, Book 3, Ch. 22).

39. Seanna Watson, in a posting at the talk.origins newsgroup on November 2, 1992. Portions of this section are adapted from Price 1980.

unbelieving biologists and geologists had actually gotten the story *correct*. The problem was they didn't realize it was *only* a story.

Why did this argument fail to attract any supporters, even among creationists? Simply because all (but Gosse) could see what extreme special pleading this was. Certainly it was all beyond disproof, but so was the Hindu claim that the world was illusory *maya* (indeed a very similar claim!). For that matter, who could prove the world had not been created a mere ten minutes ago, with Gosse recalling his formulation of a theory he had never actually formulated? Alas, such solipsism has never been very attractive, not even to modern scientific creationists. They know all too well that the argument would get them laughed out of the courts and off the debating platforms.

Yet if one carefully examines creationist polemical literature, one is surprised to find this "recessive" argument is still around. "Scientific" creationists owe an unacknowledged debt to Gosse's hypothesis. Boardman et al. provide a most obvious example in their 1973 work *Science and Creation*, zeroing in on a trouble spot while discussing astronomy and its implications for the age of the universe:

> The Biblical record places the creation of the universe at ten thousand years or less in the past; whereas, the presently accepted distance scale held by astronomers measures the universe in billions of light years. If the light rays now reaching the earth were created in transit at the time of the creation of the stellar objects, they must have been created carrying information descriptive of historical physical events (such as super novae) which never actually occurred, because we would now be observing light rays which were created in transit and never were radiated from the stars which they seem to image. [Boardman et al. 1973, 26]

Of course, the issue is not just limited to the age of the earth. As we will see below in "Peekaboo Deity," creationists of all types must accept a form of Gosse's divine deviousness to disregard the overwhelming evidence for evolution, too.

Priestly Postulations

Rudolf Bultmann distinguished between two elements of scripture that make it difficult for modern, educated people to take the Bible seriously: the ancient world picture and the mythology. Strictly speaking, they are not the same. The first is the product of ancient pre-technological observation, early science (or "natural philosophy"). The second is sacred narrative in which supernatural beings interact with humans.

If we are to understand the Priestly creation account of Genesis 1:1-2:4a, we must grasp that, unlike the J account of the Garden of Eden in Genesis 2:4b through the end of chapter 3, which is pure myth, the Priestly six-day account is ancient speculative science. It is entirely improper to call the Priestly account a myth. In fact, the author has rationalized or demythologized certain elements of the earlier Combat creation myth. The Priestly writer, for example, turns the primordial sea dragon Tiamat into "the deep" (*tehom*) and into *tohu*, "formless." The land monster Behemoth he transforms into another abstraction, "emptiness" (*bohu*). Tiamat is the plural of *tohu*, while Behemoth is plural for *bohu*. This is demythologizing, a process Bultmann shows was already underway in the New Testament (Bultmann 1958), but obviously it is older than that.

Julius Wellhausen (1844-1918) explained what is going on in the Priestly account:

> In the beginning is chaos; darkness, water, brooding spirit, which engenders life, and fertilises the dead mass. The primal stuff contains in itself all beings, as yet undistinguished: from it proceeds step by step the ordered world; by a process of unmixing, first of all by separating out the great elements. The chaotic primal gloom yields up to the contrast of light and darkness; the primal water is separated by the vault of heaven into the heavenly water, out of which there grows the world above the firmament which is withdrawn from our gaze, and the water of the earth; the latter, a slimy mixture, is divided into land and sea, whereupon the land at once puts on its green attire. The elements thus brought into existence, light, heaven, water, land, are then enlivened, pretty much in the order in which they were created, with individual beings; to the light corresponds the lamps of the stars, fishes to the water, to the heaven the birds of heaven, and the other creatures to the land. [Wellhausen 1957, 297]

Peter Enns claims, "These stories were not written to speak of 'origins' as we might think of them today (in a natural-science sense). They were written to say something of God and Israel's place in the world as God's chosen people" (2012, 5). But Wellhausen dismissed that excuse before Enns was born. He saith:

> It is commonly said that the aim of this narrative is a purely religious one .... Had he [the Priestly narrator] only meant to say that God made the world out of nothing, and made it good, he could have said so in simpler words, and at the same time more distinctly. There is no doubt that he means to describe the actual course of the genesis of the world, and to be true to nature in doing so; he means to give a cosmogonic theory .... He seeks to deduce things as they are from

each other: he asks how they are likely to have issued at first from the primal matter, and the world he has before his eyes in doing this is not a mythical world but the present and ordinary one ... The graduated arrangement in separating particular things out of chaos indicated the awakening of a "natural" way of looking at nature, just as this is manifest in the attempts of Thales and his successors. [Wellhausen 1957, 297-99]

One of the major Genesis conundrums is solved right off: How can there have been light before the sun and stars were created? Even to ask the question is not only to presuppose modern physics and astronomy but also to miss the structure of gradual differentiation of specifics from generalities created just previously. As Wellhausen notes, the contemporary Ionian philosophers looked at things in much the same way, positing a primordial element from which everything gradually emerged as from a centrifuge. Thales posited water as the primal element (much like our Priestly writer). Anaximander made it "the Indeterminate Boundless," Anaximenes said it was air, and Heraclitus said it was fire.

It's not myth. It's science, ancient rather than modern. Thus one cannot say, with Bernard Ramm and others, that the Bible speaks "non-postulationally," merely employing the common language of people as they observe nature, the sun "rising," etc. How can Ramm propose such a distinction? Because he imagines the biblical authors were just like him, interested only in theology. That is an error. What we read in the six-day creation account is a natural philosopher's attempt to reason out the origin and development of the natural world. He turns out to be wrong, but one must imagine he would have heartily welcomed correction. And of course he never claims that "God told me this" (even though the Jewish apocalyptic writers did make such claims, setting forth their astronomical speculations in the exciting form of heavenly journeys of Enoch and others to whom angels vouchsafe the secrets of the heavens, of the stars and the weather).

Like the early Greek philosophers who tried to figure out the origin of the world, the Priestly writer did his best, but he was wrong. That's no surprise. He didn't have the complex geological and biological information we have at our disposal. But, come to think of it, neither did Anaximander. That didn't stop him from making some pretty intelligent guesses about human origins. Man, he surmised, "came into being from another animal, namely the fish, for at first he was like a fish" (Hippolytus, *Philosophumena*). He declared "not that fishes and men were generated at the same time, but that at first men were generated in the form of fishes, and that growing up as sharks do till they were able to help themselves, they then came forth on the dry ground" (Aristotle, *Symposium*). Observing that "man alone requires careful feeding for a long time," unlike all the other creatures who "can quickly get food for themselves," Anaximander figured that "at the beginning man was generated from all sorts of animals" (Plutarch, *Stromateis*).[40]

Anaximander just didn't know the mechanism: natural selection. That was Darwin's great insight.[41] "[T]he ancient Greeks had proposed that life originated in a primordial slime. Darwin's extraordinary achievement lay in the coherence and persuasiveness of his scientific explanation in terms of natural–rather than supernatural–causes" (Wiley 2002, loc. 1429).

We can't let the Bible off the hook here. The contemporary pagans hadn't figured it all out, but the Bible writers weren't even close to being as right as they were. There was indeed something like evolution being talked about in biblical times, and the Bible is ignorant of it.

Let's be Kind

While the Bible misses the mark by failing to mention anything like evolution when it could have, we do need to defend it in one important respect. As the Christian evolutionary biologist Joan Roughgarden[42] points out, "It's clear that the Bible isn't making any statement one way or the other about whether species can change after they've been created" (2006, 27). The biblical "kind" is not the same as a "species":

---

40. The quotations in this paragraph are from Fairbanks 1898, 11 & 13-14, copied from history.hanover.edu/texts/presoc/anaximan.htm. It must be acknowledged that Anaximander got many things wrong (e.g., "Animals come into being through vapours raised by the sun," *Philosophumena*). But it's still impressive to see him realizing that man was not created *ex nihilo* in his current form, something still denied by about half of all Americans!

41. Three main pieces of evidence led to Darwin's insight about Natural Selection: (1) breeders had used artificial selection to produce "variations useful to man"; (2) "other variations useful in some way to each being in the great and complex battle of life, should sometimes occur in the course of thousands of generations;" and (3) the sad Malthusian reality that "many more individuals are born than can possibly survive." Given that, Darwin asked, can there be any doubt "that individuals having any advantage, however slight, over others, would have the best chance of surviving and of procreating their kind? On the other hand, we may feel sure that any variation in the least degree injurious would be rigidly destroyed. This preservation of favourable variations and the rejection of injurious variations, I call Natural Selection" (1859, 80-81).

42. *Roughgarden*: What a wonderful name for one dealing with the problemmatical tale of Eden!

> As a matter of fact, a close inspection shows us that it is probably a mistake to read Genesis 1 as talking about the kinds of plants and animals in a taxonomic sense (or even as implying that the kinds are fixed barriers to evolution). Rather, the passage makes plenty of sense if we can picture the perspective of an ancient Israelite: such a person already knew full well that if you want to grow wheat or barley, you plant wheat or barley seeds; if you want more sheep, you breed them from other sheep. [Collins 2011, 110]

It's a fair point. When Genesis 1 says animals multiplied "each according to its kind," we shouldn't take that to be any biblical refutation of speciation by the "slight modification[s], which in the course of ages chanced to arise" to drive Darwinian natural selection (p. 82). Evolution is "the accumulation of concrete changes of different kinds, big and small, revealed by direct comparison of increasingly available gene and genome sequences" (Koonin 2011, loc. 907).

Given enough time, all those changes eventually change an organism enough that it is considered a new species. It's difficult to define a "species" with precision, but a time-honored and generally accepted view relies on the distinction of "reproductive isolation," advanced by Ernst Mayr. Two organisms that are incapable of mating productively with each other are considered to be of different species.

Contrary to the caricatures made by creationists, evolutionary theory claims nothing like the sudden mutation of one species into a very different one. Nobody's talking about a bird hatching out of a fish egg! Relatively gradual change from mutations occurring here and there in the vast space of the genome isn't something that Genesis 1 explicitly rules out. Indeed, it is ludicrous to imagine that the Bible writers had any idea of such things when Darwin himself didn't even understand the biological mechanism underlying his theory.

The Sliding Scale of Inerrancy

> *It is much easier to be an "inerrantist" when the intended meaning of the original author can be disregarded.*
>
> —Thom Stark, *The Human Faces of God*

James Barr once pointed out how Bible inerrantists are by no means literalists, though they would like to think they are (1977, 46). No, Barr says, inerrantists will resort to any harmonization, any figurative, any strained interpretation, in short, any non-literal interpretation of a difficult text to make the text seem inerrant. "Well, er, ah, if the writer meant this, his statement would be perfectly correct!" Sure, and if pigs had evolved wings, they could fly. Those Christian conservatives who seek some accommodation with evolution by hook or by crook are among the worst of such culprits.

The old bumper sticker reads, "The Bible said it! I believe it! That settles it!" But to describe what inerrantists actually do, it ought to read: "I said it! The Bible believes it! That settles it!" And when they do this, they are committing what they regard as the great sin of Humanists and theological Liberals: exalting the mere opinions of men into the supposedly infallible oracles of God. Our theistic evolutionists seem to be rapidly closing the gap between the Liberal position and their own. They are willing to give away the store, or at least all the merchandise, if they can only keep the empty shelves.

Deny Me Three Times

There are three related ways of discounting Scripture advocated by evangelical and Catholic writers embroiled in the evolution debates, whether conservative or liberal. First, there is the old stand-by *progressive revelation*. This position admits that the Bible presents numerous assumptions common to the ancient cultures in which it was written or revealed. The clinging bands of prescientific cosmology and barbaric morality remain on the text alongside genuinely revelatory elements for the simple reason that God could reveal new truths only very gradually. He needed to meet people where they were, which meant leaving a great deal of error uncorrected for the time being.

This approach is invoked to account for the easy tolerance of slavery and the oppression of women in scripture. Needless to say, it comes in quite handy in explaining why the Bible writers took for granted and often referred to a world picture we now know to be false. God thought moral and religious teaching to be more urgent matters, so he left the unenlightened human race to catch up on their own, scientifically and otherwise.

The fatal flaw in this theory is that progressive revelation tends to collapse into *no* revelation, since presumably *anything* new would be an affront too difficult to assimilate. On the other hand, if you are willing to accept the shocker that you cannot merit salvation by your own good deeds, it shouldn't be any big deal to learn that slavery is wrong, especially since the Sophists and Pre-Socratics had already been saying so for a long time. Was it impossible for the ancient Israelites to entertain the notion that today's species had in some manner evolved from earlier life forms? Nothing was stopping the Ionian philosopher Anaximander from teaching that.

Second, there is *hermeneutical ventriloquism*, the practice of creationists and fundamentalists who find a way to twist the biblical text in order to pretend the writers were actually describing a spherical earth or even the Big Bang, only we couldn't recognize it until secular science had figured it out. And then the apologist crows that the Bible anticipated modern science by virtue of divine inspiration. It boils down to reinterpreting the Bible according to science, which makes the Bible parrot the results of science. It is told what to say rather than telling us anything new. What sort of "revelation" is that?

We might compare this approach with the *pesher* exegesis of the Dead Sea Scrolls and the Gospel according to Matthew. *Pesher* means "puzzle solution." The technique presupposed that, while biblical prophecies had once accurately predicted historical events, now long past, the old texts were not mere relics but could be understood as bearing new meanings, hitherto-unsuspected esoteric messages disclosed by the Holy Spirit. These meanings were coded prophecies relevant not to Isaiah's and Jeremiah's contemporaries, but to future generations after the original prophets were long dead. Conveniently, they applied to events in the life of the particular sect and its heroes, like the Teacher of Righteousness at Qumran or Jesus in Matthew's gospel. With the eye of faith, or the proper key, one could see *in retrospect* that there had been an esoteric prediction and that it had come true.

A notorious example is that business in Isaiah 7:14 about a young woman conceiving and bearing a son. It was actually just a prediction of the imminent downfall of the Israelite-Syrian alliance threatening Judah. Matthew knew that, but those events were long over, and he had a Messiah to defend. Was Isaiah a dead letter then? Of course not! As inspired Scripture, the biblicist says, it must have new truths to reveal. So Matthew could see in Isaiah 7:14 a long-dormant prediction of the virgin conception of Jesus. In very nearly the same way, fundamentalists tell us that when Psalm 51:7 says, "Purge me with hyssop and I shall be clean," it was predicting the discovery of penicillin, found in hyssop. It didn't help us discover penicillin; it just enabled us to see how much the Holy Spirit knew before we did!

Third, the notion of God using ancient genres, ancient science, even overt myths, means that God "inspired" (really, just allowed) the writers to create *their* book, not *his*. The Bible is a museum of their theologies, not a revelation from God. "Some seem to expect the Bible to be a document that fundamentally transcends its setting" (Enns 2012, 143). But the less we look at it this way, the less of a revelation it is. With this third view, espoused by Lamoureux, Enns, et al., the inspiration of scripture has retreated to safety in the cleft in the rock where it is not amenable to falsification. Everything that has been debunked is automatically no longer within the scope of inspiration, though whatever is left, is. If it proves to be wrong, then such subject matter turns out retrospectively to fall outside the scope of revelation. God cared only about the specifically religious stuff.

If God inspired a Bible containing ancient, erroneous science and myths, then he was just trying to document what ancient people thought. He wanted us to discover modern physics, astronomy, medicine, etc., for ourselves. But then have we not embraced the view of old-time Deism? There had been no revelation from God because he had already endowed us with "reason, the only oracle of man," as Ethan Allen entitled his 1854 book. Such a doctrine of "inspiration," being thus unfalsifiable, is meaningless, compatible with any and every state of affairs. We can no longer specify what it means by indicating its boundaries, stipulating what would be inconsistent with it. So we don't know what to say it is claiming!

The old-time understanding of inspiration was falsifiable, based on the Bible being free of either factual errors or internal contradictions. It was a meaningful standard, but one that the Bible failed long ago. So inspiration is revised such that nothing the Bible says—no matter how factually erroneous, mythic, or self-contradictory—can be judged incompatible with inspiration. It makes inspiration into glossolalia, possibly sounding edifying to some, but not *meaning anything* anymore. This approach to inspiration usually buttresses a hermeneutic of demythologizing, a la Rudolf Bultmann, though evangelicals like Enns and Lamoureux carefully restrict it to the Old Testament, as we will see.

Something New Has Been Added!

The doctrine of progressive revelation (or of "accommodation"), focusing on the dead weight of ancient, erroneous thought preserved in the biblical texts, is still a favorite among modern Christian evolutionists. Indeed, they must have it. Peter Enns, for example, characterizes Genesis as "an ancient text designed to address ancient issues *within the scope of ancient ways of understanding origins*" (2012, 36, his emphasis). Labeling the genre of its opening chapters as "myth, legend, supra-historical narrative, story, metaphor, symbolism, archetypal, etc." is beside the point; Genesis simply does not overlap with "the modern scientific investigation of human origins." To think otherwise "is an error in genre discernment" (p. 36).

Enns is thus importing into the Bible Stephen Jay Gould's "non-overlapping magisteria" gimmick. In 1997, Gould proposed that religion and science, *understood rightly* (i.e., as he understood them), do not conflict because religion does not (i.e., should not) make any factual claims. It ought to restrict itself to the far-flung realms of theological speculation and moral calculation. It should, in other words, stay confined to Cloud-Cuckoo Land. Actually, this was nothing new. Theologians like Paul Tillich used to say the very same thing, and were pilloried for it by the rank-and-file believers, just the sort Enns, Lamoureux and their colleagues are now trying to convert to evolutionism.

This stubborn fact remains: As long as people believe that divine interventions once occurred—in the production of various life-forms, the parting of the Red Sea, the resurrection of Jesus—their religion will be making fact-claims that are not immune from disconfirmation by historical or scientific method. Their claims trespass on the territory of science and history, and they cannot expect to slink over the fence unnoticed.

Besides, even if Enns, Jerry Korsmeyer, and Lamoureux have correctly pegged the underlying larger theological agendas beneath the stories of Genesis, this hardly means that the writers of those stories did not think things happened in the way they described. What does it mean for these apologists for theistic evolution to say that the Genesis writers' intent was not to instruct us as to how human and other life originated? We may not want to hear it anymore, but that does not mean they didn't intend to tell us. "You can't *mean* that!"

Another theistic evolutionist (albeit one who did not think to brand the approach with a new name like Lamoureux's "evolutionary creationism" or Francis Collins's snappy "BioLogos"), Kenneth Miller, thinks himself privy to the thoughts of God and what he thought best to reveal and to keep hidden.

> In order to reveal Himself to a desert tribe six thousand years ago, a Creator could hardly have lectured them about DNA and RNA, about gene duplication and allopatric speciation. Instead, knowing exactly what they would understand, He spoke to them in the direct and lyrical language of Genesis. [Miller 1999, 257]

But surely God needn't have gone into that kind of detail to get closer to the truth than Genesis does. Remember, it's not as if contemporary natural philosophers weren't thinking along the lines of a spherical earth, heliocentricity, and even evolution, as the Sophists and the Ionian philosophers did.

Lamoureux takes an even dimmer view of the ancient intellectual capacity: "It is logically impossible for finite minds to understand an infinite Divine Being unless He descends to our human level" (2008, 270). Does that include germ theory and heliocentricity? If so, if they are intrinsically beyond human ken, then how is it we can understand these things today?

> When revealing to the early Hebrews that God created the world and their community, the Holy Spirit descended to their level of understanding and employed their scientific and historical categories in order to communicate as effectively as possible. [Lamoureux 2008, 19]

Communicate *what*? Misinformation? Lamoureux indulges himself in a bit of Reader Response criticism in order, he thinks, to vindicate his approach. "The power of the eternal Truths in these chapters [Genesis 1-11] is proven by the countless lives that have been impacted and changed in every generation" (p. 270). But surely it was not the blurry abstractions to which Lamoureux reduces these tales that so affected the generations, but rather the particulars of the stories which they believed true and he does not. Moreover, one must suspect that Lamoureux is really thinking of the overarching story of Christian salvation into whose service the stories are pressed. Nobody is getting baptized in the Holy Ghost, receiving Jesus as savior, or even accepting the existence of God by reading the stories of Cain, Abel, Lamech, and Enoch.

Theistic evolution "accommodationists" repeatedly argue that to have corrected the assumptions of the ancients would have produced hopeless confusion. Yes, Lamoureux says of the Bible's references to human infertility (always the woman's fault), *of course* it was "within the Holy Spirit's power to reveal the actual cause of barrenness. When applicable, he could have stated that flagella motility in sperm cells was poor." But, he claims, no ancient person would have understood that. "Instead, God reveals at the comprehension level of His people in order not to confuse or distract them" (2008, 141). Similarly, the "Lord 'used' the common idea of a causal connection between demonic activities and being deaf, mute, or crippled" (p. 143).

But this just trades the problem of ancient information overload for another one. Think of the total chaos that has resulted lo these many years from the misleading preservation of all that ancient nonsense as part of scripture, when it supposedly was never the point in the first place. God wanted to spare the ancients confusion—but not us!

All that stuff was about "God answering prayer and fulfilling His promises," the "Message of Faith," Lamoureux says. "No Christian doctor today studies the Word of God" to understand his patients' infertility problems (p. 141). Well, then, why should that doctor study "the Word" to understand his *salvation* problems, which he wouldn't even be worrying about if it weren't for the condemnations of humankind made by the very same book?

And, come to think of it, why should God or Christ *not* have revealed a good bit about modern medicine and its administration to his ancient, suffering people? And while he was at it, if he had really wanted the Great Commission fulfilled lickety split, why not give the apostles the blueprint for the printing press, as he once gave the plans for the ark to Noah? He could have jump-started the Reformation by more than a thousand years, just like that. These are serious questions showing how the rationalizations of the evangelical evolutionists, as with all apologists for the Bible, are mere *ad hoc* hypotheses to cover the biblical butt.

For example, Francis Collins simply declaims that the "intention of the Bible was (and is) to reveal the nature of God to humankind." His basis for that is not some superior insight into the likely intentions of ancient writers. It is simply what he needs to be true, the excuse that allows him to ask, "Would it have served God's purposes thirty-four hundred years ago to lecture to His people about radioactive decay, geologic strata, and DNA?" (Collins 2006, 175).

If the goal of simplicity and clarity were the real issue rather than just the reductionistic bias of the Christian biologists, why did God encrust this simple message amid endless tedium about whether to use the hides of manatees versus seals in the curtains of the Tabernacle? Why all the picky stipulations of which insect species were on the kosher menu? Whether the Passover lamb had to be eaten roasted, boiled, or raw? To ask such questions is by no means to make fun of the niceties of the Torah. As Jacob Neusner has shown, all such elements form part of an exhaustive cultural-philosophical system. But for evangelical evolutionists who imagine the whole Bible was written to lead readers to a personal relationship with Christ, it is all equally trivial.

Consider what the reaction of Christians across the whole spectrum of literalism and liberalism would have been if the Bible had actually said that life forms evolved from each other. No one would have had a problem with it! Indeed, believers would be cheering the scientific validation of their holy book, and we wouldn't be seeing the foot-shuffling excuses being made for God, like Lamoureux's assertion that the "Lord comes down to the level of ancient peoples by employing their view of medicine" (2008, 144). He finds it "reasonable to suggest that God, as a loving Father, came down to the level of the ancient Hebrews and spoke using the concepts of science and history that they understood" (p. 13). But then the Bible fails to serve as a revelation in any real sense; it takes *and leaves* people right where they are.

And if Lamoureux is right, wouldn't people be even more obtuse when it comes to ethics and theology? Did God leave them where they were there, too? Where is there any "revelation"?

The Bible as Ventriloquist Dummy

Theistic Evolutionist (or "evolutionary creationist," as he prefers to call himself, perhaps hoping to avoid readers' suspicions associated with the old brand name) Lamoureux correctly discerns what we are calling hermeneutical ventriloquism on the part of Bernard Ramm and other neo-evangelicals:

> Progressive creationists embrace a two-way relationship between Scripture and science ... The events in Gen 1-11 shape the explanation of biological data, in particular human origins; and discoveries in cosmology and geology direct the interpretation of these opening Biblical chapters. [Lamoureux 2008, 27]

The approach, however, is a generation or two older than that, for even the old Princeton theologians already advocated it. Charles Hodge (1797-1878) wrote:

> The proposition that the Bible must be interpreted by science is all but self-evident. Nature is as truly a revelation of God as the Bible, and we only interpret the Word of God by the Word of God when we interpret the Bible by science.... For five thousand years the Church [including Israel] understood the Bible to teach that the earth stood still in space, and that the sun and stars revolved around it. Science has demonstrated that this is not true. Shall we go on to interpret the Bible so as to make it teach the falsehood that the sun moves round the earth, or shall we interpret it by science and make the two harmonize? [from Noll and Livingstone 2003, 66]

But it is older even than that, having been espoused by no less than Galileo himself. According to Giberson and Collins, "Galileo's task was . . . to use the new science of his day to *remove a misunderstanding* about what the Bible was teaching about the motion of the earth" (2011, 88). They "suggest that Darwin's theory of evolution . . . offers the same sort of help in understanding the Genesis creation story" (p. 89). The task, as they see it, is "of developing dramatic new understandings of Scripture in response to the advance of science" (p. 91).

Bernard Ramm certainly proved himself an able ventriloquist; his Bible received its inspiration not so much from the Holy Ghost as from modern science.

> [I]t is certainly obvious that if there is any truth at all in the time processes indicated by modern geology then creation at 4004 B.C. is an impossibility and we either must admit a straight contradiction between Genesis and geology or (i) try to overturn geology completely or (ii) seek a different interpretation of Genesis. We believe the evidence for the antiquity of the earth from all geological methods of measurement to be overwhelming, and we certainly seek another interpretation of Genesis. [p. 123]

But what good is such a Bible? You have to wait on scientists to tell you what it *means*?

Ramm defended theistic evolution as a legitimate position for evangelicals to hold. He himself preferred progressive creation: the belief that one species did not evolve into another, taking great amounts of time to do it, but rather that God specially created, presumably out of thin air, new species at great intervals. Is there any apparent reason he might want to take so long? Yes, of course: to enable Ramm to believe in special creation despite the vastly long history of the earth as revealed in the geological record! Ramm waited to hear what science had to say, then pretended that the Bible had said the same thing all the time. We failed to recognize what it was saying until science cleared everything up for us.

Such hindsight hermeneutics are alive and well in today's crop of theistic evolutionists. The track record of science forces them to uneasily acknowledge that we can't quite take the Bible at face value:

> It may be that we should refine our reading of the Bible writer in light of the scientist's opinions; but sooner or later we will have to decide whether the Bible can actually refer to real persons and events or not. [Collins 2011, 108]

> In the light of what scientists can tell us about human origins, modern Catholic biblical scholars recognize that the origin stories in Genesis 1-11 are not meant to be understood as historical fact. [Korsmeyer 1998, 120-21]

> [S]cience-savvy Christians, who hope to find scientific support for the Bible, are faced with the task of continually reinterpreting texts to make them consistent with the latest scientific discoveries. [Cunningham 2010, 152]

Lamoureux accepts this sliding scale, too, explicitly stating that "science has *hermenutical primacy* over Scripture in passages dealing with the structure, operation, and origin of the physical world" (Lamoureux 2008, 161, our emphasis). At least on the face of it, he is willing to acknowledge that science is the rock on which any Church of Continued Bible Relevance must be built.

Limited Inerrancy

In a number of ways, the approach of today's evangelical evolutionists recalls certain controversial trends in the thinking of the "limited inerrancy" debates of the 1970s and 80s. For a detailed review of that topic and the crisis of biblical authority that has continued apace for Evangelical Christianity in many areas other than just evolution, we refer you to *Inerrant the Wind: The Evangelical Crisis of Biblical Authority*, a book written by one of your co-authors (Price 2009).

One of the main characters in that drama was Daniel P. Fuller, who raised a storm of pious outrage when he suggested that the Bible need be deemed inerrant only when its statements bore on (unverifiable) matters of revelation and religion, not on (verifiable) matters of history and science (Price 2009, 86-89). Thus there's no problem with whether the mustard seed or the orchid seed is the smallest. Sometimes our theistic evolutionists, e.g., Denis Lamoureux, seem to echo that approach. At other times, they sound like Clark H. Pinnock who rejected but only slightly modified Fuller's approach.

Pinnock argued that any and every biblical *assertion* must be judged inerrant and infallible, but not so the background *assumptions* of the ancient writers. Jesus (or Mark) might have incorrectly assumed the mustard seed to be the smallest, but he was not *asserting* it. "The Lord's purpose in this parable is not to teach botany," Lamoureux assures us, but rather to use "the science-of-the-day in order to reveal an inerrant prophecy about the kingdom of God" (2008, 137).

And yet the two theories of Fuller and Pinnock reduce to the same thing in the final analysis, since Pinnock certainly was not willing to entertain the possibility that the biblical writers' *religious* assumptions might be erroneous, e.g., that God loved the human race. No, only secular subject matter might have been negotiated away. Listen now to Lamoureux:

> [P]assages in the Bible referring to the physical world feature both a *Message of Faith* and an *incidental ancient science*. According to this interpretive principle, biblical inerrancy and infallibility rest in the Divine Theology, and not in statements referring to nature. [2008, 110]

Here is the confusion we find in Pinnock and Fuller: Lamoureux confuses "affirmation" [or Message] versus "incidental ancient scien[tific assumptions]" on the one hand with religious versus natural subject matter on the other. The biblical writers, e.g., the apocalyptic seers, mapping out the heavens, and the Priestly writer, recounting the stages of creation, surely did mean to make assertions about nature. We cannot credit them, though, so we must (as an *ad hoc* maneuver) restrict the scope of inerrancy to the (unverifiable) religious ("revelational") assertions.

But Fuller and Pinnock are not the only limited inerrantists Lamoureux manages to parallel: "In a subtle way, the Bible itself through this contradiction offers evidence that its purpose is not to reveal scientific facts about the earth" (2008, 113-14). Indeed, this rule of thumb is absolutely central to Lamoureux's approach (and he is hardly alone in it). How does he claim to know what in the Bible, though stated there, forms no part of what God and his writing staff really intended us to learn from the Bible? Simply that, if something in the Bible turns out to be mistaken, then *ipso facto*, that cannot be one of the things God wanted us to believe!

This notion was formulated with surprising bluntness by yet another 1970s limited inerrantist, evangelical philosopher Stephen T. Davis. He said we ought to believe whatever the Bible says until we find it is mistaken. And then we go on believing again till we hit the next bump in the road. Apparently he did not accept the maxim "Fool me once, shame on you; fool me twice, shame on me."

But the evangelical evolutionists are probably a bit closer still to Dewey M. Beegle, with whom Clark Pinnock used to debate. Peter Enns suggests that "when we allow the Bible to lead us in our thinking on inspiration, we are compelled to leave room for the ancient writers to reflect and even incorporate their ancient, mistaken cosmologies into their scriptural reflections" (Enns 2012, 95). Beegle, in his notorious book *The Inspiration of Scripture*, urges us to interpret the "claims" of scripture in light of the "phenomena" of scripture, or in other words, to interpret what scripture *says* in terms of what scripture *does* (Beegle 1963).

Beegle was trying to get out from under the completely deductive approach to biblical inspiration and inerrancy propounded by Princeton theologian Benjamin B. Warfield, who had said that the whole thing was as simple as lining up the Bible verses where New Testament writers declare the Old Testament writings to be inspired or "God-breathed," to be perfect, never to be "broken" (like a legal infraction), etc. If "what Scripture says, God says," and God cannot lie, then there can be no errors of any sort in the pages of the Bible. So when we seem to find errors or contradictions, we must strive to explain them away, because we simply cannot be seeing what we think we are seeing. This might be called a "Christian Science" approach to the Bible, dismissing manifest marks of biblical error as an illusion just as Mary Baker Eddy declared all bodily ailments to be illusions. One tries to just wish the problems away.

Beegle observed that, before we go off pontificating, we might want to check our inferences. Are we so sure we know precisely what God was doing when he inspired his Bible? How can we know until we examine the Bible he inspired? And if it seems to contain contradictions and errors, then we had best get busy and frame a doctrine of inspiration that can accommodate these things. And if we are so sure (as Warfield was) that we can read the mind of God, then why have a Bible at all? Who needs it?

What Enns, Lamoureux, and company seem to be doing is emergency Bible budget cutting, trying to discard line items that have too high an intellectual cost while maintaining a non-negotiable core of infallibility. How do we know what statements are infallible in the inspired Bible? Many are called infallible, but few are chosen for belief. All that stuff about ancient science and pre-science in scripture? Well, it's wrong, so it can't be intended for us to believe: "These stories were not written to speak of 'origins' as we might think of them today (in a natural-science sense). They were written to say something of God and Israel's place in the world as God's chosen people" (Enns 2012, 5).

So why should the rest of it, the stuff we conveniently can never empirically test out anyway, still be believed implicitly? There is no good reason, except that failing to do so means giving up one's claim to be a Christian.

Scripture's "Mega-purpose"

A slight variation on the theme of a Bible that is reliable only when it comes to "revelational" subjects, matters we could never discover by ourselves, is the "central message" or "mega-purpose" approach according to which it is not *every* religious statement that we must receive as infallible, but only those strictly germane to the broader plan of salvation.

Korsmeyer figures that "what the authors of scripture were inspired to write were human stories, history, poetry and prose, sagas, legends, myths and laws, which contain that which God wishes them to contain for the purpose of our salvation" (1998, 49). But is that all God wanted to reveal in scripture?

There are plenty of religious assertions in the New Testament that would seem to fall outside the tunnel-vision scope of "the message of salvation." To board that heaven-bound train, do you really need to know (or, more to the point, *believe*) that Satan and the archangel Michael disputed over possession of Moses' corpse, as in Jude 9? Or that Elisha caused a sunken axe-head to float back to the surface (2 Kings 6:6)? Or that Jesus walked on the surface of the lake? Only if you preach the gospel of inerrancy. But it's supposed to be Jesus who is your savior, not the Bible.

This much might be painful for Christians to admit, but we can tighten the screw a turn more. Do you really need to believe in the Second Coming of Christ to be saved? The Virgin Birth? The Real Presence in the Eucharist? Must you believe in justification by faith in order to be justified by faith? How about the distinctions between your denomination and everybody else, specific doctrines that often don't even have much support in the Bible?

Obviously the problem gets even more acute when we factor the Old Testament into the equation. Does it contain *any* Christian message of salvation *at all*, without having one read into it? Isn't the Old Testament taken up with questions of prosperity, blessing, and health in this present life? Israelite independence on this side of the grave? If there is any kind of an afterlife, doesn't everyone land in the shadows of *Sheol*?

Well, according to Korsmeyer, "the Bible does not teach science or the origin of life, but does teach about our relationship to God and how to live. Once one has grasped the nature of Scripture, and let go of the idea that all its stories are literally true, the biggest problem, for Christians, is removed" (1998, 71-72). He only thinks so because it is a fallback position since the Bible turned out to be so wrong on these other issues, which it certainly does seem to have wanted to address! This is making virtue of necessity and grossly oversimplifying the Bible. And these people call Richard Dawkins and Daniel Dennett "reductionistic"?

Lamoureux agrees with Korsmeyer:

> Instead of offering the facts-of-history, the Holy Spirit through these ancient historical paradigms reveals simply that God created men

and women in His Image, they have all sinned, He judges them for sin, and Israel is His chosen vessel for blessing humankind. Therefore, *Gen 1-11 does not reveal actual past events in the creation of humanity, the entrance of sin into the world, the judgment of the first sinners, or the origin of the Hebrews and all other nations.* [Lamoureux 2008, 255]

So why accept the supposed attendant "theological truths"? It's like getting all the problems wrong but still getting an "A" on the exam.

Lamoureux shows his tin ear for the music of the Torah: Since he doesn't give a damn about such "trivia," neither can God. "The purpose of these passages in the book of Leviticus is not to reveal taxonomy. Rather, the practice of dietary laws was part of God's plan in separating the Hebrews from the pagan cultures around them" (Lamoureux 2008, 136). This argument is precisely the same as that used by ancient Diaspora Jews like the author of the *Epistle of Aristaeus*, who is a bit embarrassed about the micro-stipulations of Jewish *kashrut* and claims the regulations on food, etc., were entirely arbitrary, just aimed at building higher walls between Jews and their paganism-infected neighbors. This utilitarian approach is aimed at conceding to the scornful Gentile the silliness of the Jewish laws on one level and trying to rationalize them on another. Same with Lamoureux, who is certainly giving short shrift to what centuries of scribes and rabbis deemed pretty important. It seems unimportant only to a born-again Christian oblivious of these matters.

According to Peter Enns, "Many Christian readers will conclude that a doctrine of inspiration does not require 'guarding' the biblical authors from saying things that reflect a faulty ancient cosmology" (2012, 94). You bet! What else are they going to do? It's a strategy of cutting one's losses, nothing more.

Such interpreters are intellectually guilty fundamentalists looking for clever ways of discounting what the Bible says, things that they don't want to hear. This is the whole debate over limited inerrancy in a nutshell. The scope of what the Bible "infallibly" teaches is reduced (conveniently!) to just those areas safe from proof or disproof, just like the supposedly resurrected Jesus appearing only in closed rooms and on remote sea shores to small groups of his friends. This tactic removes faith from disconfirmation but at the cost of making it completely arbitrary.

The Island of Doctor Lamoureux

As one of the most important of the 1950s neo-evangelicals, Bernard Ramm could never in his wildest nightmares have anticipated the direction taken by today's professed evangelicals Denis Lamoureux and Peter Enns. A "typical religious liberal," he wrote, "would negate most of the theological structure of the New Testament on the grounds that its theological structure is that of the world view of ancient antiquity." Ramm said those liberals "write too much off as cultural" (1954, 53), and offered the example of "Bultmann who says that the entire New Testament (in his case) is prescientific and therefore mythical" (p. 81).

Ramm himself cut the pie into two slices with the following "general guide." First, "Whatever in Scripture is in direct reference to natural things is most likely [expressed] in terms of the prevailing cultural concepts." Then, "whatever is directly theological or didactic is most likely trans-cultural" (p. 53). How convenient! The distinction is simply an apologetic tactic, to allow him to shorten the line of defense. There is no other rationale; he just needs it to be so.

Lamoureux, Enns, and their fellows go way beyond this, paralleling Bultmann's Bible-mythicism in startling detail. According to Lamoureux, we need to cultivate "an awareness of the ancient intellectual categories in Gen 3, and practice in looking beyond these to what the Holy Spirit intended to reveal." The Spirit "uses the intellectual categories of the ancient Hebrews and neighboring nations to reveal the reality, manifestation, and consequence of human sin" (p. 205). In fact, Lamoureux disclaims, "there never was an idyllic paradise at the beginning of time" (p. 201). Likewise, "the fall of humanity into sin did not occur as stated in the Bible, and the cosmic fall never happened" (p. 205).

Pinnock at least made the biblical thought forms the theological fire wall. Lamoureux and Enns are going significantly farther than this in the direction of Bultmann. Conrad Hyers, too:

> The intent of the use of the seven-day week in discussing origins is not to provide numerical, chronological, and historical information—in which case these materials might be said to be in conflict with modern scientific accounts—but to make the religious affirmation that the totality of the universe has its origin in God, who is the one supreme power behind and within the universe, and whose works are (as the text concludes) "very good." The number that corresponds to this affirmation of totality and "very goodness" is the number seven. [Hyers 1984, 25]

So is the ostensibly more conservative C. John Collins, who avers that, like the Mesopotamian stories of prehistory, "Genesis aims to tell the true story of origins." But the parallels between them "also implies that there are likely to be figurative elements and literary conventions that should make us very wary of being too literalistic in our reading" (Collins 2011, 58). Is he not having his cake and eating it, too? And awarding himself a license for rewriting the story?

It would appear to be the same sort of move as that made by the old Protestant Rationalists. They, paradoxically, sought to vindicate the stories of the Bible by removing the supernatural, miraculous causation. They wanted to supply a more scientifically acceptable version that would preserve the basic events as inerrantly accurate. Jesus was really crucified and really appeared alive subsequently to his disciples, but after having merely passed out on the cross. C. John Collins says:

> The historicity of Adam is assumed in the genealogies of 1 Chronicles 1:1 and Luke 3:38. Similarly, although the style of telling the story may leave us uncertain on the exact details of the process by which Adam's body was formed, and whether the two trees were actual trees, and whether the Evil One's actual mouthpiece was a talking snake, we nevertheless can discern that the author intends us to see the disobedience of this couple as the reason for sin in the world. [2011, 66]

The whole argument is like that of Preterism, an approach to reading the apocalyptic technicolor of Mark 13 and the Book of Revelation in such a way as to dissolve (really, to demythologize) the terms into mere metaphor so as to salvage the bare "event." Oh yes, the Parousia of Christ *did* occur within the lifetime of Jesus' contemporaries, but it was just a colorful way of talking about the fall of Jerusalem. Yeah, that's the ticket. Same with the Garden of Eden: Well, I guess there wasn't really a talking snake, heh heh, but . . .

C. John Collins tests the water with a Rationalistic toe, but Enns, like Lamoureux, takes the Bultmannian high dive: "The fact that biblical authors wrote these things [e.g., the multiple heavens cosmology] down does not mean they are accurate descriptions of physical reality. Rather, they simply reflect ancient ways of thinking" (Enns 2012, 94). In fact, Enns's approach to scripture exactly parallels his approach to evolution: God "inspired" a book that looks the same as if he *hadn't* inspired it (as witness the use of legends, accommodation, ancient thought categories), just as theistic evolution has God "creating" the world by a process that appears to have been without direction.

Consider an analogy to the frustrating "peace process" between the State of Israel and her Arab tormentors. It is a constant see-sawing of offers and counter-offers, and the question is: how much territory are you prepared to surrender for the sake of peace? In Lamoureux's case that peace amounts to reduction of cognitive dissonance. Young-earth creationists are like the PLO or Hamas in that they won't give up anything. The terrorists just want Jews dead and Israel wiped off the map. Old-earth creationists like Bernard Ramm are like conservative elements in the Israeli government, unwilling to make genuine sacrifices for an accord. But "evolutionary creationists" are like the suicidal Israeli liberals, willing to give away almost the whole store for empty and ultimately doomed promises of peace from an implacable enemy still eager to devour them.

In plain words, this means that Lamoureux is willing to admit that the first eleven chapters of Genesis are poetry and embody an ancient and untenable understanding of history. It is all mere parable seeking to communicate the "theological," "spiritual" truths that the Creation is good, that humans bear the "image" of God, and that we are fallen and in need of redemption.

Are We Not Bultmenn?

Lamoureux and Enns are doing with Genesis precisely what Bultmann was doing with the gospels: ruthless demythologizing. What's the difference? Both argue that the revelation was imparted in terms of the outmoded worldview of the ancients, and that for modern Christians to require an embrace of that superannuated worldview would be to set up a "false stumbling block" in the path to the gospel. Lamoureux laments that "a stumbling block has been placed between [unbelieving scientists] and Jesus by scientific creationists" (2008, 25). Of course, he refers to evolution deniers like Henry Morris, Duane T. Gish, and Ken Ham.

Note the evangelistic-apologetic agenda here. Theistic evolutionism is an example (though not one he would have endorsed) of what Francis A. Schaeffer used to call "pre-evangelism." But can we imagine secular scientists finding this pathetic "me too" appeal any more attractive?

At any rate, it is hardly a surprise when we recall that Ramm had exactly the same concern: "It is our purpose to show that there is nothing between the soul of a scientist and Jesus Christ save the disposition of the scientist himself" (1954, 43). For him, the evangelical attitude toward science was basically just a function of public relations, making evangelical Christianity respectable so sophisticated unbelievers could take it seriously and get saved. Not exactly the usual motivation for scientific research.

But it is exactly the motive underlying Bultmann's demythologizing program. It is easy for conservatives to forget that Bultmann's primary concern was evangelistic and apologetic. He knew the war was lost when it came to maintaining and defending both the ancient world-picture and supernaturalism (itself a kind of pre-science, an attempt to explain natural phenomena by reference to unseen personal wills). And he asked whether there might be some remaining gospel truth relevant to modern man. And he concluded that there was. Christianity offered an existential posture of authentic self-understanding *vis a vis* the world and the transcendent God. When we interpret the ancient stories of the New Testament by demythologizing them, this becomes apparent, for all myths are narrative representations of the culture's self-understanding.

Here is a famous passage from Bultmann's seminal essay, "New Testament and Mythology":

> Can Christian preaching expect modern man *to accept the mythical view of the world as true?* To do so would be both senseless and impossible. It would be senseless because there is nothing specifically Christian in the mythical view of the world as such. It is simply the cosmology of a pre-scientific age .... For all our thinking to-day is shaped irrevocably by modern science. A blind acceptance of the New Testament mythology would be arbitrary, and to press for its acceptance as an article of faith would be to reduce faith to works ... The only honest way of reciting the creeds is to strip the mythological framework from the truth they enshrine ... No one who is old enough to think for himself supposes that God lives in a local heaven. There is no longer any heaven in the traditional sense of the word. The same applies to hell in the sense of a mythical underworld beneath our feet. And if this is so, the story of Christ's descent into hell and of his Ascension into heaven is done with. We can no longer look for the return of the Son of Man on the clouds of heaven ... It is impossible to use electric light and the wireless and to avail ourselves of modern medical and surgical discoveries, and at the same time to believe in the New Testament world of spirits and miracles ... Again, the biblical doctrine that *death is the punishment of sin* is equally abhorrent to naturalism and idealism, since they both regard death as a simple and necessary process of nature ... what a primitive mythology it is, that a divine Being should become incarnate, and atone for the sins of men through his own blood! [Bultmann 1961, 3-7]

Now, Lamoureux does not hesitate to stick his neck out (like Lamarck's giraffes!) and use the word "myth" for certain biblical tales: "Genesis 6:1-4 features a motif that was part of the intellectual make-up of the ancient Near East. Dare I use the word 'myth' to describe the literary genre of this passage?" (2008, 216). Like Bultmann, too, he is happy to abandon the ancient world picture. He admits that Paul "uses the 3-tier universe in one of the most significant passages in the New Testament—the Kenosis of Jesus (Phil 2:5-11)" (p. 274). But then why not admit with Bultmann that divine incarnation was fully as much of a piece of ancient window-dressing as the three-decker universe—and as negligible?

Ditto for the business about an eschatological judgment of souls: "On the day of judgment, God will call on us to give an account for our conduct" (Lamoureux 2008, 283). Why is this not to be dismissed as a piece of ancient apocalyptic conceptuality? Just because Lamoureux wants to go so far and no farther. Biblical protology can be thrown to the wolves because it gets in the way of what means more to Lamoureux, namely accepting the facts of science. But biblical eschatology must be maintained because he wants that ticket to heaven. That is even more important to him than science.

Again like Bultmann, Lamoureux realizes he does not live in the spirit-and-miracle world of the New Testament when he assures his readers that, *of course*,

> no Christian doctor, nurse, or other healthcare professional believes that evil spirits are the cause of epileptic seizures, the loss of sight or hearing, and the inability to talk or stand upright. For the most part, medical science has proven that these disabilities and diseases have physical mechanisms that can be treated by physical means. Spiritual causes, such as demonic activity within a person, are not considered in medical diagnosis and treatment, and Christian medical staff in no way feel they are compromising their faith. [pp. 143-44]

Backtracking

We look over all this with a twinge of "been there, done that" understanding. We've both done our time in evangelical and fundamentalist Christianity, warding off the cognitive dissonance that arises with the realization of so much error in the founding text of one's cherished faith. As the yawning abyss of apostasy looms closer, one desperately alternates between stomping on the intellectual brakes and searching for an as-yet hidden way out. Daniel Dennett articulates what many doubting believers will not:

> Without ever being frankly aware that a cherished ideal is endangered in some way, people may be strongly moved by a nameless dread, the sinking sense of a loss of conviction, a threat intuited but not articulated that needs to be countered vigorously. [2006, 205]

The discomfort is as evident for institutions as for individuals, except that it is a rare church that will ever say, "You know what? We were wrong about all this." That would open the door for the everyday faithful to ask some uncomfortable questions about their ecclesiastical leaders. How can you claim infallibility or the guidance of the Holy Spirit when you were so demonstrably wrong about the very nature of human existence, and remained so for such a long time? And why should we listen to you now, as you fuss about things like contraception and which movies we are allowed to watch?

Backtracking is a tricky business in theology, even without the unsettling prospect of taking a red pencil to the inerrant (or at least "inspired") Word of God. It must be done slowly and as quietly as possible, preferably one funeral at a time.[43] "Like sausage-making and the crafting of legislation in a democracy," Dennett observes, "creed revision is a process that is upsetting to watch too closely, so it is no wonder that the fog of mystery descends so gracefully over it" (p. 205).

White offers a revealing comparison in his survey of "the retreat of the church after its victory over Galileo" (1895, Ch. 3, §6). For Protestant theologians, he says, the retreat

> was not difficult. A little skilful warping of Scripture, a little skilful use of that time-honoured phrase, attributed to Cardinal Baronius, that the Bible is given to teach us, not how the heavens go, but how men go to heaven, and a free use of explosive rhetoric against the pursuing army of scientists, sufficed. [§6]

The Catholic church had a harder time of it, having to deal with the fact that "papal infallibility was committed fully and irrevocably against the double revolution of the earth." So, the

> apologetic army, reviving an idea which the popes and the Church had spurned for centuries, declared that the popes *as popes* had never condemned the doctrines of Copernicus and Galileo; that they had condemned them as men simply; that therefore the Church had never been committed to them. [§6]

Nice try.

---

43. This is borrowed from a famous line by Max Planck: "Science advances one funeral at a time." Apparently, creed revision can be an unsettling prospect for some scientists, too. Fortunately, the overall process of science is quite dedicated to the practice.

Branch II: The Creature

When I consider thy heavens, the work of thy fingers, the moon and the stars, which thou hast ordained; What is man, that thou art mindful of him? And the son of man, that thou visitest him?

—*The Book of Psalms*

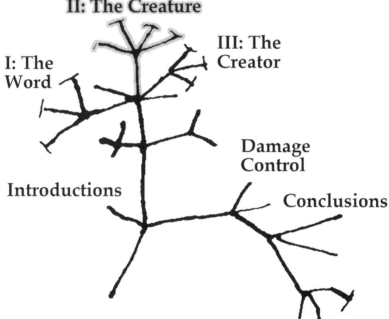

Apex or Ex-Ape?

> *I said to myself concerning the sons of men, "God has surely tested them in order for them to see that they are but beasts." For the fate of the sons of men and the fate of beasts is the same. As one dies so dies the other; indeed, they all have the same breath and there is no advantage for man over beast, for all is vanity. All go to the same place. All came from the dust and all return to the dust. Who knows that the breath of man ascends upward and the breath of the beast descends downward to the earth?*
>
> —The Book of Ecclesiastes (NASB)

The writer of Psalm 8 looks at man's lowly place in the cosmos and is filled with wonder that God would take any thought of him, much less "crown him with glory and majesty" and "put all things under his feet." The Psalmist had it half right. We are mere specks in the universe, even in the biosphere that has evolved on this warm, wet planet we call home. After millions of years of evolution from small, furry tree dwellers, our brains and now our culture have finally given us a unique ability to introspect about ourselves. Now we look in the mirror and see God's image and a crown of exaltation over nature, when the true picture is of a naked ape, an organism containing trillions of cells locked in a symbiotic embrace for the propagation of their genetic legacy—human, viral, and bacterial.

Aren't We Special?

Ever "since Darwin published his world-shattering treatise on the theory of evolution, the case for man's primate origins has been bolstered a thousandfold." So says the neuroscientist V.S. Ramachandran in the opening words of his *The Tell-Tale Brain*, adding that it is now "impossible to seriously refute this point: We are anatomically, neurologically, genetically, physiologically apes" (2011, loc. 344). We may know that on an intellectual level, but it doesn't seem to sink in very deep. "We feel like angels trapped inside the bodies of beasts, forever craving transcendence" (loc. 5369). That feeling comes from the same brain whose advanced capabilities convince us of how special we are. No wonder it's a hard thing to get past, cognitively.

We can't really act like we are just evolved primates, any more than we can avoid thinking in terms of mind-body dualism. That kind of thinking is just part of who we are, as another neuroscientist, Robert Burton, points out.

> A full exposition of the underlying brain mechanisms won't prevent us from seeking larger meanings any more than understanding the big bang theory stops us from wondering what surrounds the universe or came before the beginning. It is our fate. [Burton 2008, loc. 1946]

Sam Harris agrees, though he seems somewhat open to the possibility of extra-neural consciousness and some kind of post-mortem existence (2005, 208), which would be a surprising view for an atheist neuroscientist. Nearly every one of us, he says,

> experiences the duality of subject and object in some measure, and most of us feel it powerfully nearly every moment of our lives. It is scarcely an exaggeration to say that the feeling that we call "I" is one of the most pervasive and salient features of human life: and its effects upon the world, as six [now seven!] billion "selves" pursue diverse and often incompatible ends, rival those that can be ascribed to almost any other phenomenon in nature. [p. 214]

Ramachandran is practical and clear-headed about our evolved and material state, but that does not keep him from experiencing a sense of wonder about "the puzzle of our own unique and marvelous brain" (2011, loc. 362), which

> is made up of atoms that were forged in the hearts of countless, far-flung stars billions of years ago. These particles drifted for eons and light-years until gravity and chance brought them together here, now. These atoms now form a conglomerate—your brain—that cannot only ponder the very stars that gave it birth but can also think about its own ability to think and wonder about its own ability to wonder. [loc. 364]

Its evolution was mostly a gradual one, over many millions of years. During that time, before the "great leap forward" of cultural selection began some tens of thousands of years ago, natural selection's tinkering was

> gradual and piecemeal: a dime-sized expansion of the cortex here, a 5 percent thickening of the fiber tract connecting two structures there, and so on for countless generations. With each new generation, the results of these slight neural improvements were apes who were slightly better at various things: slightly defter at wielding sticks and stones; slightly cleverer at social scheming, wheeling and dealing; slightly more foresightful about the behaviors of game or the portents of weather and season; slightly better at remembering the distant past and seeing connections to the present. [loc. 521]

Even the physical structure of our brains indicates a long evolutionary development, with "the neo cortex, with its capacity for thought" being built "right on top of other structures in the brain." In pointing this out, biologist Paul Ehrlich is not surprised by it, because all

> the old control systems must remain in place and functional while new ones, providing new capabilities, evolve. Regulation of limb movement and blood chemistry remains as essential for our proper functioning as it was for early amphibians and for our rodent-size forebears who ate the eggs of dinosaurs. [2000, 110]

Some of our theistic evolutionists seem to think that we were an inevitable end result of all this, that evolution was part of God's plan to come up with humans who would worship him. We are no accident, says Lamoureux: It "was the Lord's plan and purpose to make people" (2008, 283). "For long centuries," the God of C.S. Lewis "perfected the animal form which was to become the vehicle of humanity and the image of Himself" (from Collins 2006, 208). Giberson and Collins quote Simon Conway Morris's bold surmise that "something like ourselves is an evolutionary inevitability" (2011, 204). Their own thoughts on the subject are more nuanced:

> Scientific evidence can be viewed as compatible with and even supportive of the traditional Christian belief that human beings—or creatures like human beings—are a fully intended part of creation. Since Christians believe that God upholds all of creation from moment to moment, God is the ground or basis for the myriad and subtle nuances of nature responsible for the convergences that give rise to human beings. [p. 205]

Kenneth Miller is circumspect, too, describing a God who was confident that intelligence would emerge and decided to reveal himself when it did (1999, 252-53). The man who was Miller's pope (until his startling resignation and replacement by Pope Francis) said it quite poetically in 1968:

> The clay became man at that moment in which a being for the first time was capable of forming, however dimly, the thought "God." The first Thou that—however stammeringly—was said by human lips to God marks the moment in which spirit arose in the world. Here the Rubicon of anthropogenesis was crossed. [from Horn and Wiedenhofer 2008, loc. 148]

But Professor Ratzinger wasn't being humble about humanity's place in the universe. He said man is *entitled* "to regard himself as the point of reference, at least for the question about himself." Why? Because of "the special relation that man assumes with respect to all the rest of reality" (loc. 96). If our dogs and cats could communicate such an idea to us, we would laugh and scratch them behind the ears, shaking our heads. Miller can't avoid the human conceit, either, referring to God and "His creatures" (1999, 252).

Really, though, no other viewpoint is compatible with the special relationship that Christianity posits between humanity and a God who had his second person at the ready from the beginning of time to save it. Humans were just inevitable, either as part of a carefully crafted plan, or as an expectation that would emerge somehow. And that gives God a lot to do, even working within the constraints of a naturalistic universe and evolution. "If, as Christianity and other religions teach, God created the universe with a special place and plan for humanity," Stenger observes, "then he would have had to step in countless times along the way—every time there was a mutation on the path to *Homo sapiens*, to make sure we evolved" (2009, 230).

It was a long and winding path. Dedicated paleoanthropologists have been uncovering hominid fossils for decades to painstakingly build a picture of our gradual evolution over at least six million years (Gibbons 2007, loc. 3685). As we would expect from what we know about how evolution operates, the fossil evidence indicates a process "of trial and error rather than a 'straight-ahead slog' from primitiveness to modern humans" (loc. 3743).

The relics providing some of the most irrefutable evidence for our evolutionary origins have been found not in the sediment layers of Africa, but in the DNA within the cells of our own bodies. Just **1.2%** of the nucleotides in that DNA are different from the corresponding ones in chimpanzee DNA (Roach et al. 2010, 638). The people with the most genetic variation have their ancestry right where you'd expect to find it: Africa, where the stew of human evolution has been simmering the longest.[44] That is where our ancestors remained, the ancestors of every last one of us, until at least 100,000 years ago (Wilcox 2003, 242).

---

44. "Populations with African ancestry contributed the largest number of [DNA sequence] variants and contained the highest fraction of novel variants, reflecting the greater diversity in African populations. For example, 63% of novel SNPs [single nucleotide polymorphisms] in the low-coverage project and 44% in the exon project were discovered in the African populations, compared to 33% and 22% in the European ancestry populations" (The 1000 Genomes Project Consortium 2010, 1064). Ehrlich says that "the Xhosa people of southern Africa . . . alone retain 90 percent of human genetic variability" (2000, 98).

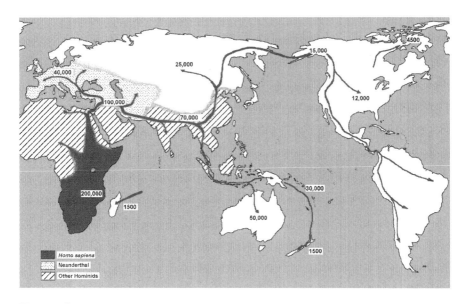

FIGURE 9: Descent and Spread of *Homo sapiens*.

"If human beings were a completely separate creation," asks Kitcher, "why did the creative force find it apt to form our species in the chromosomal image of the great apes?" (2007, 54). We have nearly the same DNA, and we have it packaged much the same way, too. With just a few exceptions, the "chromosomes from humans, chimpanzees, gorillas, and orangutans are highly similar and can be aligned with one another" (Fairbanks 2007, 20). This is remarkable evidence of our common ancestry with the great apes. But it is one of those exceptions, human chromosome 2, that clinches the case.[45]

We have one fewer chromosome than the great apes. Our second longest one matches up with two of theirs, set end to end. And that's exactly what happened; the two chromosomes linked up and formed a single one in us sometime after our lineages diverged. We can tell from an "ancient telomere at the fusion site," which "is now a nonfunctioning relic of evolution embedded in the middle of the chromosome" (Fairbanks 2007, 27).

---

45. The other nine exceptions are inversions of DNA segments between the two species' chromosomes. The "segment's orientation is reversed in one species when compared to its corresponding segment from the other species" (Fairbanks 2007, 89-90). That type of DNA rearrangement isn't as dramatic as the fusion of two chromosomes to form a single human one, but it's also compelling evidence of evolution from a common ancestor.

A telomere is a DNA tip at both ends of all the chromosomes in their species and ours. It has a characteristic sequence of six DNA base pairs repeated 50-100 times. At the exact point on our chromosome 2 where the telomere of one great ape chromosome ends and the telomere of the other chromosome begins, guess what we find? "In this region is a DNA sequence with 158 copies of the tandem repeat found in telomeres" (p. 23). The match isn't perfect, but this is another case where the mismatch is positive evidence, not negative. The differences are mutations, "precisely what we expect if the fusion happened long ago in the remote ancestry of humans. After the fusion event, the repeats no longer functioned as telomeres, so mutations (changes in the DNA sequence) in them had no harmful or beneficial effect" (p. 27).

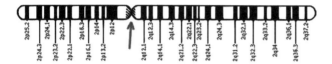

FIGURE 10: Diagram of human chromosome 2 with its telltale fusion site.

Disputing our evolutionary origins in view of this evidence is like coming across two cars smashed together and pretending that there hasn't been an accident. "Well, isn't this an odd eight-wheeled vehicle with a crunched-up midsection! Detroit works in mysterious ways!"

Monkey Business

Much of the opposition to Darwin's new theory arose from the all-too-obvious applicability of evolution to human beings along with all other forms of life. He glossed over the topic with a single sentence in *Origin* ("Light will be thrown on the origin of man and his history," 1859, 490), but faced the topic head-on with his 1871 *Descent of Man* (2nd ed. 1888). One of the main objections he dealt with was, as he titles *Descent*'s third chapter, the "comparison of the mental powers of man and the lower animals." He gives one example after another of non-human animals exhibiting similar behaviors to those we cherish as "human," especially the other primates:

> All have the same senses, intuitions, and sensations,—similar passions, affections, and emotions, even the more complex ones, such as jealousy, suspicion, emulation, gratitude, and magnanimity; they practise deceit and are revengeful; they are sometimes susceptible to ridicule, and even have a sense of humour; they feel wonder and curiosity; they possess the same faculties of imitation, attention, deliberation, choice, memory, imagination, the association of ideas,

and reason, though in very different degrees. The individuals of the same species graduate in intellect from absolute imbecility to high excellence. [Darwin 1888, 120]

There have been some impressive examples of such "human" intellectual attributes in our primate relatives, including the ability to reason. Darwin cautions us not to make too much of our supposedly innate abilities in that department, noting "how little can the hard-worked wife of a degraded Australian savage, who uses very few abstract words, and cannot count above four, exert her self-consciousness, or reflect on the nature of her own existence" (1888, 127-28). He relates one case from Johann Rengger, "a most careful observer." Rengger gave monkeys lumps of sugar wrapped up in paper, sometimes with "a live wasp in the paper, so that in hastily unfolding it they got stung." The monkeys were quick learners; "after this had *once* happened, they always first held the packet to their ears to detect any movement within" (p. 118).

A more recent example is of a bonobo who captured a bird and was then urged by her keeper to let it go. She did so by climbing as high as she could up a tree, unfolding the stunned bird's wings and spreading them open, and "throwing the bird as hard as she could" (from de Waal 2006, 31). This shows a special kind of reasoning at work: adopting the viewpoint of another being (p. 30). "Having seen birds in flight many times," this bonobo "seemed to have a notion of what would be good for a bird" (p. 31). It is also evidence of another attribute we are all too eager to claim for ourselves, empathy.

All of this was demonstrated by an old bonobo at the San Diego zoo when a moat had been drained for cleaning.

> After having scrubbed the moat and released the apes, the keepers went to turn on the valve to refill it with water when all of a sudden the old male, Kakowet, came to their window, screaming and frantically waving his arms so as to catch their attention. After so many years, he was familiar with the cleaning routine. As it turned out, several young bonobos had entered the dry moat but were unable to get out. The keepers provided a ladder. All bonobos got out except for the smallest one, who was pulled up by Kakowet himself. [from de Waal 2006, 71]

Kakowet understood that the young folks were about to drown, and he was not going to let that happen. With a heroic action that involved some impressive abstract reasoning, he shows us that one does not need to be human to be humane.

Heroic concern for others is also evident in a study done in the 1960s. The subjects conducted themselves a lot more nobly than their human experimenters, who

> found that rhesus monkeys refuse to pull a chain that delivers food to themselves if doing so shocks a companion. One monkey stopped pulling for five days, and another one for twelve days after witnessing shock delivery to a companion. These monkeys were literally starving themselves to avoid inflicting pain upon another. [de Waal 2006, 29]

Francis Collins cites the "inner voice that causes me to feel compelled to jump into the river to try to save a drowning stranger, even if I'm not a good swimmer and may myself die in the effort" (2006, 28). Laudable, yes, but no different from the chimpanzees observed by Jane Goodall:

> Chimpanzees cannot swim and, unless they are rescued, will drown if they fall into deep water. Despite this, individuals have sometimes made heroic efforts to save companions from drowning—and were sometimes successful. One adult male lost his life as he tried to rescue a small infant whose incompetent mother had allowed it to fall into the water. [from de Waal 2006, 33]

As Darwin did with the limitations of human reasoning powers, it is important to take an honest look at ourselves when it comes to morality. Collins claims that "the concept of right and wrong appears to be universal among all members of the human species" (2006, 23). In response to that, the blood of Abel cries out from the ground, along with countless other innocents who have been butchered over the sordid history of humankind. Just listing a few names from recent history (the ancients were by no means any gentler) should be enough to make the point: Armenia, Buchenwald, Cambodia, Rwanda.

Do Unto Others

Despite warning his fellow believers against the God of the Gaps, Collins maintains that "DNA sequence alone, even if accompanied by a vast trove of data on biological function, will never explain certain special human attributes, such as the knowledge of the Moral Law and the universal search for God" (2006, 140). Actually, science already has provided ample explanation of those attributes, without even needing to peer into the DNA that codes for them.

The "universal search for God" certainly is unique to humans, but that proves less about God's existence than Collins's personal wish to feel like a special creation. Only humans have evolved the mental and vocal capacity that is needed to communicate such an abstract concept as "searching" for an invisible entity. One could just as easily postulate that there is no God because only members of the one species that evolved religion feel any need for a divine connection. After all, the God that Collins defends was so concerned about having his name declared throughout all the earth as to inflict plagues on an entire nation for that express purpose (Exod. 9:16). Why wouldn't he bother to induce some animals to show some sign of fealty to him, perhaps by digging holes for their dead instead of leaving them to scavengers? It's no more specious an argument than the one advanced by Collins.

Even among humans, we have counter-examples to this supposedly universal sense of divine longing. Darwin says he and his companions who traveled to Tierra del Fuego on the Beagle "could never discover that the Fuegians believed in what we should call a God, or practised any religious rites" (1888, 145). The Pirahã Indians of the Amazon don't even have a word for God, according to the linguist and former missionary (now atheist) who lived among them, Daniel Everett (2008, loc. 4720). None of them has ever been known to convert, despite more than two centuries of missionary efforts (loc. 4814). "They have no craving for truth as a transcendental reality. Indeed, the concept has no place in their values. Truth to the Pirahãs is catching a fish, rowing a canoe, laughing with your children, loving your brother, dying of malaria" (loc. 4879).

In "Peekaboo Deity," we discuss our evolved propensity to see agency behind natural events. Isn't that what's really behind any innate drive we have to "search for God"? What lends the whole enterprise its air of transcendent duty is religion, and we will have something to say about the evolutionary origins of that later, too, in "The Memes Shall Inherit the Earth."

Now, let's turn to this "Moral Law," our supposedly God-given sense of right and wrong. This is strongly associated with altruism; when we talk about doing the right thing, we usually mean doing right by others. As a social species, humans have a taboo against selfishness, which is

> almost the definition of vice. Murder, theft, rape and fraud are considered crimes of great importance because they are selfish or spiteful acts that are committed for the benefit of the actor and the detriment of the victim. In contrast, virtue is, almost by definition, the greater good of the group. Those virtues (such as thrift and abstinence) that are not directly altruistic in their motivation are few and obscure. The conspicuously virtuous things we all praise—cooperation, altruism, generosity, sympathy, kindness, selflessness—are all unambiguously concerned with the welfare of others. [Ridley 1996, 38]

Thus Collins got a strong feeling of "why it seemed so right" for him to be selflessly serving a young African patient, and in his mind, "nothing about the evolutionary explanations for human behavior could account for" it (2006, 217). But there *is* an explanation, a simple one provided by the relentless calculus of natural selection: "Evolution favors animals that assist each other if by doing so they achieve long-term benefits of greater value than the benefits derived from going it alone and competing with others" (de Waal 2006, 13). Cooperation happens all the time: between species, within species, and—this may be a surprising realization—even within organisms. It's happening inside you, right now.

Evolutionary theory says that all of the body's structures and functions are ultimately intended for the survival of its genes in succeeding generations. "All phenotypic features of organisms—indeed, cells and organisms themselves as complex physical entities—emerge and evolve only inasmuch as they are conducive to genome replication" (Koonin 2011, loc. 6703). Your own cells started differentiating into separate roles within a few days of conception, and for almost all of them, that put a stark limitation on their future as individual replicators. Absent some artificial intervention like cloning or DNA testing, the genes *in every cell but your gonads* will never go beyond your body. Their chance at immortality lies only in the success of their identical twins that make it aboard new chromosomes formed in sperm or eggs.[46] "Our own genes cooperate with one another," Dawkins points out, "not because they are our own but because they share the same outlet—sperm or egg—into the future" (2006, 245).

It's all about genetic replication. For reasons Dawkins explains thoroughly (pp. 259-63) and we won't go into, evolution works well partly because it forces that replication into a single starting point for each new organism.[47]

> No matter how many cells there may be in the body of an elephant, the elephant began life as a single cell, a fertilized egg. The fertilized egg is a narrow bottleneck which, during embryonic development,

---

46. An individual gene has about a 50-50 chance of its twin making the cut, literally, due to the recombination of paternal and maternal chromosomes that occurs when a new sperm or egg is formed. (We discussed this in "Cast of Characters" as one of several sources of random variation that drive evolution.) It's not as drastic as it sounds; there is so little genetic variation between any two human beings that the alternative gene probably has the same DNA, anyhow.

47. One fascinating exception is the *chimera*, a single organism formed from multiple fertilized eggs. Yes, there are human chimeras. Their bodies are a patchwork of tissue bearing two separate sets of DNA. They get different results from genetic testing depending on where they take their cells from! This raises an interesting issue for "life begins at conception" theology, which is worth mentioning but otherwise beyond the scope of this book: From which fertilized egg does a chimeric person obtain his or her soul? What happened to the soul in the other one?

widens out into the trillions of cells of an adult elephant. And no matter how many cells, of no matter how many specialized types, cooperate to perform the unimaginably complicated task of running an adult elephant, the efforts of all those cells converge on the final goal of producing single cells again—sperms or eggs. [p. 259]

Without all that cooperation, there would be no new elephants. Or you, for that matter, without it happening in the bodies of your parents. The cells in their livers dedicated their existence to keeping those two people alive, but the real purpose was making it possible for *you* to live. And now your liver is working toward the same goal, for the chance of you producing children with livers, too. With cells working together in organisms, "the intricate mutual co-evolution of genes has proceeded to such an extent that the communal nature of an individual survival machine is virtually unrecognizable" (p. 47). "The division of labour is what makes a body worth inventing" (Ridley 1996, 41).

This symbiosis occurs between separate organisms, too. The bacteria in your gut are an example; we give them a warm, wet place to live and they help us digest the food we send down there. Dennett points out that, of the perhaps one hundred trillion cells in our bodies, *nine out of ten* are not human. "Most of these trillions of microscopic guests are either harmless or helpful; only a minority are worth worrying about. Many of them, indeed, are valuable helpers that we inherit from our mothers and would be quite defenseless without" (Dennett 2006, 86).

So it should be no surprise that there is cooperation *within* species. An extreme example is what happens in colonies of social insects—ants, bees, wasps, termites—where most individuals are effectively slaves to the good of the genes they share with everyone else in the colony.[48] Their society

---

48. Some species of ants have actually figured out how to enslave other species, as Darwin observed firsthand: "One day I fortunately chanced to witness a migration from one nest to another, and it was a most interesting spectacle to behold the masters carefully carrying . . . their slaves in their jaws" (1859, 223). The slavers take advantage of the "built-in nervous programs" in their slaves, who go about "doing all the duties that they would normally perform in their own nest. The slave-making workers or soldiers go on further slaving expeditions while the slaves stay at home and get on with the everyday business of running an ants' nest, cleaning, foraging, and caring for the brood" (Dawkins 2006, 177).

There is cooperation going on here, too, even if one party to the transaction is an unwilling one. The slaves get enough sustenance and protection for their genes to survive, even though it is not their own kin that they were serving—like cows who line up in rows to have their udders suckled by milking machines, feeding humans rather than calves, or the "aphid species that are much cultivated by ants" for sucking juice from plants (p. 181).

"achieves a kind of individuality at a higher level"—sharing food as if there were a communal stomach, information as if there were a nervous system, and repelling intruders as if there were an immune system (p. 171). The queen and her drones "are the analogues of our own reproductive cells in our testes and ovaries," and the "sterile workers are the analogy of our liver, muscle, and nerve cells" (p. 172).

Most of the individuals in the colony will never reproduce—the sterile workers, the "honey-pot" ants with their "grotesquely swollen, food-packed abdomens, whose sole function in life is to hang motionless from the ceiling like bloated light-bulbs, being used as food stores by the other workers" (p. 171). They are dead-enders, but no more so than your liver cells. It makes sense, evolutionarily, for your liver or a whole worker ant to devote itself to the reproduction of DNA that is identical or very nearly so. We are all "gigantic colonies of symbiotic genes" (p. 182). Is it any wonder, then, that we see humans engaging in selfless behavior for the good of their own kin?

There are many nuances to kin selection, and Dawkins explains them in fascinating detail in *The Selfish Gene*: the importance of siblings as well as offspring, our imperfect mechanisms for identifying close relatives, the balancing of investment a mother makes in current versus future children. Even suicidally altruistic behavior can make evolutionary sense, when the kin-altruism gene, "on average, tends to live on in the bodies of enough individuals saved by the altruist to compensate for the death of the altruist itself" (Dawkins 2006, 93). A mother who sacrifices her life to save four of her children is rewarded by having twice as many of her genes (1/2 the DNA of four people) remain walking around than they would have if she had just saved her own lousy hide. Or, to put it more precisely, the genes that instill the deep-seated trait of maternal altruism would survive. The only sense in which evolution is really the "survival of the fittest" is when you are talking about the survival of genes. "The currency used in the casino of evolution is survival, strictly gene survival," though "for many purposes individual survival is a reasonable approximation" (p. 55).[49]

---

49. Our reference to maternal altruism as the example, rather than fathers who certainly are motivated to protect their children as well, is not just a whim. Females invest much more into their offspring than do males, who can walk around impregnating available females as much as their rivals and society permit. It would be no surprise to see the instinctive kin-selection motive being stronger for a mother, who has devoted much of her life to being pregnant with and tending small children, than for a father who, until recently in human society, can envision many more opportunities for reproduction. Besides the investment factor, we can "expect that there will normally be some evolutionary pressure on males to invest a little bit less in each child, and to try to have more children by different wives. By this I simply mean that there will be a tendency for genes that say 'Body, if you are male leave your mate a little bit earlier than my rival allele would have you do, and look for

It is possible that some of what we perceive as selfless altruism toward people who aren't related to us is actually driven by limitations in our abilities to instinctively recognize others as kin, limitations that Dawkins discusses at length (pp. 98-109) for humans and other animals. These are not necessarily conscious decisions. "What really happens is that the gene pool becomes filled with genes that influence bodies in such a way that they behave *as if* they had made such calculations" (p. 97, our emphasis). The "'true' relatedness may be less important in the evolution of altruism than the best estimate of relatedness that animals can get," which explains why a mother might be more concerned about the welfare of her children than the children are about each other. The amount of DNA sharing (50%) is exactly the same between a mother and child as it is between siblings (p. 105), but she is more certain of the relationship.

Collins and others mention the example of Oskar Schindler, a German who is justly lauded for saving many Jews during the Holocaust. Why might he do such a thing when he is clearly not related to them? It is easier to understand people helping others whom they have reason to believe might be relatives. Conversely, "racial prejudice could be interpreted as an irrational generalization of a kin-selected tendency to identify with individuals physically resembling oneself, and to be nasty to individuals different in appearance" (Dawkins 2006, 100). Ehrlich notes that

> the universality and persistence of systems for keeping track of relatives and determining behavior toward them (kinship systems) in human societies can be seen as an argument for a general influence of kin selection on the genome of *Homo sapiens*. After all, what other reason could there be for human societies to have developed complex systems of terminology to allow individuals to keep track of their genetic relationships to one another? [2000, 39]

But as we will see, Ehrlich has many reservations about how much the genes really influence human behavior. We hesitate to speculate further, as this is starting to get us entangled with the controversies of evolutionary psychology. That field has come in for a lot of criticism, especially when it starts speculating about our behaviors being the legacy of cavemen fighting over women and food. The primatologist Frans de Waal finds the objectives of evolutionary psychology laudable, and remains optimistic about it. But he notes that it

---

another female', to be successful in the gene pool. The extent to which this evolutionary pressure actually prevails in practice varies greatly from species to species" (Dawkins 2006, 146-47).

is unfortunately better known for a few narrow theories about why women fall for rich guys, why stepfathers are not to be trusted, and how rape is only natural. Moreover, in the promotion of these ideas, theoretical convictions have often been more conspicuous than data. [de Waal 2002, 187]

Regarding Dawkins's point about physical resemblance, for example, one recent study seems to provide little support. Experimental subjects played a "two-person sequential trust game" with online participants who were represented by a facial image. The subjects were no less likely to make "selfish betrayals of the partner's trust" when their partner's purported face was a digitally morphed version of their own face than under control conditions, even though the participants did trust the former class of partners more (DeBruine 2002).

There still can be some genetic payback to altruism, though, even without relatives real or imagined. Reciprocal altruism is doing something for others out of the expectation, perhaps a subconscious or societally constructed one, that someone else will help you out, too. If you are expecting the same individual to pay you back, that is direct reciprocity. If you are in a society where the payback might come from others, it's *indirect* reciprocity (Nowak 2012, 37-38). This is a bit further advanced than the unthinking symbiosis between you and your gut bacteria, though the principle is the same. Dawkins devotes a chapter of *The Selfish Gene* to reciprocal altruism (2006, 166-88), accounting for suckers, cheaters, and grudgers, and drawing from his zoological knowledge to offer interesting examples of cooperation in the animal kingdom. But he also acknowledges the limits of his explanations for this, as well as for kin selection.

Yes, we "may well have spent large portions of the last several million years living in small kin groups," and "kin selection and selection in favour of reciprocal altruism may have acted on human genes to produce many of our basic psychological attributes and tendencies." Yet these ideas "do not begin to square up to the formidable challenge of explaining culture, cultural evolution, and the immense differences between human cultures around the world," Dawkins finds (p. 191). This is where he joins up with Ehrlich in looking to evolution beyond the gene, though Ehrlich is not enthusiastic about Dawkins's specific destination, memetics. (Again we reference the upcoming chapter, "The Memes Shall Inherit the Earth.")

Ehrlich says that humans, uniquely it seems, exhibit *true* altruism, which goes beyond just gene replication and mutual back-scratching. The "answers are partial at best" about altruism, even the genetically self-interested kind that we have just discussed (2000, 311-12).

> Virtually nothing is known about any genes that altruistic human beings might promote by helping relatives, and even less is known about the possible reproductive differentials involved. Moreover, empathy and altruism often exist where the chances for any return to the altruist are nil. Indeed, careful psychological experiments suggest that much of human helping behavior is divorced from any real prospect of reproductive or other reward. [p. 312]

Ridley's engaging book *The Origins of Virtue* addresses these questions, as you might expect from its subtitle, *Human Instincts and the Evolution of Cooperation*. Society, he says, "evolved as part of our nature. It is as much a product of our genes as our bodies are" (1996, 6). He downplays the analogies with social insects and the effect of kin favoritism among humans, noting that nepotism is a dirty word (p. 40). It is on the talking and trading aspects of human society that he focuses: "Exchange for mutual benefit has been part of the human condition for at least as long as *Homo sapiens* has been a species" (p. 200). He discusses game theory to show how individual benefit can be maximized through pro-social behavior, points out how society features mutually advantageous divisions of labor, and notes the benefits of sharing. No man can eat a whole mammoth, he notes in an epigraph (p. 103), and says that sharing of food "is a very effective way of reducing risk without reducing overall supply" (p. 102).

In none of this does Ridley make recourse to the much-maligned concept of group selection, which has "fatally weak assumptions" behind it (p. 179). He says that "group selection will drown the effects of individual selection" only "if groups have generation times as short as individuals, only if they are fairly inbred, only if there is relatively little migration between groups and only if the whole group has as high a chance of going extinct as the individuals within it." Otherwise, "selfishness spreads like flu through any species or group that tries to exercise restraint on behalf of the larger group." He summarizes the group selection problem with this memorable sentence: "Individual ambition always gets its way against collective restraint" (Ridley 1996, 179).

The Harvard researcher Martin A. Nowak includes group selection in his list of "five mechanisms governing the emergence of cooperation" (2012, 37-38). The others are direct reciprocity, clustering of cooperation ("spatial selection," where there are "clumps of cooperators and defectors"), kin selection, and indirect reciprocity. Nowak acknowledges the debate about group selection, but says his own mathematical modeling and that of others "has helped show that selection can operate at multiple levels, from individual genes to groups of related individuals to entire species" (p. 38).

Even looking at things *as individuals*, de Waal notes that we "have much to lose if the community were to fall apart, hence the interest in its integrity and harmony" (2006, 54). A civil and fair society serves *us*, not the other way around. It provides a setting for Nowak's indirect reciprocity: "I'll scratch your back and *someone* will scratch mine" (Nowak 2012, 38, our emphasis).

Now, despite being an eminent scientist in his own field, Collins seems to look past all of this science to see an unavoidable spiritual dimension to humans. He pays little heed to the explanations of kin, reciprocity, and culture, claiming that "selfless altruism" is a major challenge for the evolutionist, "a scandal to reductionist reasoning" that "cannot be accounted for by the drive of individual selfish genes to perpetuate themselves" (Collins 2006, 27).

It turns out that he doesn't pay much attention to the depressing view the Bible and Protestant theology have about human goodness, either, where "we are all as an unclean thing and all our righteousnesses are as filthy rags" (Isa. 64:6). If man is born unto trouble, as the sparks fly upward (Job 5:7), if there is none righteous, not even one (Rom. 3:10), then why is Collins claiming otherwise? He is trying to oppose the godless materialists with a Bible that seems to agree more with them than him on this point.

For whatever reason, Collins seems to backtrack when he writes on the subject with Giberson some years later. They graciously note that "many atheists and agnostics have created ethical systems that encourage generosity and promote fairness," and "are themselves persons of great integrity, demonstrating in their own lives that they are able to find an adequate grounding for their personal morality." Giberson and Collins also acknowledge that many of their fellow "Christians are self-indulgent and enjoy great wealth they have no intention of sharing." They still suggest, tepidly, that "it often seems that a moral order with God in it is more robust than one without," but they "caution against making too much of this argument" (2011, 141). The problem is, of course, that Collins himself had made the argument into a large portion of his own earlier book. Without it, there's not much left for him (or BioLogos) to talk piously about.

Gloating in Genesis

Traditional Christianity maintains the anthropocentric view that man is the pinnacle of creation, distinct from all other species. (We discussed the biblical basis for this in "Eden Disorder.") Nowadays the claim is made by denying or rationalizing the science, but at first it was done with ignorant sincerity. No other creatures were walking around and talking about themselves, and it just made sense that we were something very different, in kind rather than just degree.

Two examples of such thinking, both involving our now-unique trait of walking upright, come to us from the first centuries CE. In his *First Apology*, Justin Martyr saw the distinction as a symbolic one, and a very Christian symbology at that. He claimed that

> the human form differs from that of the irrational animals in nothing else than in its being erect and having the hands extended, and having on the face extending from the forehead what is called the nose, through which there is respiration for the living creature; and this shows no other form than that of the cross. [Ch. 55]

Minucius Felix saw God especially revealed as the artificer of man by the "very beauty of our own figure: ... our upright stature, our uplooking countenance, our eyes placed at the top, as it were, for outlook; and all the rest of our senses as if arranged in a citadel" (*Octavius*, Ch. 17).

As we have seen, the Bible begins all the self-congratulatory strutting right in the first chapter of Genesis, where man is supposedly made in the image and likeness of God.[50] In discussing this verse, Luther acknowledged that the "beasts greatly resemble man. They dwell together; they are fed together; they eat together; they receive their nourishment from the same materials; they sleep and rest among us. Therefore if you take into account their way of life, their food, and their support, the similarity is great." But there is, thought Luther like generations of Christians before and after him, "an outstanding difference between these living beings and man ... that man was created by the special plan and providence of God. This indicates that man is a creature far superior to the rest of the living beings that live a physical life, especially since as yet his nature had not become depraved" (*Lectures on Genesis*, Ch. 1, v. 26).

Throughout the Bible, Old and New Testaments alike, the Creator of the Universe is occupied with the concerns, thoughts, and deeds of this one species of mammal. Romans 8:17-23 goes so far as to attribute a sense of "anxious longing" to the entire creation as it "waits eagerly for the revealing of the sons of God, i.e., the glorification of human beings who are "heirs of

---

50. Actually, "gods," since it says, "let *us* make man in *our* image, after *our* likeness" (Gen. 1:26). That slip is a residue of the polytheism we discussed back in "The World the Biblical Writers Thought They Lived In." Unless of course you are persuaded by Luther's assertion that "'Let Us make' is aimed at making sure the mystery of our faith, by which we believe that from eternity there is one God and that there are three separate Persons in one Godhead: the Father, the Son, and the Holy Spirit" (*Lectures in Genesis*, Ch. 1, v. 26).

God and fellow heirs with Christ" (NASB). Really, though, is it any surprise to see this focus? Religion is all about a deity having a relationship with *humans*, after all, not dogs and cats.

A notable exception is the Book of Ecclesiastes, which flatly contradicts traditional Christian dogma about the special status of human beings.[51] In a remarkable passage quoted in the epigraph to this chapter, the "Preacher" who narrates Ecclesiastes wishes that the sons of men would realize that they themselves are beasts (3:18) and asserts that the fate of both is the same:

> As one dies so dies the other; indeed, they all have the same breath and there is no advantage for man over beast, for all is vanity. All go to the same place. All came from the dust and all return to the dust. Who knows that the breath of man ascends upward and the breath of the beast descends downward to the earth? [Eccl. 3:19-21, NASB]

---

51. Ecclesiastes is quite a doctrinal headache for inerrantists, as pointed out in Suominen 2012, §6.20, from which this note has been adapted. "The Preacher" who narrates Ecclesiastes says "the earth abideth for ever" (1:4), which is incompatible with the idea of Judgment Day, and that one fate befalls both the wise man and the fool who walks in darkness (2:14), which dispenses with any post-mortem judging at all. He looks back on his life and wisdom, saying "there is no remembrance of the wise more than of the fool for ever; seeing that which now is in the days to come shall all be forgotten. And how dieth the wise man? as the fool" (2:16). He concludes that there "is nothing better for a man, than that he should eat and drink, and that he should make his soul enjoy good in his labour. This also I saw, that it was from the hand of God" (2:24).

The Preacher doesn't look forward to any eternal correction of injustice for the oppressed. Rather, he says he congratulated the dead more than the living, who had no comforter in their oppression. But better than both of them is the one who has never existed, who has never seen the evil that is done under the sun (4:1-3). This is quite a contrast to how we often comfort ourselves about one who has "gone to his reward!" The reward of a man, the Preacher contends, is "to eat, to drink, and enjoy oneself in all one's labor in which he toils during the few years of his life which God has given him" (NASB, 5:18). When you're dead, it's lights out: The dead "know not any thing, neither have they any more a reward; for the memory of them is forgotten. Also their love, and their hatred, and their envy, is now perished; neither have they any more a portion for ever in any thing that is done under the sun" (9:5-6).

Thom Stark finds it remarkable that either Ecclesiastes or Job, another book he finds notably "subversive," was "kept within the sacred curriculum." Ecclesiastes, he says, was disputed among the rabbis "because of its hedonistic and skeptical message, but it was ultimately salvaged with the help of some earlier editorial emendations, and by virtue of a tradition (wrongly) attributing the book to King Solomon" (Stark 2011, 12).

Regardless of that canonical speed bump, Christians have never accepted themselves as being just another species of primate. Christian apologists from Luther to Lamoureux reverently trace their fingers along the lines of Genesis, look up, and conclude that humans are special *because another human said we are*, in a text that still other humans have designated as the Word of God:

> [N]o matter which interpretation of the image of God we prefer, we must, if we are to be careful, acknowledge that it implies that there is something about human capacities that is different from those in any other animal. [Collins 2011, 95]

> For long centuries, God perfected the animal form which was to become the vehicle of humanity and the image of Himself. [C.S. Lewis, *The Problem of Pain*; from Collins 2006, 208]

> Whichever scenario of origins Christians embrace, we can agree that God is the creator of all and that humans are unique because they partake of God's spirit. [Hurd 2003, loc. 2623]

> Men and women are unique and distinguished from the rest of creation because they bear the Image of God and have fallen into sin. [Lamoureux 2008, 31]

Even Peter Enns, who accepts so much of modern biblical criticism, insists on an anthropocentrism tied to the "image of God" statement of Genesis 1:26. He claims that language denotes the dominion over the creation that God assigned to his favorites, the human race (Enns 2012, xiv-xv & 49). Enns disputes the common understanding of "image" as being "the soul, God-consciousness, or other qualities that make us human" (p. xiv). Rather, it "refers to humanity's role of ruling God's creation as God's representative" (p. xv). Is there a distinction in that difference?

Catholic Confusion

Two of the theistic evolutionists who seem to most distance themselves from Genesis 1:26 are Roman Catholics: Kenneth Miller and John Haught. Before discussing their views of humanity's place in an evolved biosphere, we think it worthwhile to ponder if they also are willing to distance themselves from their recent Pope, who hardly seemed an advocate of evolution.

After hearing an entirely science-grounded talk on the topic in 2006 by Peter Schuster, president of the Austrian Academy of Sciences, Benedict XVI acknowledged that he had been "impressively shown the logic of the developing theory of evolution." But, acting like a moth who swoops around the light yet cannot quite stand the heat of landing on it, Benedict noted "the internal corrections (especially to Darwin's ideas) that were discovered along the way" and "the questions that remain open." (Not that he was trying to "cram the dear Lord into these gaps," he hastened to add.) He threw in a bit of NOMA, finding it "important to underscore that the theory of evolution implies questions that must be assigned to philosophy and that in and of themselves lead beyond the internal scope of the natural sciences" (Horn and Wiedenhofer 2008, loc. 1945-50). And hey, maybe it isn't really true, after all!

> In particular, to me it is important, first of all, that to a great extent the theory of evolution cannot be proved experimentally, quite simply because we cannot bring 10,000 generations into the laboratory. That means that there are considerable gaps in its experimental verifiability and falsifiability due to the enormous span of time to which the theory has reference. [loc. 1950]

Perhaps he would have been more fully persuaded by one of his own flock. Writing *Finding Darwin's God* several years before Benedict's silly argument, Kenneth Miller called it unfortunate that "there is a school of thought that rejects the very idea that any theory about the past can be scientific" (1999, 22). He points out how a "police detective would scoff at the notion that crimes can be solved only when they are witnessed directly," and it isn't any different for science. "We may not be able to witness the past directly, but we can reach out and analyze it for the simple reason that the past has left something behind" (p. 23).

Schuster was invited to lecture the Pope despite being an agnostic. But even if the faithful Catholic Miller were "to lay out graphs and charts and diagrams, to cite laboratory experiments and field observations, to describe the details of one evolutionary sequence after another," the results would be limited. To "the true believers of creationism," apparently including the Pope in his heart of hearts, "these would all be sound and fury, signifying nothing. The truth would always be somewhere else" (Miller 1999, 173). Benedict seems to cling to the comforting possibility that a more palatable truth might still be out there, somewhere:

> A second thing that was important to me was your statement that the probability is not zero, but not one, either. And so the question arises: How high is the probability now? This is important especially if we want to interpret correctly the remark by Pope John Paul II: "The theory of evolution is more than a hypothesis." When the Pope said that, he had his reasons. But at the same time it is true that the theory of evolution is still not a complete, scientifically verified theory. [Horn and Wiedenhofer 2008, loc. 1953]

A bigger issue for Catholic acceptance of evolution, particularly its implications for human origins, is the Church's demand that the faithful believe in monogenism, the descent of all people from a first human couple. Regarding acceptance of the alternative, the "conjectural opinion" of polygeneism, Pope Pius XII wrote in *Humani Generis* that

> the children of the Church by no means enjoy such liberty. For the faithful cannot embrace that opinion which maintains either that after Adam there existed on this Earth true men who did not take their origin through natural generation from him as from the first parent of all or that Adam represents a certain number of first parents.

Wiley provides this quote from the 1950 encyclical in her book *Original Sin* (2002, loc. 1504), and notes the limited freedom that it gave to Catholic scientists and theologians. Yes, they could "inquire into human origins and physical evolution" (loc. 1502). But they were kept on a leash, one that all of our Catholic theistic evolutionists seem to have slipped off, as they must in order to maintain their scientific integrity: The faithful were, and still are, "prohibited from acceptance of a scientific theory of human origins proposing the descent of humankind from more than a single pair of ancestors" (loc. 1503).

Miller seems to brush off such theological details. "As a scientist," he knows that "the appearance of living organisms was not sudden, but gradual" (1999, 257). This statement comes after a scant few paragraphs he offers about Adam and Eve, where he mentions the "obvious conflict" between the two Genesis accounts (pp. 256-57). Thus the reader is left to infer (wink, wink) that *Homo sapiens* gradually appeared, too, and Miller avoids a direct confrontation with the dogma of his church. "As a Christian," he assures, "I believe that Genesis is a true account of the way in which God's relationship with the world was formed" (p. 257). Well, as critical thinkers, we believe he is just trying to have it both ways.

Haught floats above mere biblicism and anthropocentrism with a lofty "drama of life" theme, which we will discuss along with other aspects of Process Theology in "Passive-Aggressive Creationism." As with all the troubling aspects of evolution for Christian theology, he is admirably candid in presenting the difficulty of human origins: "What could it possibly mean therefore to claim, in keeping with Christian tradition, that human beings are created in the image and likeness of God?" (2010b, 43). The "totality of nature and its long evolutionary history are God's creation, and not our own." Most of the universe's "countless surprising evolutionary developments" have little or nothing to do with us. And it "may well hold the promise of outcomes in the future that, like many of those in the past, are not narrowly definable in terms of its simply being a home for our own species" (Haught 2000, 159).

Nonetheless, even Haught cannot quite avoid a few lapses, dressed up though they are in fancy language:

> The fact that we are informationally discontinuous with other kinds of life at least makes it conceivable that a theological affirmation of special human dignity poses no conflict with science. [2010b, 51]

> [O]ur sense of dignity is inseparable from our capacity to open ourselves to the future. Theologically, I want to connect our sense of self-worth and the place of humanity within the universe to the biblical idea of a God of promise. The fundamental biblical theme of promise encourages theology to think of God as creating the universe and shepherding life's evolution—without coercion—by inviting the entire universe toward an ever-new future within the timeless milieu of God's unbounded, compassionate love. The God of evolution is none other than the God who calls Abraham into a refreshingly new future.... The process of life's evolution and the story of human existence on earth are infused with special value here and now by virtue of their being open to being taken into the everlasting embrace of the God who may be thought of as the world's Absolute Future. At least in some measure, our own special significance or value consists of our inherent openness to the "power of the future" that goes by the name God. [pp. 51-52]

He finds his distinctions in our "informational discontinuity" with all other life—chimps don't write books about their special place in the Universe, after all—and our "openness" to God. But Haught still has the deity fussing over us with "love" and "calling." We aren't standing next to Yahweh haggling with him about righteous souls in Sodom like Abraham was (Gen. 18). No, we're being "grasped, renewed, and dignified by the inexhaustible Future we call by the name God" (p. 52). Hey Yahweh, how many fancy words will it take to save our self-image from the fire and brimstone of materialism? Will three of them do? Or do you need a few more?

When testifying in *Kitzmiller v. Dover*, Haught said that he's not a Process theologian. "People have called me that, but I've never identified myself as such. I use ideas from many, many different kinds of theology, including process theology."[52] However, we fail to see much distinction between his

---

52. Trial transcript from talkorigins.org/faqs/dover/day5pm2.html. In that testimony, Haught also indicated he believes that Jesus would not have been captured on video if a camera were running in the upper room after the resurrection. And this man claims to be a Roman Catholic!

ideas and those generally associated with Process Theology, where all things, after a fashion, supposedly have souls, sparks of subjectivity. This is the point of Whitehead's concept of "prehension," as we will see in "Groping for Gaps." Everything, even a lousy rock, "experiences" whatever befalls it. The humble pebble experiences the water eroding it. It may not be *aware* of experiencing it, but we can imagine, or at least say, that *from its own vantage point*, the pebble is the subject suffering the erosion. Nor is this some bizarre, unintended implication of a doctrine that seeks to make a different point. It has to be true if we can speak of God "luring" and "wooing" all creation toward some greater fulfillment. The Process God is a sort of pan-psychism (or Vitalism, or Animism), a theoretical atavism roundly blasted by Jacques Monod in his scathing discussion of Teilhard de Chardin (Monod 1974, 38-40).

Whether such a scenario makes any sense is highly dubious, but if one thinks it does, one has no business making human beings unique in their capacity for fellowship with the Divine. Of course, maybe Haught and his allies believe human beings will be superceded by future products of evolution in comparison to whom we will recede into the shadows along with the Pterodactyl and the Trilobite. Then we can be like John the Baptist, content to prepare the way for our betters.

Gimme that Old Time Anthropocentrism!

What seems to put all theologians out of business, Haught included, is the materialist reality that we are just "survival machines—robot vehicles blindly programmed to preserve the selfish molecules known as genes," as Dawkins puts it (2006, preface). Our bodies are complex and wonderful, yes, but this is the unvarnished truth of the matter when you strip away all the comforts of mythology and philosophy: *We are what our genes evolved us to be*, for their sake rather than our own. One might do a reversal of the incredulity of Psalm 139: "Such knowledge is too low for me; I cannot attain unto it."

And we're just not that different from the other myriad forms of life that the genes have constructed for their propagation. Consider the simplified "tree of life" diagram of Figure 11.[53] It is drawn to a phylogenetic scale at which

---

53. Actually, the "Tree of Life" idea is something of an oversimplification. Horizontal gene transfer occurs between microorganisms after they have branched from each other, a phenomenon that neither Figure 11 nor the tree metaphor accounts for. The continued relevance of the metaphor is a subject of debate among evolutionary biologists (Koonin 2011, loc. 2471-80), and "forest of life" might be a better one (loc. 2548-769). No "single tree can fully represent the evolution of complete genomes and the respective life forms," says Koonin, but "the realistic picture of evolution

you can barely distinguish mammals from reptiles, much less humans from other primates.

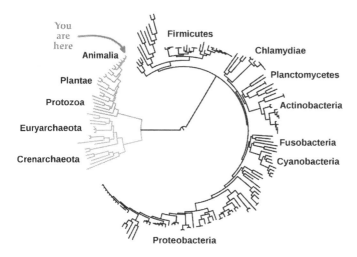

FIGURE 11: Phylogenetic "Tree of Life."

It's understandable for us as humans to be less than thrilled by this. "Darwin himself saw clearly that if he claimed that his theory applied to one particular species, this would upset its members in ways he dreaded, so he held back at first" (Dennett 1995, 335). Ruse quotes E.O. Wilson about "man's ultimate nature":

> We keep returning to the subject with a sense of hesitancy and even dread. For if the brain is a machine of ten billion nerve cells and the mind can somehow be explained as the summed activity of a finite number of chemical and electrical reactions, boundaries limit the human prospect—we are biological and our souls cannot fly free. If humankind evolved by Darwinian natural selection, genetic chance and environmental necessity, not God, made the species. [*On Human Nature*, 1978, from Ruse 2000, 107]

The yearning for some special essence of humanity isn't just expressed by religion, either. "Virtually every high priest of the humanities, of anthropology and of psychology preaches the same old, defensive sermon of human uniqueness that theologians clung to when Darwin first shook their tree" (Ridley 1996, 155). But there is nobility to be found in our evolutionary heritage, Ramachandran says from an entirely scientific

---

necessarily combines trees and networks." The resulting "forest of life" metaphor refers to "the complete collection of phylogenetic trees for individual genes" (loc. 2750).

perspective: "We are the first and only species whose fate has rested in its own hands, and not just in the hands of chemistry and instinct. On the great Darwinian stage we call Earth, I would argue there has not been an upheaval as big as us since the origin of life itself" (2011, loc. 353).

Prehistoric Propitiation

Hurd begins his informative essay "Hominids in the Garden?" with a detailed account of the fossil record, laying out his facts like the knowledgeable anthropologist that he is. Then he comes to the hard part: figuring out what to do with all this as a professor at a Christian college. "How might we explain theologically the fact that some early fossil populations do not resemble modern apes or modern humans?" Most "were not human ancestors," but like us they "used stone tools, walked bipedally, and had a humanlike tooth structure. Unlike modern humans, they had small braincases, greater arm/leg length ratios, and relatively simple cultural practices" (Hurd 2003, 222-23).

Chimps can smash nuts with rocks, soak up water with leaves, and poke sticks into termite mounds to extract the goodies, but that's about the extent of their skills with making or using tools (Zimmer 2001, 265-66). It's not a trivial task to make stone tools, as one of us (Ed) appreciates from having watched his neighbor make an arrowhead using a neolithic toolkit of leather apron, obsidian core, and antler billet. The flint knapper must identify and prepare a "platform" on the core from which to remove the next flake, and then whack the core just right with his stone, wood, or antler billet to form a new edge while leaving a ridge down the middle of the blade. It requires patience, visualization in three dimensions, and the ability to learn from others—unless you think the technique behind every one of those stone tools was the independent invention of a different lone artisan.

The individuals who made the hand axes scattered around the world "from Africa to Asia starting 1.5 million years ago" (Wilcox 2003, 236) were well advanced of any non-human ape alive today. Those tools gained some sophistication—symmetry and smooth edges—over a million years ago (Ramachandran 2011, loc. 2307). Around 500,000 years ago, hundreds of stone tools were left along the shore of Lake Elmenteita in what is now Kenya, "most made of glassy black obsidian, the rock that comes from volcanic lava. There were tear-shaped hand axes, two-sided flakes, and even triplets of round stones that look like black billiard balls" (Gibbons 2007, loc. 496). These "were among the first hard evidence of a sophisticated ancient Stone Age culture in eastern Africa" (loc. 497), and it was a human culture. Not yet *Homo sapiens*, but human.

So let's take a moment to shine a flashlight on the never-mentioned implications of the belief of many theistic evolutionists, Protestant and Catholic, that the human soul appeared somewhere in the evolution of *Homo sapiens* from these uncouth hairy ancestors. At some point, when the genetic stew had sufficiently simmered, God injected a soul into two (or more) of the big lunks. One of these, or perhaps their representative, was assigned the name Adam.

These first ensouled humans were spiritual orphans, apparently. Their parents would have looked and acted exactly like them, with only a handful of DNA mutations distinguishing them, biologically. But this theology views them as mere brutes, just like the South American Indians were treated by Europeans colonizing the newly discovered Americas, like Israel treated the nuisance Canaanites. The Pope reassured the Conquistadors that they were free to kill the natives rather than evangelizing them because these copper-tanned folks were not really folks at all. They didn't have any souls to be concerned about.

Not too many centuries after this, Nazi theologians like the mad monk Jörg Lanz von Liebenfels (author of the shunned and abhorrent *Theozoologie*, 2001) argued that Jews, Africans, and other "undesirables" were the product of blasphemous miscegenation between Adam and female beasts who caught his wandering eye in Eden. You see the pattern: disqualify ethnic groups one despises as "illegal immigrants" from sub-human stages of evolution. Human, because we can reproduce with them, but not quite human enough to suit us. They lack a soul, as the Tin Woodsman lacked a heart and the Scarecrow wanted for a brain.

To his credit (the man knows his science), John Haught has "long thought there is something evasive, artificial, and theologically shallow about" the idea that our "human souls are what make us special" (2010b, 46). Indeed, "the question now arises whether there is any clear distinction between life and lifeless matter, let alone between one species and another" because

> evolutionary biology presents life as an incessant river of genes, flowing continuously from one generation to another. And ... this same continuum of genes, along with the proteins they configure, can be chemically subdivided into large molecules, and the latter into atomic units ... [p. 45]

As we saw earlier, Haught finds human uniqueness in "the arena of information," where we "stand out as discontinuous with the rest of life" (p. 51). On that point, aside from all of Haught's accompanying obscurantist mumbo-jumbo, fair enough.

For the rest of Christianity, where souls and salvation are of paramount importance, what about those ancestors hundreds of thousands of years ago? Even the members of our own species who lived over a hundred thousand years ago in Africa pose a dilemma for traditional Christianity. What kind of salvation was available to these people, before even the most primitive forms of Israelite sacrificial worship? The "oldest known human-made religious structure," was built about 12,000 years ago, and is decorated with graven images of animals.[54] That would be prohibited by Exodus 20:4; people weren't worshiping the God of Israel there.

Obviously, in one sense all this is moot since there are no longer any cave men left around being "discriminated" against. It's like complaining that the money-grabbing Ferengi are being caricatured on *Star Trek*. But it still seems worthwhile to point out the insidious thinking involved in this theistic evolutionist gambit. It seems innocent only because the scapegoat ("soulless") forebears are all gone, unable to confront us except in the occasional skeleton or fossilized skull fragment.

---

54. So says Wikipedia about Göbekli Tepe, in a well-written article that includes many external references. You may have noticed by now that we are big fans of Wikipedia, despite all the valid academic warnings not to cite it. It contains some really good stuff, and we refer to it, cautiously, along with the usual slew of books, magazines, and peer-reviewed journals.

Justifying Jesus

> *The idea that we are descended from "beasts" is one reason why many people have been repelled by evolutionary theory. And the idea that Christ would share that relationship is especially shocking to many Christians.*
>
> —George L. Murphy, in
> *Perspectives on an Evolving Creation*

The figure of Jesus is also problematic given the reality of evolutionary science. First, why is a virgin-born, death-cheating miracle worker any more plausible than an *ex nihilo* created first human being? Science has clearly disproven the latter, so why exempt the former?[55] Second, there is the issue of Jesus' own human nature, given that at least half his DNA came from a human woman. He was tempted in all points, like us (Heb. 4:15), and now we know that he must have been formed from the same flawed stuff as we are.

Is Jesus Safe from Adam's Fate?

If Jesus is the second Adam, and we demythologize the Genesis Adam, will we wind up doing the same to his New Testament counterpart? That is, of course, the great worry of fundamentalists who believe in either of these Adams purely on the say-so of the Bible. Not the Bible as a historical sourcebook, but the Bible as a supernatural Aladdin's Lamp of infallible doctrines. Their reasoning, based on this premise, is quite sound, not that Jesus *must* be as mythic as Adam but that he *might* be, and there would no longer be any way to be sure. It is not that there is no come-back. This is the point at which fundamentalists will reply, "You ask me how I know he lives? He lives within my heart," or maybe, "I talked with him this morning."

Later we will see how this response is alive and well—and just as futile as ever. But on the basis of it, you know what comes next: Subjective experience and its bodyguard, the will to believe, work retroactively, guaranteeing that the Edenic Adam was just as real as the non-negotiable Jesus. Not that one has a "personal relationship" with Adam, but if what we say of Adam we must also say of Jesus, then it follows that what we say of Jesus must be equally true of Adam: both are as historical as your old Aunt Fanny. But our sophisticated Christian evolutionists hope to opt out of this

---

55. There is much to say about the lack of historical evidence for Jesus, and Robert M. Price has written extensively on the topic. But this book discusses Christian theology, not historical Jesus studies.

loop. They have no choice anymore about rejecting a historical Adam. He is gone with the wind. Lamoureux, Enns, and the rest have eaten of the Tree of the Knowledge of Paleontology and Genetics. But they don't want to cast Jesus under a cloud of suspicion, so their next move is just to draw a line in the sand beyond which they will not pursue consistency. Not that Jesus would *have* to be mythical like Adam, you understand, but he theoretically *might* be, and who wants to have to argue that case, too? Some take a deep breath and decide to try. But not very hard.

As we will see, they rely on second-hand, discredited historical apologetics, the historiographical equivalent of Henry Morris and Duane Gish. Okay, that's where we're headed. Now let's back up.

"Adam and Eve might be ancient vessels that transport divinely inspired Messages of Faith," Lamoureux will go so far as to suggest. The message is that "humans are created in the Image of God, they have fallen into sin, and God judges them for their rebellion" (2008, 20). But then surely Jesus might as easily be a symbolic fiction to "transport" the "Messages of Faith" that God forgives and renews sinners. Why not? Lamoureux echoes Bultmann in "New Testament and Mythology" again when he pronounces:

> Ancient peoples experienced and embraced the notion of causality. Their science is evidence of this fact. However, their understanding of cause-effect relationships in nature differs markedly from that of today. Ancient causation is often agentic [*sic*] and interventionistic. That is, willful agents or personal beings are causal factors that act dramatically in the origin and operation of the world. In the ancient mindset, the notions of "supernatural" and "natural" are intertwined and not demarcated sharply . . . . And like the ancient cultures around them, Israel and the early Church believed that demons and evil spirits caused sufferings, disabilities, and diseases. [Lamoureux 2008, 154-55]

He acknowledges that "obvious concerns arise in light of the historical shift from ancient to modern" views of causality. "Are all statements of divine action in Scripture, including the miraculous healings of Jesus," just cases where the ancients believed in false causes (p. 155)? No, don't worry, he hastens to assure us. The gospels are eyewitness accounts of miracles (p. 156), so we can suspend the whole logic of our theory and leap into the very "piecemeal supernaturalism" that we were trying to avoid![56] Sure, Genesis 1 "features an ancient origins science," but the "bodily resurrection of Jesus is a literal and historical event" (p. 160).

---

56. See our discussion of "Refined versus Piecemeal Supernaturalism" in the chapter on Process Theology, "Passive-Aggressive Creationism."

Maybe it is a "reasonable concern" to wonder, "if statements in Scripture about nature are not scientifically, historically, and literally accurate, then neither are the miracles and resurrection of Jesus." But, hilariously, Lamoureux offers a set of "hermeneutical brakes" (his term!) that can stop the concerned Christian from sliding down the slippery slope "to liberal theology, and maybe even a loss of faith" (p. 162). See if this passes your own road safety inspection:

> [F]irst-century individuals were certainly capable of knowing whether or not water had been turned into wine, a paralytic had walked away, and a man born blind could now see. And for a generation that saw many crucifixions, it was well within the scope of cognitive competence of witnesses to the death and resurrection of Jesus to know these events had actually happened. [p. 162]

Our author's horizons are much too narrow. He simply assumes that, where historical events after Genesis 11 are concerned, the record is historically accurate, that the gospels are eyewitness accounts, and that archaeology confirms the accuracy of both Testaments. Clearly he is unaware of recent biblical archaeology that puts Bible-believers in the same sinking boat with the red-faced Mormons and their discredited idea of American Israelites. Worse yet, it simply does not occur to him that the issues are the same in the gospels as in Genesis. He faults young-earth creationists for arbitrarily rejecting uniformitarianism, the inevitable working assumption that natural processes have always operated in the same way we see them operating now, an axiom without which no scientific research is even possible. But he does not see that the very same principle forbids the historian taking seriously the gospel miracles.

Lamoureux happens to be worried only about science, not history; only about Genesis, not the gospels. He is able to dictate and declaim that the early chapters of Genesis do not need to have "concord" with the facts simply because this would allow him to accept evolution, which he wants to accept. But the gospels he likes just fine as they are. No conflict with evolution there.

Why not treat the supernaturalism of the New Testament like the prescientific cosmology of the Old? Roman Catholic evolution-apologist John Haught says we should:

> **Q:** What do you make of the miracles in the Bible—most importantly, the Resurrection? Do you think that happened in the literal sense?
>
> **A:** I don't think theology is being responsible if it ever takes anything with completely literal understanding. What we have in the New Testament is a story that's trying to awaken us to trust that our lives

make sense, that in the end everything works out for the best. In a pre-scientific age, this is done in a way in which unlettered and scientifically illiterate people can be challenged by this Resurrection. But if you ask me whether a scientific experiment could verify the Resurrection, I would say such an event is entirely too important to be subjected to a method which is devoid of all religious meaning. [Haught 2010a, 97-98]

Obviously, Haught is happy to go the whole way with Bultmann, though his demythologized gist is far more banal than Bultmann's. Jerry Coyne writes eloquently and bitingly about the phenomenon of what he calls "making religious virtues from scientific necessities" (see, e.g., Coyne 2011a). That is exactly what Haught does when he says it would somehow degrade his theological platitude if one were to lay the profane hands of scientific research upon it. It's "entirely too important" for such treatment. Tut, tut. Haught wants this special pleading of his to be respected rather than written off as the desperate measure it really is.

He has long since arrived at the point Lamoureux is desperately hoping to avoid. Lamoureux is filled with false bravado when it comes to his personal savior. He claims that "the Bible offers a trustworthy account of Jesus of Nazareth," that archaeology "is also in accord with the New Testament. Notably, the historical reality of a man named Jesus in first-century Palestine stands firmly established" (2008, 15).

Lamoureux knows what his conservative readers (whom he seems to be trying mightily to court) are thinking: Where does the believer "draw the line? Or asked more directly, if Adam is not historical, then is this also the case with Jesus and His crucifixion and resurrection?" He argues that it isn't, because of some purported "foundational interpretive principle of literary genre" in which, arbitrarily, the "early chapters of Scripture and the Gospels are completely different types of literature." All of his fancy footwork about Genesis "is *not* applicable to the New Testament and the record of the Lord's ministry" (pp. 373-74). And why not? Because

> Genesis 1-11 is built on recycled ancient Near Eastern motifs that are ultimately a retrojection into the distant past of an ancient phenomenological perspective on the world. In sharp contrast, the New Testament is based on the testimony of people who actually encountered Jesus. [p. 374]

Well, the very same scholarly life preserver that Lamoureux throws at Genesis to rescue it from drowning in irrelevance informs us about the gospels, too. It shows that those stories about an itinerant teacher, healer, and exorcist are actually not eyewitness accounts of things that really happened, but rather analogous recyclings of ancient myth patterns retrojected into the imagined past to serve as a foundation legend for the community of faith. Exactly like Genesis!

Even if people did witness such things (the miracles somehow escaped even Paul's attention), even if some did have visions of a postmortem Jesus, why not assume that the historical Jesus became utterly mythicized into an Incarnate Son of God whose sacrifice atoned for sins? After all, Lamoureux finds the Holy Spirit's revelations quite accommodating about ancient categories of thought and belief. So maybe all that stuff about a resurrection was myth, too, just another way in which "the Lord comes down to the level of ancient peoples," which Lamoureux thinks happened when it came to "employing their view of medicine" (p. 144). The kid writhing on the ground foaming at the mouth really had epilepsy, not an evil spirit, despite what Mark 9:17-26 says. This kind can come out only by anticonvulsants and neurosurgery, not prayer and fasting. Lamoureux knows that. And this is in the New Testament, an event in "the Lord's ministry" for crying out loud, which is supposedly off-limits to demythologizing. The fact is that he must avert his gaze from the central story of his faith, lest in really seeing it he makes it disappear.

The evangelical evolutionists are like the Gentile "God-fearers" of the ancient world who admired Judaism but did not want to go the whole way to become official proselytes. They didn't want to embrace Jewish dietary customs and get circumsized. So they were allowed to make pilgrimage to the Jerusalem temple, but could approach only so near. They had to remain in the Court of the Gentiles, warned by a big plaque to come no closer on pain of death.

Denis Lamoureux, Francis Collins, Peter Enns, and the rest have seen enough of science to know they want to go all the way into its temple, bringing the Bible with them. The issue is certainly not one of scientific knowledge: Dr. Collins is the former head of the Human Genome Project and current director of the National Institutes of Health, and one of Lamoureux's three earned doctorates is in biology. But still they can't get past that dire threat, whose posted warning is not about ritual purity but scientific consistency. Its intellectual rulebook applies as much for historiography as it does for biology and genetics.

The requirement is simply *uniformitarianism*: One cannot factor alleged ancient miraculous interventions into the natural order on the mere possibility that they once occurred, much less because one wishes they had. But Francis Collins balks at this necessary operating principle, protesting that "there are clearly parts of the Bible that are written as eyewitness accounts of historical events, including much of the New Testament" (2006, 175). Writing with his co-author Karl Giberson, he says their BioLogos brand of theistic evolution "accepts the possibility of God-ordained miracles, of course, but looks for situations where no natural explanation could possibly apply, like the resurrection of Jesus Christ" (Giberson & Collins 2011, 71). Robert John Russell says we "should avoid an interventionist argument as far as possible," but there is still that non-negotiable subject matter: "obvious exceptions arise in the incarnation and resurrection of Christ" (Russell 2003, 343).

George C. Cunningham sees through this screen of pious obfuscation. He agrees that "Collins is correct in pointing out the dangers of this quest for scientific support." But Collins, he says, "defends his faith by claiming that the supernatural is beyond the scope of science." That isn't "a demonstration of their compatibility," just "a separation of scientific knowledge and religious faith." Collins recognizes the danger that

> you can "eviscerate the real truths of faith" and will need to decide "where to place a sensible stopping point." Unfortunately, Collins's stopping point is the use of flawed and unreliable written reconstructions of alleged eyewitnesses to the life of Jesus. [Cunningham 2010, 152-53, quoting Collins 2006, 209]

Peter Enns can almost discern the facts of the matter, but they seem to him as trees walking. The "foundational stories of Genesis seem to fit so well among other—clearly ahistorical—stories of the ancient world, in what sense can we really say that Israel's stories refer to fundamentally unique, revealed, historical events?" (Enns 2012, 37). He is talking about the Principle of Analogy. Why not use it on the gospels, too? (And this is the truth of the warning that if Adam be dismissed as mythical, so will Jesus.) In fact, Enns comes remarkably close to doing just that:

> [F]or many scholars of Christianity and ancient religions in general, the resurrection of Christ is every bit as mythical as Adam, given the commonality of resurrection stories in the ancient world—not only in Paul's time but also in other religions throughout antiquity. No student of early Christianity can afford to brush this aside. [p. 125]

However, he refuses to consider that possibility seriously:

> It is the recent event that Paul claims to bear witness to along with more than five hundred others who saw the resurrected Christ.... [T]he resurrection is not a cultural assumption that Paul makes about primordial time, as he does with Adam. It is for Paul a present-time reality, an actual historical event. [pp. 125-26]

But that is beside the point. It is hardly a present event for us. The discourse of 1 Corinthians 15 makes it seem recent, but then so was the slaying of the

Bull of Heaven for the writer of *Gilgamesh*. We are not Paul's contemporaries, and it is altogether circular to assume that we are reading an eyewitness account. Why does Enns stop where he does? He admits that "accepting the resurrection of Christ is truly a matter of faith" (2012, 126). Then let him stop making it substitute for historical reasoning.

Jesus Christ Superchimp?

We have seen (and *boy*, have we seen) the difficulties that modern genetics poses for traditional theology's understanding of Adam and Eve as the parents of the whole human race. Our intent is not to embarrass the ancient texts but, on the contrary, to defend their right to stand as monuments of ancient belief and thought without warping and twisting them in order to make them seem compatible with modern knowledge. In the same way, we must inquire as to the implications of scientific genetics for the genesis of the Second Adam, Jesus Christ. As with the more sophisticated forms of creationism, the conflict of science is not precisely with the Bible but with certain doctrines derived from a selective reading of it, with theology more than the biblical text itself. Will we find such conflicts in the case of the genetic origins of Jesus?

Christian doctrine has historically held that Jesus Christ was miraculously conceived of the Holy Spirit (the third person of the Godhead) in the Virgin Mary, with no contribution from Joseph or any mortal father. The resultant child, Jesus, was already as an infant both God and man. He was a single person (this personhood being derived from God and divine at its core), supporting two full natures, divine and human. He was not a demigod, half human and half divine. He was not God wearing a fleshly scuba suit. He was not even two-thirds human (possessing human body and soul, but no human spirit) yet fully God (the divine Logos substituting for the human spirit). As the Nicene Creed has it, Jesus Christ, God's Son, was "begotten, not made."

Theologians have seldom if ever contended that all these careful nuances were explicitly taught upon the biblical page. Had that been so, it should never have taken three centuries for the underlying issues and questions to arise. The credal definitions were hammered out by convocations of ancient bishops and theologians. Their goal was twofold. They wanted to frame a Christology compatible with the various biblical hints as to the nature of Jesus Christ. And they wanted to stay consistent with a doctrine of salvation by divinization (*theosis*), the infusion of God's immutable immortality into humans as if by a kind of transfusion from God (as Jesus Christ). Especially for the latter, the savior had to have been fully God, not some mere angel, prophet, or wisdom teacher.

As if all this were not difficult enough, new conundrums arise when we superimpose our understanding of genetics. For we have to ask: What is the nature of the genetic material that God must have caused to appear inside Mary and to unite with her set of chromosomes? Did God create them *ex nihilo*? If so, does that conflict with the Nicene assertion that the Son was begotten, not made? Presumably that claim refers to the eternal begetting of the Logos, the Father's eternally present generation of the Word. Was the divine Logos already united with this created set of human chromosomes when they merged with Mary's? It would seem so; otherwise we are entertaining Nestorianism, the heresy that the divine and human in Jesus never truly united, leaving two personhoods, as well as two natures, in Jesus Christ. Just as Cyril of Alexandria insisted that Jesus Christ had already been God even in diapers, orthodoxy would have to affirm that the newly created DNA in Jesus was already God as well as human.

Theistic evolutionists might wish to drop the virginal conception doctrine, winking and nodding that *of course* Joseph was not just an adoptive father, perhaps admitting what the Talmudic writers said about the Roman Pandera. Jesus' nuclear DNA was 100% human, all 46 chromosomes of it. Otherwise, as Paul Tillich warned, the virgin birth makes Jesus into a demigod. But then they would be stuck with a kind of adoptionism: a human Jesus who at some point was given divine honors. Such a scenario would seem to imply that the divine Logos "descended upon" the apeish chromosomal configuration, which had originally been alien to it. Where is the divinity?

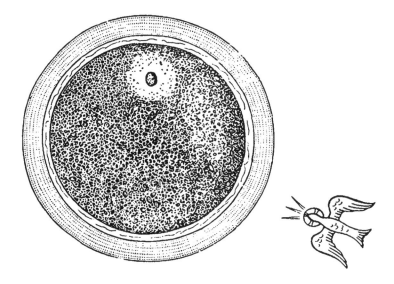

FIGURE 12: Mary's egg was fertilized by . . . what, exactly?

Let's look at what would be required for evolutionary compatibility with the more traditional view of the Incarnation: the union of divinely created DNA with the heavenly Logos. When the Holy Ghost came upon Mary and the power of the Highest overshadowed her (Luke 1:35), it brought a set of freshly constructed chromosomes, which fused with hers. Perhaps a single God-incarnate sperm cell was constructed near her egg to burrow into it and get the job done? This is not mockery, but an honest contemplation of how our Christian naturalists might attempt to bridge—just this once—the chasm between natural and supernatural conception.[57]

When the gospel stories were written, the Incarnation didn't seem so outrageous. Perhaps a century later (using the traditional dating), Tertullian thought it quite reasonable that Christ's flesh "came not of the seed of a human father," because

> Adam himself received this flesh of ours without the seed of a human father. As earth was converted into this flesh of ours without the seed of a human father, so also was it quite possible for the Son of God to take to Himself the substance of the selfsame flesh, without a human father's agency. [*Anti-Marcion*, On the Flesh of Christ, Ch. 16]

Mary seems relegated to being a mere sperm receptacle and warming oven with nutrient supply, as Eve was in Acts 17:26: God "made from *one man* every nation of mankind to live on all the face of the earth." In the fourth century, Ambrose made it quite clear that this is how he viewed Mary's role in bearing Jesus. Augustine quotes him approvingly in his *Treatise on the Grace of Christ and on Original Sin* against Pelagius, as follows:

> It was no cohabitation with a husband which opened the secrets of the Virgin's womb; rather was it the Holy Ghost which infused *immaculate seed* into her unviolated womb. For the Lord Jesus alone of those who are born of woman is holy, inasmuch as *He experienced not the contact of earthly corruption*, by reason of the novelty of His

---

57. Of course, there are theological befuddlements aplenty. To restate with uncomfortably direct language what the Nicene Creed actually posits, God has a first part and a second part that was somehow "eternally begotten" by the first part, as well as a third part that "proceeds from" the first part and is what made the second part "incarnate," by impregnating the virgin Mary. If it isn't weird enough for the monotheistic God of Israel ("Hear, O Israel: The LORD our God is one LORD," Deut. 6:4) to have three "persons," you can ponder the idea of the third part proceeding from the first part while the co-eternal second part is incarnated into a man by the third part. The thoughtful Christian has a lot more headaches to contend with than those arising from evolution, issues that one of your co-authors (Robert M. Price) has written about extensively in numerous other books.

immaculate birth; nay, He repelled it by His heavenly majesty. [Book II, Ch. 47, our emphasis]

The idea persisted in Augustine's admirer Luther, who mentioned the passage in Acts when commenting on the wonder of human reproduction; "man is brought into existence from a droplet of blood, as the experience of all men on the entire earth bears witness." The "woman receives semen," which "becomes thick and, as Job elegantly said (Job 10:10), is congealed and then is given shape and nourished until the fetus is ready for breathing air" (*Lectures on Genesis*, Ch. 2, v. 21). But that can't have been a universal view whenever the idea of the Immaculate Conception came along to nix concerns about Mary passing along the stain of Original Sin. (It wasn't made a Catholic dogma until 1854, with the Papal bull *Ineffabilis Deus*.) If she was just an incubator, what's the worry? Cooties?

Of course, we have now left the realm of science far behind. But let's continue considering this fusing of Marian and made-to-order chromosomes, because it raises some interesting questions. They arise from the fact that the divinely produced DNA would need the *appearance* of having come from an evolved human father. If those genes were somehow qualitatively different, how could they be compatible with what Jesus must have inherited from Mary? The paternal and maternal chromosomes have to line up:

> A single inversion on a chromosome inherited from one of the two parents causes a slight loss of fertility. So does a fusion. In fact, single chromosome rearrangements like inversions and fusions are a major cause of infertility in humans. Although one chromosome rearrangement is usually insufficient for complete loss of fertility, multiple rearrangements can result in complete sterility, as in mules. [Fairbanks 2007, 91]

It is not a question of a miracle to which we might appeal in order to leap the gap. If we say chromosomes are incompatible, we can't even imagine a miracle helping; that would be like saying a miracle can make a square peg fit into a round hole, or like saying that a miracle can produce a four-sided triangle. It's just meaningless.

So we are stuck with another case of the omphalos absurdity we saw in "Everybody's Working for the Weekend." The Holy Spirit injected a ready-made Y chromosome into Mary (along with 22 others from falsified meiosis in a non-existent human father), complete with endogenous retroviruses, fossil genes, and other hallmarks of evolution. Gosse had to protest that there was no deception intended by Adam's useless navel, nor concentric tree-rings in the first created trees, nor growth lines in the first shells (1857, 347). God just couldn't have made plants and animals without their "retrospective marks," or else we wouldn't have recognized them (p. 349). Today, all but brain-dead creationists see how ridiculous and dishonest that is, but how is this one special case of the Incarnation any different? "Man would not have been a Man without a navel," Gosse says (p. 349). Neither would Jesus be one without a Y chromosome *faked* to look like it had been passed down, with occasional mutations, from an endless line of human paternal ancestors.[58]

Then there is the darker side of our human nature. As we'll be discussing in "Sinful Selection," we owe the genetic component of that—the selfish, violent, and lustful potentialities—to ancestors who had what it took to survive and reproduce in a harsh and brutal world. Those who accept evolution, Christians included, know that they cannot look back wistfully on some pre-fallen humans who were free of such unsavory traits. There was no innocent, pre-Fall Adam.

Let's pause a moment to acknowledge the role of cultural evolution, which is responsible for much of what we consider human nature. It is true, as Ehrlich reminds us, that we must not give our genes a greater burden than they can bear. They "cannot incorporate enough instructions into the brain's structure to program an appropriate reaction to every conceivable behavioral situation or even a very large number of them" (p. 124). And natural selection is a blunt instrument, affecting seemingly unrelated features of the organism. A single gene may be used several times, its expression into the proteins of different body features regulated by separate genetic switches (Carroll 2005, 112-13).[59]

---

58. Matthew 1:2-17 and Luke 3:23-38 each provide a listing of such ancestors, with fatal discrepancies between the two. But of course, Joseph was merely "the husband of Mary, *of whom* was born Jesus" (Matt. 1:16), as Luke puts it, Jesus "being (*as was supposed*) the son of Joseph." The *Syriac Sinaitic*, a 4th century manuscript, renders Matthew's statement as follows: "Joseph, to whom was betrothed Mary the Virgin, *begat* Jesus, who is called the Christ" (from Agnes Smith Lewis, ed., 1910, *Old Syriac Gospels*, emphasis added). One wonders if the "of whom" and "as was supposed" disclaimer language is an interpolated product of the high Christology of later decades, which could no longer accommodate the adoptionism associated with a human paternal line. The genealogies are like fossil genes lingering in the DNA of the text, mutated into uselessness and contradiction.

59. The term "gene" is used, somewhat confusingly, to refer to both a unit of heredity and a particular segment of DNA that codes for a protein—or different proteins as regulated by different segments of non-coding DNA. Carroll's main thesis in *Endless Forms Most Beautiful* is that the other DNA is of critical importance in evolutionary development, the formation of organisms from embryos using a genetic toolkit that has itself been painstakingly evolved. In human DNA, 2-3 percent of "our dark matter contains genetic switches that control how genes," the segments of protein-coding DNA, are used (Carroll 2005, 112).

Ehrlich offers the example of fruit flies behaving differently after being selected in the lab for DDT resistance (p. 18). He is skeptical "that many human behavioral traits claimed to be the direct result of natural selection actually are" (p. 19). But he still recognizes a "substantial genetic influence," finding it "obvious that a large amount of genetic evolution was involved in developing the human capacity to have an extensive culture, and that evolution has resulted in a variety of human behavioral predispositions" (p. 123).

For Jesus to be a real man rather than an angel or docetic apparition, the divine part of his DNA would have needed to code for all those predispositions. The result would be *exactly the same* as if the DNA came from a mortal man having all those traits himself. Mustn't we say that Jesus, as a genuinely human being, was defined by the same animalistic genetic heritage as ours? With Holy-Spirit DNA that is in a sense defective by design, Jesus would not be immune from these inclinations.

Well, perhaps the "fallenness" of human nature entails only the *liability* to be tempted. After all, Christian doctrine freely admits that Jesus was "tempted at all points, even as we are" (Heb. 4:15). Does fallen nature necessarily mean that you cannot avoid sinning sooner or later? If it only means "temptability," then perhaps there is no problem and one need not posit that God equipped the incarnate Word with a set of "fallen" chromosomes. It's not far from Ehrlich's view of human nature itself: largely the product of cultural evolution built on what genetic evolution had first supplied.

But now we may be getting in deep over our heads with the old Christological riddle: "Was Jesus incapable of sinning, or capable of not sinning?" If the former, he was just accomplishing what would be expected of him, like the son of a family-owned business rising to a prominent executive position there. If the latter, Jesus was working (however admirably) with a fallen nature no better inherently than our own. In that case, his achievement would seem far greater, and this is just the way the fourth-century Arians viewed it, those who believed that the Logos was not of the very same impassible nature as the Father. There are issues for traditional Christian doctrine either way.

The Soggy Foundation of Original Sin

> *Now, if the book of Genesis is an allegory, then sin is an allegory, the Fall is an allegory, the need for a Savior is an allegory, and Adam is an allegory—but if we are all descendants of an allegory, where does that leave us? It destroys the foundation of all Christian doctrine—it destroys the foundation of the gospel.*
>
> —Ken Ham, Answers in Genesis

Historians of doctrine seem agreed that the belief in the atonement of Christ on the cross requires belief in Original Sin. If Adam did not fall and take the human race down with him, there is nothing to be saved *from*. If it were simply a matter of a generous God forgiving our trespasses, as we forgive one another, there would be no need for the cross. Unless Adam introduced a fatal and virulent toxin into the human race, no such radical surgery as the atonement provides would be needful. It is not so much a case of making the punishment fit the crime as of making the crime fit the punishment, for otherwise "Christ died in vain" (Gal. 2:21). But the reality of evolution erases all traces of a historical Adam, and therefore of any historical Fall. Christian evolutionists know that. And it sends them hastening back to the drawing board to rewrite the Eden story and salvage as much of it as they can.

It is clear that these biologists are winging it, making things up as they go. They do not have the appearance of researchers seeking to uncover the facts. They seem rather to be spin doctors or defense attorneys desperately trying to concoct a good story. And that, as we will see, is just what they are doing.

Visiting the Iniquity

"Original Sin" is the doctrine, usually attributed to Augustine, that the disobedience of Adam—eating the fruit of the Tree of the Knowledge of Good and Evil—destroyed the virtue with which he was created. It introduced a taint of both guilt and depravity that Adam and Eve would pass on to their offspring, ultimately to the entire human race. It's a grotesque extension of the inherited guilt that Yahweh threatened against those who might dare break his law given on Sinai, "visiting the iniquity of the fathers upon the children, and upon the children's children, unto the third and to the fourth generation" (Exod. 34:7). At least that version seemed to have a statute of limitations for great-great-great grandchildren.

Ingersoll put the matter thus: "This God waiting around there, knowing all the while what would happen, made [Adam and Eve] on purpose so it would happen; and then what does he do? Holds all of us responsible; and we were not there." Our focus is on the problems evolution raises for this idea (and they are formidable), but Ingersoll's acerbic summary of the theological problems is well worth quoting:

> What did Adam do? I cannot see that it amounted to much anyway. A god that can create something out of nothing ought not to have complained of the loss of an apple. I can hardly have the patience to speak upon such a subject. Now, that absurdity gave birth to another—that, while we could be rightfully charged with the rascality of somebody else, we could also be credited with the virtues of somebody else; and the atonement is the absurdity which offsets the other absurdity of the fall of man. Let us leave them both out; it reads a great deal better with both of them out; it makes better sense. [*Lecture on Orthodoxy*, 1884]

We have all inherited the guilt of Adam and Eve, the story goes. (Eastern Orthodox and Protestant Christians are not uniform in sharing this feature of the doctrine.) It's based on the first chapters of the Hebrew scriptures, but the idea comes entirely from Christianity. In her comprehensive and refreshingly objective study of Original Sin, Tatha Wiley notes the origins of the doctrine in "the early Christian era," when "the story of expulsion from the garden became the primary revelatory text for why the forgiveness of Christ is universally necessary" (2002, loc. 403). The Jews certainly showed no interest in it, failing to mention Adam or Eve in any other writings they included in the Hebrew canon (loc. 420).[60]

Wiley shows how the "theology of Original Sin developed incrementally in the patristic writings." It was "a response to a broad range of questions—the relation of God to evil, human nature, the reason for divine redemption, the necessity of Christ, the practice of infant baptism, and the role of the church in God's plan of salvation" (loc. 471). The few Christian writings preserved from the first two centuries C.E.—the *Didache*, *First Clement*, *Barnabus*, Ignatius, *The Shepherd* of Hermas, Justin Martyr—did not have the doctrine

---

60. Wiley cites two apocryphal, noncanonical writings that do make a brief mention of Adam or Eve: "In the Apocalypse of Moses (first century C.E.), Eve says to God, 'I have sinned, I have sinned, God of the universe, I have sinned against you . . . against the angel elect . . . against the cherubim; and through me all sin entered creation.' In Baruch 54:19 (early second century C.E.), Adam is the focus: 'But each of us has been the Adam of his own soul'" (Wiley 2002, loc. 421).

fully formed (loc. 475-85). The "conception of sin as an inherited condition was not yet on the horizon" (loc. 485).

Justin, whose writings are traditionally dated at around 160 C.E., "referred to the sin of Adam, to fallen humanity, and to the power of sin and death" in his *Dialogue with Trypho*, "but without further explanation" (loc. 526). He "brings out the relation between sin and death found in the ancient Hebrew writers and in Paul. By their sin, human beings brought death upon themselves." But Wiley believes Justin "was not concerned with developing an explicit principle explaining what he took as fact, the universality of sin." Rather, Adam and Eve's sin was "the prototype of personal sin" (loc. 530).

Around the end of the second century, as a proto-orthodoxy was finally starting to show some dominance in the cultural evolution jungle of competing Jesus cults, Irenaeus came along with his *Against Heresies*. He fended off opposing views of Jesus at two extremes: the claim of docetism that Jesus had only the illusion of humanity, and the adoptionist view that Jesus began his life as the fully human offspring of both Mary and Joseph. His position, which became the orthodox one, was that Jesus was born of a virgin, and somehow both human and divine. To defend it, he brought up Adam and Eve:

> For the Lord, having been born "the First-begotten of the dead," and receiving into His bosom the ancient fathers, has regenerated them into the life of God, He having been made Himself the beginning of those that live, as Adam became the beginning of those who die. Wherefore also Luke, commencing the genealogy with the Lord, carried it back to Adam, indicating that it was He who regenerated them into the Gospel of life, and not they Him. And thus also it was that the knot of Eve's disobedience was loosed by the obedience of Mary. For what the virgin Eve had bound fast through unbelief, this did the virgin Mary set free through faith. [Book III, Ch. 22]

It's a mouthful, but the point for our discussion is this: Perhaps for the first time since Paul, a Christian writer made the Fall binding on everybody who followed the original perpetrators. We all became tied up in the "knot of Eve's disobedience." He also mentions God's "condemnation, which had been incurred through disobedience," which made Jesus necessary as a "recapitulation" (Ch. 23). This was Irenaeus's interpretation of Paul, says Wiley, with an emphasis on obedience:

> Christ took up all things since the beginning into himself as the recapitulation of Adam. His obedience destroyed the effects of Adam's disobedience, reclaimed humankind from the devil, restored God's plan of salvation, and reestablished the process of divinization begun in Adam but interrupted by sin. [2002, loc. 507]

Three other important writers of the pre-Nicene period would at least mention the events in Eden: Clement of Alexandria, Tertullian, and Origen. Clement of Alexandria was an acestic sourpuss who wrote page after page restricting the most innocent pleasures of life: married sex, bathing, laughing, even smiling. He places the blame for our condition on "that wicked reptile monster [who], by his enchantments, enslaves and plagues men even till now." That's not Godzilla he's talking about, but the Edenic Serpent. And that's how Clement makes a connection with the first parents: shared experience with the "seducer," who "is one and the same," rather than ancestry. He "that at the beginning brought Eve down to death, now brings thither the rest of mankind" (*Exhortation to the Heathen*, Ch. 1).[61]

Tertullian mentions the Fall, but in a way that shows the absurdity of Original Sin as clearly as any New Atheist writing today: "Man is condemned to death for tasting the fruit of one poor tree, and thence proceed sins with their penalties; and now all are perishing who yet never saw a single sod of Paradise" (*Anti-Marcion*, Book 1, Ch. 22). It's not that he thought the Fall didn't happen or was irrelevant. It merely gave an inclination toward sin, for which infants needed no baptism of forgiveness (Wiley 2002, loc. 563).

Of the three, it is Origen who might be considered the founder of the Original Sin doctrine. (Perhaps it should be called "*Origenal* Sin"? Sorry.) He

> correlated infant baptism with sin and explicitly named the sin "original sin." He answered the obvious theological question raised by offering infants the forgiveness of baptism: What kind of sin could exist in an infant? "All are tainted with *the stain of original sin* which must be washed off by water and the spirit." The premise of original sin shaped Origen's understanding of the purpose of redemption and the role of the church. "Certainly," he wrote, "if there were nothing in infants that required remission and called for lenient treatment, the grace of baptism would be unnecessary." [loc. 571]

But the name that will always be associated with Original Sin, alongside Adam, is Augustine of Hippo (354-430 C.E.). He famously argued with Pelagius about human nature and divine grace (loc. 721). Wiley thinks his thoughts on the subject preceded that conflict. She notes that, in his *Confessions* written a decade earlier, Augustine distinguished between his

---

61. Though he doesn't explicitly mention "Satan," Clement seems to commit the same error as Irenaeus (and modern Christians) of making the Serpent out to be more than he actually was. Luther offered the nuance that Satan was hidden *within* the serpent (*Lectures on Genesis*, Ch. 3, v. 14).

personal sins and what he said was "the chain of original sin by which 'in Adam we die,'" citing 1 Corinthians 15:22 (Wiley 2002, loc. 725).

Unquestionably, though, it is Augustine's dispute with Pelagius that generated the earliest fully formed writings on Original Sin. Over and over again he cites Romans 5, doing a detailed exegesis of the passage in Book II of his Letter to Valerius:[62]

> Let these objectors tell us how it can be "by one offence unto condemnation," unless it be that even the one original sin which has passed over unto all men is sufficient for condemnation? Whereas the free gift delivers from many offences to justification, because it not only cancels the one offence, which is derived from the primal sin, but all others also which are added in every individual man by the motion of his own will. "For if by one man's offence death reigned by one, much more they which receive abundance of grace and righteousness shall reign in life by One, Jesus Christ. Therefore, by the offence of one upon all men to condemnation; so by the righteousness of one upon all men unto justification of life." Let them after this persist in their vain imaginations, and maintain that one man did not hand on sin by propagation, but only set the example of committing it. How is it, then, that by one's offence judgment comes on all men to condemnation, and not rather by each man's own numerous sins, unless it be that even if there were but that one sin, it is sufficient, without the addition of any more, to lead to condemnation,—as, indeed, it does lead all who die in infancy who are born of Adam, without being born again in Christ? [Ch. 46]

Augustine wonders why his opponent searches "for a hidden chink when he has an open door." It is all very clear to him:

> "By one man," says the apostle; "through the offence of one," says the apostle; "By one man's disobedience," says the apostle. What does he want more? What does he require plainer? What does he expect to be more impressively repeated? [Ch. 47]

The charge was made even then that Original Sin was Augustine's idea, and he denied it: "It was not I who devised the original sin, which the catholic faith holds from ancient times; but you, who deny it, are undoubtedly an innovating heretic" (Ch. 25). And that heresy was not to be tolerated: "Now,

---

62. Attributing this work is very confusing. We are working from a Kindle book entitled *The Collected Works of 46 Books by St. Augustine*, one of which is *On Marriage and Concupiscence*, edited by Philip Schaff (1887). That, in turn, consists of a single Letter Addressed to the Count Valerius made up of three "Books," the second of which we cite.

whoever maintains that human nature at any period required not the second Adam for its physician, because it was not corrupted in the first Adam, is convicted as an enemy to the grace of God" (*Grace of Christ*, Book II, Ch. 34).

The Eastern Orthodox only eat part of the Augustinian apple, saying that what we have inherited are Adam's corruption and mortality, not his guilt from what many of them prefer to call *ancestral* sin (Shingledecker 2013). But for Augustine and the bulk of Christianity in the 1600 years since, the bony finger of accusation is pointed at us from the moment we emerge from the womb. If mere imitation of Adam's mistake were all that Paul had meant, Augustine says Paul "would have said that sin had entered into the world and passed upon all men, not by one man, but rather by the devil." Instead, he "used the phrase 'by one man,' from whom the generation of men, of course, had its beginning, in order to show us that original sin had passed upon all men by generation" (Letter to Valerius, Book II, Ch. 45). The rot goes so deep that an infant is born condemned, and would head straight to hell if it died before being baptized.

> It may therefore be correctly affirmed, that such infants as quit the body without being baptized will be involved in the mildest condemnation of all. That person, therefore, greatly deceives both himself and others, who teaches that they will not be involved in condemnation; whereas the apostle says: "Judgment from one offence to condemnation," and again a little after: "By the offence of one upon all persons to condemnation." [*Merits and Forgiveness of Sins*, Book I, Ch. 21]

We doubt that even the most fundamentalist of our current Bible-thumping creationists seriously entertains the absurdity of newborn babies being tortured for eternity. Augustine himself came across those "who think it unjust that infants which depart this life without the grace of Christ should be deprived ... of eternal life and salvation." Rather than address that for the outrage that it is, Augustine simply cites the broader outrage of Christian exclusivity. He asks why his critic is "not similarly disturbed by the fact that of two persons, innocent by nature, one receives baptism, whereby he is able to enter into the kingdom of God, and the other does not receive it, so that he is incapable of approaching the kingdom of God?" (Ch. 30). The only response to the monster that theology had created was, as it is now, to create another one: the denigration of our own sense of morality. Just shrug your shoulders and blame God for the horrors and cruelty you've imputed to him: "How utterly insignificant, then, is our faculty for discussing the justice of God's judgments" (Ch. 29).

Even on its face, Original Sin is a pathological and offensive doctrine. It is an oddity of religious cultural evolution that first mutated into existence with Paul, then into several competing forms. One of those, Augustine's, was culturally selected over centuries to provide a rationale for an atoning Jesus. Most of the Christian creeds that are being propagated in churches today include versions of it, with elaborate camouflage further evolved to conceal the underlying absurdity: Limbo for unbaptized infants (a doctrine recently gone extinct), an ill-defined "age of accountability," even universalism. But now, with the certain knowledge that there was never any Original Sinner, the doctrine of Original Sin stands fully exposed, unable to hide any longer.

Paul and Adam

Our Christian evolutionists try to avert their eyes and pretend to find some redeeming qualities to Original Sin, because they have to. A fundamental aspect of Christianity—the act of a universal sinner required a universal redeemer—is at stake, though some of them try to deny or minimize that. And, regardless of the modern efforts at explaining away an ancient and ill-informed doctrine, they still need to figure out what to do with Paul.

Enns and Lamoureux struggle manfully with the problem of having to take Paul seriously. As the major Christian apostle, what he says about Adam and the origin of sin is even more important to Christian thinking than anything found in Genesis. And Paul has plenty to say (perhaps too much, given the trouble it causes!) about Adam's sin. As is well known, Romans 5:12-19 draws a typological connection between the First and the Last Adam. Adam damned all, while Christ saves all, each through a pivotal deed with universal ramifications:

> Therefore, just as through one man sin entered into the world, and death through sin, and so death spread to all men, because all sinned —for until the Law sin was in the world, but sin is not imputed when there is no law. Nevertheless death reigned from Adam until Moses, even over those who had not sinned in the likeness of the offense of Adam, who is a type of Him who was to come.
>
> But the free gift is not like the transgression. For if by the transgression of the one the many died, much more did the grace of God and the gift by the grace of the one Man, Jesus Christ, abound to the many. The gift is not like that which came through the one who sinned; for on the one hand the judgment arose from one transgression resulting in condemnation, but on the other hand the free gift arose from many transgressions resulting in justification. For if by the transgression of the one, death reigned through the one,

much more those who receive the abundance of grace and of the gift of righteousness will reign in life through the One, Jesus Christ.

So then as through one transgression there resulted condemnation to all men, even so through one act of righteousness there resulted justification of life to all men. For as through the one man's disobedience the many were made sinners, even so through the obedience of the One the many will be made righteous. [Rom. 5:12-19, NASB]

The passage is repetitive, so much so in fact, that it is tempting to suspect it has suffered from scribal glosses as copyists added explanatory notes in the margins, notes which were subsequently mistakenly copied into the text the next time a new copy was needed. But, assuming this is all the work of a single author, we see that he did envision Adam's sinful deed introducing death as each human being thereafter actualized his inherited sinfulness by his own acts of sin. The phrase "because all sinned" by itself might imply a Pelagian understanding that each man is his own Adam, each woman her own Eve, starting out fresh but soon repeating the Fall. But this seems to be ruled out by the later qualification that all were "made sinners" by the primordial disobedience of Adam. So it does seem fair to say that the Romans passage teaches at least the basic notion of Original Sin as a will-warping taint inherited from Adam.

Paul makes the same basic point in 1 Corinthians 15:21: "For since by man came death, by man came also the resurrection of the dead. For as in Adam all die, even so in Christ shall all be made alive." Augustine recognized this; the "statement which the apostle addresses to the Romans ... tallies in sense with his words to the Corinthians" (*Treatise on Forgiveness*, Book III, Ch. 19). Paul intended no difference in meaning by referring to death instead of sin in the latter passage, says Augustine, since his discourse there was focused on resurrection rather than righteousness (Book I, Ch. 8).

Naturally all this creates a major problem once one realizes there was not only no Adam, but no First Man—no matter what his name was. It seems that *the very purpose* of Pauline Christianity is to solve a supposed problem that is rooted in myth rather than fact.

Protected from theological nuance by their ignorance and denial of the scientific proof for evolution, the fundamentalists have unwittingly become some of the best expositors of the problem. Ken Ham's protest about allegory is memorable (and, we think, accurate) enough to appear as the epigraph to this section. Similarly, Albert Mohler of the Southern Baptist Convention has this to say:

> The denial of an historical Adam and Eve as the first parents of all humanity and the solitary first human pair severs the link between Adam and Christ which is so crucial to the Gospel. If we do not know how the story of the Gospel begins, then we do not know what that story means. Make no mistake: a false start to the story produces a false grasp of the Gospel. [Mohler 2011]

Luther took the story's theological significance for granted, steeped as he was in Paulinism and innocent of any plausible alternative to the idea of a first human pair. The Fall created the need for the savior:

> The Son of God had to become a sacrifice to achieve these things for us, to take away sin, to swallow up death, and to restore the lost obedience. These treasures we possess in Christ, but in hope. In this way Adam, Eve, and all who believe until the Last Day live and conquer by that hope. [*Lectures on Genesis*, Ch. 3, §15]

A former liberal Protestant minister phrased the issue as a "prime example" of the core doctrines of Christianity that evolution challenges:

> The role of Christ as the Second Adam who came to save and perfect our fallen species is at the heart of the New Testament's argument for Christ's salvific significance. St. Paul wrote, "Therefore, just as one man's trespass led to the condemnation of all, so one man's act of righteousness leads to salvation and life for all" (Romans 5:18). Over the centuries this typology of Christ as the Second Adam has been a central theme of Christian homiletics, hymnody and art. More liberal Christians might counter that, of course there was no Adam or Eve; when Paul described Christ as another Adam he was speaking metaphorically. But metaphorically of what? And Jesus died to become a metaphor? If so, how can a metaphor save humanity? Really, without a doctrine of original sin there is not much left for the Christian program. If there is no original ancestor who transmitted hereditary sin to the whole species, then there is no Fall, no need for redemption, and Jesus' death as a sacrifice efficacious for the salvation of humanity is pointless. The whole *raison d'etre* for the Christian plan of salvation disappears. [Aus 2012]

If some among the first myriad of humans sinned and others did not, would that mean that the Sethian Gnostics were right in believing in two human races, the "kingless race" of Seth and the sinful race of Cain? On the other hand, one might admit that a few rotten apples would provoke sin in the whole barrel, but that would be Pelagianism again, and it is not clear an atonement would be either necessary or effective in repairing the damage. So the Pauline problem is twofold. First, he was wrong about there having been a First Adam; second, his ascribing to Adam a universal taint of sin running so deep as to require not merely "spot forgiveness," sin by sin, but radical metaphysical surgery through the cruciform atonement goes up in smoke. "Then Christ died in vain" (Gal. 2:21).

Faced with this dilemma, some of our Christian evolutionists perform a bit of radical surgery of their own. They are quite willing to admit Paul was wrong about Adam and to resort to "limited inerrancy," saying that Paul's assertion is correct (Christ does eradicate Original Sin), but the terms in which that assertion is cast are negligible. We are to see in the Romans 5 argument a parallel to Jesus' insignificant inaccuracy in the Mustard Seed parable: The mustard seed turns out not to be "the smallest seed on the earth" (Mark 4:31), but so what? How does that effect the point of the parable? It's small enough! But do we really have a true parallel here? Paul's appeal to Adam and his imagined sin is not some illustrative window dressing but rather the premise of an argument. How can we jettison the logical and scriptural arguments that led Paul to his theological conclusions yet retain his conclusions as if they transcended their supports?

Enns claims that "Paul's handling of Adam is *hermeneutically* no different from what others were doing at the time: appropriating an ancient story to address pressing concerns of the moment. That has no bearing whatsoever on the truth of the gospel" (2012, 102). But how can it not? If we kick away the whole Adam-Christ typology as mere first-century trappings, why did Christ have to die? The atonement of Christ is not only left hanging in mid-air, but it is even evacuated of all meaning: What is it he supposedly did, then, besides get executed?

And what a telling choice of words when Enns refers to "what remains of Paul's theology" (2012, 123). Why draw a magic circle around what's left? Why not acknowledge that all the terms of his message, and therefore his message itself, are at home only in a way of thinking that we have long since discarded? Isn't the whole notion of a dying and rising savior appeasing a judgmental deity just as obsolete as the idea of a first human progenitor suddenly popping into existence—complete with endogenous retroviruses, fossil genes, vestigial traits, and a belly button?

Natural Election

Catholic theologian Ansfridus Hulsbosch (*God in Creation and Evolution*, 1966) prefers a different understanding of Adam's fall and Christ's redemption. Irenaeus and Athanasius both advocated the Recapitulation theory. According to it, Adam did not fall from a position of righteousness but rather swerved off the course of maturation and perfection that God intended for him. Adam sidetracked the human race. Jesus Christ comes to put us back on course, pressing on toward perfect maturity. Hulsbosch contended that this model fits in marvelously well with evolution. We need not worry about a taint and how it came to infect humanity. It is rather a lack of maturity, and Christ helps us to mature.

Hulsbosch reminds us of the difference anthropologists draw between natural selection and cultural selection. Natural selection would eventually eliminate near-sightedness after enough near-sighted people got run over by cars! No more descendants to spread their defective occular genes! It is the game of chance between mutation and environment.

But cultural selection means that we can turn the course of the river of evolution by conscious decision. We decide that individual lives mean as much as species survival, so we invent eyeglasses. They save us from getting killed, but now the species will never be rid of near-sightedness. Okay, we say (especially those of us wearing glasses): It's a fair trade.[63]

You are what you are biologically. On this, at least, Freud was right: Biology is destiny. As Jesus said, "You cannot by trying add a single cubit to your height" (Matt. 6:27). Natural selection is over for you. But cultural selection is not. You can evolve further that way. You *can* cultivate the genes natural selection gave you for altruistic, or at least cooperative, behavior. Genes whisper suggestions rather than shouting commands, as Ehrlich puts it (2000, 7). This is what Christian theology calls sanctification. This is what Colossians 3:9-11 and Ephesians 2:11-15 mean when they urge you to put on the new humanity because Christ has made obsolete the old human species and produced a new humankind in his image. That's cultural selection.

How, pray tell, did this upgrade get accomplished? One might infer that a Pelagian understanding of Adam and Christ would be the most natural next step. In opposition to Augustine with his doctrine of inherited guilt and taint and evil inclination, Pelagius held that Adam and Eve had "merely" built a society on the basis of, and according to the rules of, selfishness and sin. Born into such a world, and socialized into its ways, no human being can escape becoming a sinner, too. What Jesus Christ did was essentially to start a new, alternative community in which people might come up in a society founded upon righteousness as taught by its founder, Jesus. God's grace is not some saving and sanctifying power mysteriously at work in those who partake of the Church's sacraments. It is simply God's generosity in providing this wholesome matrix in which one may thrive apart from the sinful world outside. Nothing mysterious about that.

Take Alcoholics Anonymous as an analogy, or as an example, since it is a kind of scaled-down version of the Christian Moral Rearmament movement

---

63. These first three paragraphs are adapted from Price 2010.

(Buckmanism). There is nothing particularly spooky about it. AA delivers alcoholics from their besotted hell by means of strong group accountability and peer support. So with the Church according to Pelagius, himself a monk who lived in a monastic community. And, as Augustine argued, there is no real role for Christ's death in that scenario. He is more of a teacher than a redeemer. This is why the Liberal Protestants of the nineteenth and early twentieth centuries revived a form of Pelagianism that placed its hopes not on evangelism but on education and social reform. Having rejected the atonement as "butcher shop religion," they preferred a social, collective approach creating a new, environmental conception of sin and its remedy via the Social Gospel.

The evolutionist theologian Korsmeyer, despite being a Catholic, seems to advocate Pelagianism as the natural outcome of his abandonment of Augustinian Original Sin. He applies some nuance to the word "generation" that the Council of Trent failed to convey.

> Insights from the social sciences have been helpful in consideration of the transmission of original sin, said by Trent to occur by generation, not imitation. But Trent did not discuss the meaning of "generation." For modern theologians it is considered to apply to the whole process of socialization that occurs when humans enter the world and are assimilated into the family, the local community, and the wider social sphere. The principal idea is that humans find themselves thrust into a world where sinful people, activities and structures are already in place, and influencing and training the mind long before it is capable of making fully free moral decisions. We are contaminated by evil just by being born, by our generation into the human race, and not just by imitating our ancestors. [1998, 57]

That's not Pelagianism? Doesn't Korsmeyer really mean that the modern social sciences have proven Augustine wrong and Pelagius right? That heresy has evolved into orthodoxy? He'd be closer to the truth in saying that, of course, but if he does, his book can kiss the *Nihil Obstat* and *Imprimatur* goodbye.

Gardener's Guilt

Despite Jewish indifference and a slow evolution of the doctrine by early Christians, Original Sin—the Fall of Man—is supposedly a clear teaching of Genesis 2-3. How could anyone not see it there, C. John Collins wonders. Where, pray tell, does he profess to behold it? Well, it is not quite seeing either the forest or the individual trees. It is more like seeing Bigfoot sneaking between the trunks out of the corner of your eye.

> When we ask what a Biblical author is "saying" in his text we are not limited to the actual words he uses. For example, we will note that Genesis 3 never uses any words for sin or disobedience; but it would be foolish indeed to conclude that what Eve and Adam did was not "sin." The author wants us to see that it was indeed, and to be horrified. [Collins 2011, 25-26]

He finds it absurd to suggest that "the text does not 'teach' that Adam and Eve 'sinned,'" just "because there are no words for 'sin' or 'rebellion' in Genesis 3." God's question in Genesis 3:11 ("Hast thou eaten of the tree, whereof I commanded thee that thou shouldest not eat?") is "as good a paraphrase of disobedience as we can ask for." What about the fact that "the text of Genesis does not say that humans 'fell' by this disobedience"? Both "objections stem from a failure to appreciate that Biblical narrators tend to prefer the laconic 'showing' to the more explicit 'telling,' leaving the readers to draw the right inferences from the words and actions recorded" (p. 61).

But what are these "right" inferences? For Collins, they are the ones that align the text with Christian theology. The far more adventurous Lamoureux dares to say, "Significantly, the category 'original sin' is not in the Bible. This concept was formulated by St. Augustine" (2008, 291). The punch line, for him, is merely that "all humans are sinners" (p. 292). Nonetheless, Original Sin may yet be necessary to the orthodox theology that he, like C. John Collins, wants to retain. This, obviously, is why he insists on interpolating it into the Eden story.

Robin Collins considers several possible interpretations for Original Sin, which seem to us like a number of emergency fire exits arrayed around a room such that one will be nearby no matter where you're sitting. The one that aligns with science is the "biological interpretation," where Original Sin is "nothing more than biologically inherited propensities, such as aggressiveness and selfishness, that help the individual or one's kinship group survive but typically do not promote the flourishing of the larger community" (Collins 2003, 495).

This stands slightly apart from the main point of Original Sin. We are not just damned at birth, but also depraved, having inherited a tendency to heap our own sins on top of the first one. John Calvin, that cheerful theological thinker, called it "total depravity." It doesn't mean everyone is as bad as they could be all the time, that we are a species made up of Mansons and Dahmers. It just means, a la Freud, that our every action, no matter how apparently noble, has a selfish motivation, which may be hidden from oneself and others. Jonathan Edwards (another ray of sunshine; he gave the infamously sadistic sermon "Sinners in the hands of an angry God") explained it like this: We do have a kind of free will, but the range of choices open to us are only sinful ones. Ingersoll did not spare this aspect of the doctrine his scathing wit in his *Lecture on Orthodoxy*, either: "A god that cannot make a soul that is not totally depraved, I respectfully suggest, should retire from the business. And if a god has made us, knowing that we would be totally depraved, why should we go to the same being for repairs?"

This goes to the question of what "sins" are, really, when looking at ourselves as evolved primates. We'll discuss that further in the next chapter, but for now let's consider what the concept of sin would have meant to the Bible writers. Then we may be better able to decide whether such sin is even the sort of thing theologians claim to find in the Eden story.

The concept seems to have originated as a function of ritual taboos, having nothing, at first, to do with moral misdeeds. No one ever supposed you were doing anything "immoral" by eating a ham sandwich. The offense was rather a ceremonial transgression, violating the food taboos implied in the cultural taxonomy (just as we don't eat roaches or dogs though some cultures do). Thus "sins" were specifically incurred against God, not other mortals. The latter is where the categories "immoral" and "criminal" applied (Otto 1924).

As of the Prophets (especially Isaiah), wrongs done against neighbor and employee began to be considered to offend God, too. Isaiah excoriates those who sashayed into the temple ready to impress God with their piety but who had engaged in unjust business dealings, etc., out in the everyday world (Isa. 1:11-18). God didn't want their stinking worship as long as they were exploiting others. This seemingly had never occurred to most people. They figured God was only concerned with sacrifice formulas, ritual propriety, etc., as when he kills Nadab and Abihu for ritual carelessness, offering an improperly mixed incense ("strange fire," Lev. 10:1-3), or commands the execution of some poor jerk for gathering kindling on the sabbath (Num. 15:32-36), or atomizes poor Uzzah for touching the Ark of the Covenant, trying to steady it when it teetered on the brink of the ditch (2 Sam. 6:6-7). Isaiah and the others were part of the phenomenon of interiorizing and moralizing religion world-wide over a period of a thousand years that Karl Jaspers dubbed "the Axial Age" (Jaspers 1953). Henceforward, it was believed that moral wrongs committed against one another were also to be deemed as offenses to God, thus moralizing the concept of "sin."

Hand in hand with this development went another: The concept of divine "holiness" came to be moralized, rationalized. Originally, "holiness" denoted "separation" between God and the world as well as the "Wholly Other" character of the divine, its dreadful, frightening uncanniness. God was beyond good and evil, strictures which pertained only to puny human affairs, just as our pets are bound by rules irrelevant to us. "I create peace, and I create evil" (Isa. 45:7). "Is there evil in the city and Yahweh has not done it?" (Amos 3:6). Once God became a moral being, the evil had to be sloughed off onto the back of a different supernatural being, and this is why Jewish thinkers borrowed the Zoroastrian Ahriman, the evil anti-god, with whom Satan (hitherto God's deputy) was merged to give us a devil.[64]

The Only Expiation I Can See

Sacrifices "expiated" the pollution caused by ritual "sins" by "washing" it away in the shed blood of the animal's slit throat. It looks as if the sacrificial interpretation of the death of Jesus began as a way for Jewish Christians to explain how God could accept believing Gentiles who, though they now believed in Jesus, had never either offered Jewish temple sacrifices nor even observed the dietary laws of the Torah. God hadn't condemned them unless they were immoral, not even for non-observance of Jewish law since the Torah applied only to Jews. But if God was now to accept Jesus-believing Gentiles on a par with Jews, they had to be "cleaned up" from their long-standing ritual "uncleanness." This Jesus' sacrifice supposedly did (Williams 1975).

The rather different notion of the "substitutionary atonement" or "penal substitution" extended Jesus' sacrifice to wipe away the guilt of *immorality* insofar as it offended God. Repentance was no longer enough to save you, as it had been in the teaching ascribed to Jesus in the Synoptic Gospels. But why? Why isn't God's willingness to forgive the repentant (as illustrated in the Prodigal Son parable) enough? Because early Christians needed some reason for Jesus to have died. Individual sins, even a great number of them, might be simply forgiven, but to justify Jesus' death, something much worse had to be at stake, something requiring radical surgery.

Some of our Christian evolutionists are quite open about the seeming fact that the Original Sin doctrine was reverse engineered from belief in the salvation wrought by Jesus Christ on the cross. The solution begat the problem. For example, Jerry Korsmeyer says, "Original sin *explains* the need for Christ's coming, his death and resurrection. No Adam, no fall; no fall, no atonement; no atonement, no Savior. One can understand why this

---

64. "Satan" was at first a title, not a proper name, and it belonged to one of the sons of God (Job 1:6) who served as God's special assistant to monitor and test the mettle of God's favorites, like Job (Job 1:9-11), David (1 Chron. 21:1) and Joshua, the favored candidate for the high priesthood (Zech. 3:1). Isaiah 14 refers to the humiliation of the minor deity Halal (the planet Venus), while Ezekiel describes the expulsion of the cherub-guarded Adam from Eden; neither has to do with a fallen angel named Lucifer or Satan. Nor was the Serpent in Eden identified with Satan until much, much later, in apocryphal books like *The Life of Adam and Eve* and *The Secrets of Enoch*.

version of Christianity requires original sin" (1998, 22). Wiley concurs: "In the early Christian era, the story of expulsion from the garden became the primary revelatory text for why the forgiveness of Christ is universally necessary" (2002, loc. 403).

The important thing is Jesus' atonement, they seem to be saying, not the various ancient rationalizations for it. Can we not retain the saving death on the cross without the ancient guesses as to why it was necessary? What they do not seem to see is that the death of Jesus cannot be understood as a solution unless there is a prior problem to solve. And evolution replaces the whole Adam "problem" with a scientific explanation.

> The story of salvation by the cross makes no sense against a background of evolutionary naturalism. The evolutionary story is a story of humanity's climb from animal beginnings to rationality, not a story of a fall from perfection. It is a story about recognizing gods as illusions, not a story about recognizing God as the ultimate reality we are always trying to escape. It is a story about learning to rely entirely on human intelligence, not a story of the helplessness of that intelligence in the face of the inescapable fact of sin. [quoted in Fairbanks 2007, 161]

That cogent summary of the theological dilemma was not written by some atheist critic of Christianity. Its author was Phillip Johnson, founder of the (ostensibly non-religious) Intelligent Design movement. Wrong as he is about the "science" of ID, Johnson is right on the money about this: "There is no satisfactory way to bring two such fundamentally different stories together, although various bogus intellectual systems offer a superficial compromise to those who are willing to overlook a logical contradiction or two" (p. 162).

Lamoureux acknowledges that there "is no sin-death problem," since "Adam never existed, and therefore suffering and death did not enter the world in divine judgment for his transgression." But he fails to see the implications when he claims that "the divine revelation in Gen 3, Rom 5-8, and 1 Cor 15 is very simple: humans are sinners, God judges sin, and Jesus died for sinful men and women" (2008, 329). Evolution raises one glaring question in response to Lamoureux's point, which is slightly off the topic of Original Sin and will be discussed in the next chapter, "Sinful Selection": Men and women are sinful because of *what*?

Original Sin, however, is traditionally viewed as the main issue, what Enns cites as the "universal problem" of Romans 5. God's solution to it, which he finds in Romans 5:1-11, is none other than "the death and resurrection of Christ" (2012, 133). Sin and death "remain the foes vanquished by Christ's death and resurrection," even if we don't try to say they were Adam's fault

(p. 125). Well, *how* did Jesus' death remedy things? Repeating the ancient formula doesn't provide an answer. You can't just omit the means and keep the end. You might prefer a different atonement theory, but you need *some* reason for Christ's death and resurrection to be the means of deliverance from sin and death.

Korsmeyer embraces what sounds like a form of Peter Abelard's Moral Influence atonement theory. The cross does not render any actual change; it is not the *means* of redemption but only a *display* of it. And Korsmeyer's schema falls prey to the same objection historically raised against Abelard's view: Without some intrinsic linkage between plight and deliverance, the connection is arbitrary. Unless there is something actually *at stake* in the death of Jesus, it is hard to see how that death even reveals divine love, much less enables or actualizes it.

Suppose you and I are crossing the street and I suddenly spot an oncoming bus hurtling toward you so fast that I do not even have time to warn you. I lunge into you, knocking you out of the path of the runaway bus and getting run over in your place. Then you will be right to say my death demonstrated my love for you, because it led me to a concrete action that was necessary if you were to escape being run over. But suppose I see a bus coming up the street at a normal clip and posing no danger. I say to you, "Watch this!" And then I leap in front of the bus, to my messy death. Would that act signify my love for you? No, it would just signify my insanity, my death wish.

Abelard could not supply any integral connection between the cross and the salvation supposedly effected by it. So how could it even win us over to God by attesting his great love for us? This Moral Influence theory works only insofar as it piggybacks on some one of the other atonement theories whereby, say, Christ's death defeats and despoils the devil or pays the debt for our moral crimes, or eradicates our deep-rooted sinful nature, i.e., Original Sin. But is there anything for it to eradicate? We have to agree with C. John Collins on this point:

> The notions of sin as an alien invader that affects all people, and of atonement as God's way of dealing with the guilt and pollution that comes [*sic.*] from this defiling influence, depend on the story of the original family and their original disobedience. The Biblical terms for atonement, which have the associated ideas of propitiation, expiation, and cleansing, become meaningless without this part of the story. If this is so, then the death of Jesus loses a crucial aspect of its meaning as well. [Collins 2011, 134]

Robin Collins objects to the way the "biological interpretation" of the Fall reduces "original sin to certain inherited biological traits." He complains that "this sort of view tends to minimize the necessity of the atonement . . .

A bloody death on the cross does not seem as necessary" (2003, 496). And that, says atheist G. Richard Bozarth (quoted in Lamoureux 2008, 307), goes right to the heart of the problem, the reason "Christianity has fought, still fights, and will fight science to the desperate end over evolution." It "destroys utterly and finally the very reason Jesus' earthly life was supposedly made necessary. Destroy Adam and Eve and the original sin, and in the rubble you will find the sorry remains of the Son of God. It takes away the meaning of his death."

As we are reminded by Charles P. Shingledecker, author of *The Crazy Side of Orthodoxy*, the Eastern Orthodox Church is less affected by the loss of the Original Sinner because it has always relied less on Original Sin as a justification for Jesus. Adam's misdeed simply left us in our present human condition, having nothing to do with any moral taint or guilt. It is only the sin *we actually commit*, on our own, that requires an atonement. Each person's rap sheet starts out clean, with no pre-printed entry at the top for inherited sin (Shingledecker 2013). But the Orthodox are still stuck with what Pelagius *did* blame on Adam and Eve: our supposedly sinful tendencies. What's the point of calling human nature sinful if it is just how we are, and was never actually corrupted at all? That question is the subject of our next chapter.

Sinful Selection

Man still bears in his bodily frame the indelible stamp of his lowly origin.

—Charles Darwin, *The Descent of Man*

Traditional theology has told us that the Fall put the taint of sin and corruption into man from Adam and Eve's screw-up. An understanding of evolutionary human origins dispenses with such nonsense on two grounds. First, as we saw in "Eden Disorder," there was no such event; it is an ancient myth with evident pagan parallels. Second, those actions that have been labeled as "sin" do not arise from any subsequent corruption of our nature, but *from our very nature itself* as the descendants of reproductive survivors in a harsh and brutal world. All of those "sinful" traits that Christian clergy have condemned from their pulpits ever since Clement of Alexandria's joyless asceticism have evolutionary explanations that make a whole lot more sense than the story that Christian theologians dreamed up.

Adam Made Me Do It

The connection between human sinfulness and the Eden story was very clear in Luther's mind. Adam was in a state of innocence; there was no sin to be found because, Luther thought, logically enough, "God did not create sin." If only Adam had obeyed God and not eaten the forbidden fruit, "he would never have died; for death came through sin" (*Lectures on Genesis*, Ch. 2, v. 16).[65] Following this statement is a fanciful description of what the Fall- and death-free world would have been like:

---

[65] Eve seems to be forgotten in this discussion, though Luther certainly doesn't leave her out. He writes about the "curse" of childbearing in a way that glorifies it as an honor and blessing of sorts, while still acknowledging its discomforts and danger. Despite being somewhat ahead of his time in the way he treated his wife Katherine, Luther shows no shortage of sexism in his ultimate conclusion about Eve's fate after the Fall: If she "had persisted in the truth, she would not only not have been subjected to the rule of her husband, but she herself would also have been a partner in the rule which is now entirely the concern of males." Women, he claims, "are generally disinclined to put up with this burden, and they naturally seek to gain what they have lost through sin. If they are unable to do more, they at least indicate their impatience by grumbling. However, they cannot perform the functions of men, teach, rule, etc." Only in "procreation and in feeding and nurturing their offspring they are masters" (*Lectures on Genesis*, Ch. 3, v. 16).

For us today it is amazing that there could be a physical life without death and without all the incidentals of death, such as diseases, smallpox, stinking accumulations of fluids in the body, etc. In the state of innocence no part of the body was filthy. There was no stench in excrement, nor were there other execrable things. Everything was most beautiful, without any offense to the organs of sense; and yet there was physical life. [v. 16]

For Luther, our supposed sinfulness is just part of Original Sin, which "really means that human nature has completely fallen" (v. 16). Since "we all lie under the same sin and damnation of the one man Adam," he asked in *The Bondage of the Will*, his landmark rant against human goodness and free will, "how can we attempt any thing which is not sin and damnable?" (§152).

It's understandable that Christians up to Luther's time would hold this view. "Medievals could talk without self-consciousness about an historical Adam and Eve because an alternative view had not yet arisen in biblical interpretation or from a scientific understanding of human origins" (Wiley 2002, loc. 1519). Already in 1895, White was looking in the rear-view mirror of enlightenment at the "times when men saw everywhere miracle and nowhere law," when nothing was "more natural" as an explanation for man's evil than his corruption by a primordial fall (Ch. 8). But now there is no such excuse for our bad behavior, nor any inherent sinfulness to feel guilty about.

The Man of Sin

This is a realization that is both humbling and freeing: We are the genetic success stories of our ancestors' behavior as well as their bodies. Forget all the ancient nonsense about being saddled with some inherent "sin-corrupt" nature, about blaming everything on some mythical Adam. You are here because you had ancestors who did what it took to survive and reproduce in a very harsh world. It's as simple as that! What you inherited from them is not some taint of sin, but the very traits that allowed them to produce you.

All those "works of the flesh" that Christianity condemns—our lust for power and procreation, favoritism toward self and kin, propensity to hog energy in gluttony and conserve it in sloth, quick anger responses to threats and slights—are the results of millions of years of evolutionary experiment in what allows genes to replicate best in these human vehicles they've wound up in. It is all driven by the genes, Dawkins reminds us in his magnum opus *The Selfish Gene* (2006, originally published 1976), and the genes *don't care* about the morality of that behavior—they don't care about anything. They have no lofty morals, no sense of higher purpose; "the true 'purpose' of DNA is to survive, no more and no less" (2006, 45). Dawkins emphasizes "that we must not think of genes as conscious, purposeful agents," even though natural selection "makes them behave rather as if they were purposeful" (p. 196). Both genes and natural selection itself are often described with the language of agency, but the scientists know what they're talking about when they do so: "When we say 'genes are trying to increase their numbers in future gene pools', what we really mean is 'those genes *that behave in such a way as to increase their numbers* in future gene pools *tend to be the genes* whose effects we see in the world'" (p. 196, our emphasis).

Daniel Fairbanks (a Mormon whose scientific writing seems utterly unclouded by dogma) cites one genetic success story that is a moral failure of the worst degree: the "tyrannical rampage" of Genghis Khan "throughout central Asia to establish the largest land empire in history." The Great Khan's "sexual exploits throughout the region he conquered are well documented. Furthermore, his male relatives, who inherited the same Y chromosome as he, ruled much of the region even after his empire fell, and they also had power for sexual exploitation" (Fairbanks 2007, 110). Now, one particular Y chromosome is found in about 8 percent of all men living in that region: "The pattern of distribution and the time of its origin extrapolated from the data both point to the time and the area of his reign, strong evidence that thirty million men today inherited his Y chromosome" (p. 111).

The Bible doesn't provide a sterling example in the rape and conquest department, as must be acknowledged by anyone who has objectively read Joshua or Numbers 31 ("kill every male among the little ones, and kill every woman who has known man intimately. But all the girls who have not known man intimately, spare for yourselves," NASB). Thankfully though, few professed Bible believers (or anyone else) sink to that level anymore. Cultural evolution has erected societies in which we largely rise above the basest impulses that biological evolution has left us with.

According to Russell Kolts, a professor of psychology and advocate of an evolution-informed approach to anger management, we share these impulses "with organisms—like reptiles—that appeared long before we did in the evolutionary chain" (Kolts 2011, 29). The human brain, he says, is

> an evolutionary patchwork quilt—stitching together a complex and varied system of structures and functions, some of which date back almost to the beginnings of life itself. Our brains have evolved with a host of extremely powerful emotions, capable of harnessing our thoughts and attention, often without our awareness. [p. 28]

The way we act on these emotions "often seems suited to earlier times in the human story—they prepare us to fight, flee or lie down in submission when we'd be better served by pausing and taking a moment to become more mindful, and perhaps to analyze, consider or negotiate" (p. 28). Reflections on the evolutionary heritage of our brains gives us a sense of understanding, both for ourselves and others (p. 29). It's certainly a more productive approach for one's mental health than the one taught by Jesus. He equated the natural, common impulses of anger and forbidden sexual attraction with murder and adultery, and said that calling someone a fool warrants a descent into the fiery hell (Matt. 5:21-28).

Hellbound or not, we all lapse into angry outbursts from time to time. This rash tendency is entirely to be expected, says Kolts. Our threat "system has evolved so that it is activated *rapidly*, because defences that come on too slowly may be too late" (2011, 8). We have been prey more than predators, even for most of human evolutionary prehistory, and there isn't much time to react when the tiger is about to pounce. Unlike Jesus and the clergy who have frowned and forgiven ever since, Kolts wants us to understand that having a rapid-response amygdala for threat response "is *not our fault*; it is simply the way our brains work" (p. 8).

How do the theologians account for all this? Well, in the previous chapter we saw Robin Collins's attempt to address the issue of "biologically inherited propensities" as he describes the "biological interpretation" of Original Sin. These traits, e.g., aggression and selfishness,

> help the individual or one's kinship group survive but typically do not promote the flourishing of the larger community. Essentially, under this view, the doctrine of original sin, the Genesis story, and the various statements in the epistles tell us nothing more than what science tells us. Advocates of this view often assume that we are purely biological and physical beings. Hence science, not theology, becomes the primary place to look to understand the nature and origin of human beings. [Collins 2003, 495]

He admits this view to be common even among theologians, but thinks it is reductionistic, making "the voices of theology, Scripture, and Church tradition" into "a sort of fifth wheel." Instead, "it is the purported findings of science that are claimed to provide the correct understanding of human nature and the human condition" (p. 495). But isn't this just the Achilles Heel of all these theo-biological approaches?

Collins objects that humans seem to be more than merely physical creatures. Well, how could he *not* think that and still be religious? The biological interpretation also "tends to reduce evil merely to our acting on biological impulses, ignoring the particularly serious forms of evil that are made possible by our own self-awareness and transcendence—evils such as idolatry of self, viewing other people as mere objects, and the like" (p. 495). But these traits and deeds, too, are directly rooted in our survival instincts. As the anatomist and Christian Daryl Domning points out, our "sinful" human behaviors

> exist because they promote the survival and reproduction of those individuals that perform them. Having once originated (ultimately through mutation), they persist because they are favored by natural selection for survival in the organisms' natural environments. Since these behaviors are directed to self-perpetuation and succeed in a world of finite resources only at the expense of others, it is accurate to call them, in an entirely objective, non-psychological and non-pejorative sense, selfish. [Domning 2001]

Vices are not unique to us as humans, either. If our supposed sinfulness is the result of Forbidden Fruit in Eden, then innocent nonhuman animals must have gotten word to start screwing up in similar ways, since "there is virtually no known human behavior that we call 'sin' that is not also found among nonhuman animals. Even pride, proverbially the deadliest sin of all, is not absent" (2001). Either that, or there was a whole world full of animal Fall dramas being played out alongside Adam and Eve; perhaps among them, the first ancestral chimpanzee guiltily munched on a piece of the Banned Bark, and then looked down in horror to see that she *wasn't* a naked ape, after all!

Of course, that is ridiculous. But no more so than it would be to deny the "unambiguous conclusion" that Domning draws from his observation that animals are "doing things that would be sinful if done by morally reflective human beings":

> Logical parsimony and the formal methods of inference used in modern studies of biological diversity affirm that these patterns of behavior are displayed in common by humans and other animals because they have been inherited from a common ancestor which also possessed them. In biologists' jargon, these behaviors are homologous. Needless to say, this common ancestor long predated the first humans and cannot be identified with the biblical Adam. [2001]

Lustful Angels

Contrary to Luther and so many other traditional expositors of Christian doctrine, the Eden story isn't really about human nature being infected with evil. Adam and Eve did not so much defy God as see through his deception and gain knowledge for themselves with the help of their true benefactor, the Promethean Serpent. They paid a high price, true, but that doesn't mean they shouldn't have done it. Nor is God portrayed very nobly in that story; the writer leaves us with the impression that disobeying God wasn't necessarily wrong.

For a biblical explanation of human sinfulness, we need to look at the sixth chapter of Genesis rather than the third:

> And it came to pass, when men began to multiply on the face of the earth, and daughters were born unto them, that the sons of God saw the daughters of men that they were fair; and they took them wives of all which they chose. And the LORD said, My spirit shall not always strive with man, for that he also is flesh: yet his days shall be an hundred and twenty years. There were giants in the earth in those days; and also after that, when the sons of God came in unto the daughters of men, and they bare children to them, the same became mighty men which were of old, men of renown. And God saw that the wickedness of man was great in the earth, and that every imagination of the thoughts of his heart was only evil continually. And it repented the LORD that he had made man on the earth, and it grieved him at his heart. [Gen. 6:1-6]

This brief introduction to the Deluge story has a complex history. It began as an ethnological myth attempting to account for the great stature of the *Nephilim*, or *Anakim*, a tall people mentioned in Numbers 13:32; Deuteronomy 3:11; 1 Samuel 17; and 2 Samuel 21:15-22. If Goliath is representative, they towered over the shrimpy Israelites (and everybody else) at six and a half feet. Somehow this one group was freakishly tall by ancient standards, like certain basketball players today.

How to explain this? Ancient Israelites pegged them as demigods, tall because of divine ancestry. Somewhere back along the line, their mothers must have had dalliances with the gods ("sons of God") and produced half-divine offspring who were celebrated in legend as epic heroes: Goliath, Nimrod (Orion), and Gilgamesh. Originally no shame attached to this when the story circulated by word of mouth. It was just like the Greek myths where Zeus and Apollo fathered semi-divine heroes like Hercules, Asclepius, Perseus and Theseus on mortal women.

By the time we read it, though, the biblical compiler looks askance at such fraternizing between mortals and immortals, mainly because monotheism had become all the rage. And yet the story was sacred tradition; he did not feel at liberty to leave it out, for someone surely would have complained. So he inserted it here to provide narrative motivation for the Flood. If you viewed these shenanigans as a blasphemous outrage against nature (as in Jude verses 6-7), it might explain how evil entered, so to speak, into the human gene pool and corrupted humanity so that people thought only evil thoughts continually. They had become ripe for being drowned like rats. Somehow Noah escaped the general perversion, so God decided to start over with him.

This story was widespread in ancient Israel. Other versions with greater detail appear in Pseudepigraphical books like *The Testaments of the Twelve Patriarchs* and *The Book of Jubilees*, where we are told that these angels (as they were now reinterpreted) taught mortals the arts of weapon-making, warfare, and the use of cosmetics. In one version, it is the women who already knew the fine art of seduction who lured the poor angels into their grasp! These apocryphal versions would imply a Pelagian view of sin: the creation of a corrupt environment where the young are acclimated to sin as they are socialized, because what the fallen angels did was to inculcate bad habits. But the simpler version in Genesis 6 implies just the sort of physical taint Augustine ascribed to Original Sin.

It's really too bad Paul doesn't use this story in his discussion of Original Sin. It would fit modern genetics much better since the Fall happened simultaneously with a number of humans, not just a couple. And it's not that Paul didn't know the story. Of course he must have, being a Jew conversant with Hebrew Scripture. It's more than just that; Paul refers to the story directly in 1 Corinthians 7, where he warns the prophesying women of Corinth not to prophesy or pray with an unveiled head, since this might attract lustful angels. It's like Eve's "unveiled" nakedness attracting the depraved archons in the Garden of Eden, as we read in Nag Hammadi texts like *The Hypostasis of the Archons*.

But perhaps the story is better ignored. Just imagine Francis Collins and his pals trying to make angelic seduction respectable in the eyes of secular colleagues!

Branch III: The Creator

A god without dominion, providence, and final causes is nothing else but Fate and Nature.

—Isaac Newton,
Philosophiae Naturalis Principia Mathematica

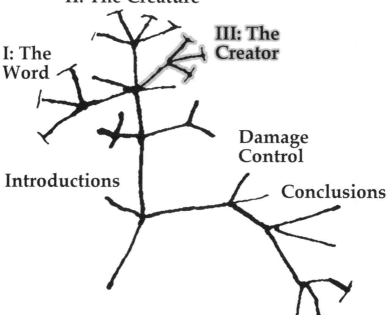

Peekaboo Deity

> *Why, we might ask with the Darwinians, do we need to posit any divine action at all if nature can create itself autonomously?*
>
> —John F. Haught, *God After Darwin*

In one episode of his *Infidel Guy* podcast, Reginald Finley complained memorably about the "peekaboo deity" who no longer shows his face as he did in the pages of the Bible. This is a God who once took an evening stroll in Eden, stood next to Abram haggling about the minimum number of righteous citizens in Sodom, and torched a pile of damp vegetables to demonstrate his superiority. He exercised day-to-day power over earthly events in everything from tipping over water jars in heaven to knitting together fetuses in the womb, and was feared not for the wrath he might show in any afterlife but in the atrocities he was committing right here on earth.

Now, however, this God seems to let himself be known only through the voices of those who claim to speak on his behalf, and his stamp of workmanship in creation is so well hidden that completely naturalistic explanations for everything from the Big Bang onward are no longer in serious dispute. Darwinian evolution, with its blind, impersonal algorithm of natural selection, "shatters the illusion of design within the domain of biology, and teaches us to be suspicious of any kind of design hypothesis in physics and cosmology as well" (Dawkins 2008, 143). Without any apparent design to the universe or even ourselves, the divine designer seems nonexistent, at the very least deceptive in making himself utterly hidden to our scientifically informed eyes.

Designus Absconditus

In the fearful ignorance of prehistory and antiquity, the only sensible explanation for the wonders around us was divinity. There was no plausible alternative explanation for the formation of living creatures than the creative work of some being who was greater than any of them. Giving credit to blind forces of nature for spreading forth the earth and giving life and breath to its people (Isa. 42:5), indeed to all things (Acts 17:25), would not only have seemed impious, but downright ridiculous.

Irenaeus defended the one true God of both Creation and Christianity (against the Marcionite heresy that considered them two distinct entities) using the design argument. The Christian should never fall from this belief, he wrote, that

> this Being alone is truly God and Father, who both formed this world, fashioned man, and bestowed the faculty of increase on His own creation, and called him upwards from lesser things to those greater ones which are in His own presence, just as He brings an infant which has been conceived in the womb into the light of the sun, and lays up wheat in the barn after He has given it full strength on the stalk. But it is one and the same Creator who both fashioned the womb and created the sun; and one and the same Lord who both reared the stalk of corn, increased and multiplied the wheat, and prepared the barn. [*Against Heresies*, Book 2, Ch. 28]

Sounding like an ancient Bill O'Reilly with his infamous "tide goes in, tide goes out, can't explain that" remark, Irenaeus asked, "What explanation, again, can we give of the flow and ebb of the ocean" (Ch. 28). Everyone admits there must be a certain cause for such phenomena, he added, and it was clear what his explanation was: God did it—the one true God, that is.

That "deity by default" view is all the more understandable when you consider the fact that it arose from minds evolved to see "agency wherever anything puzzles or frightens us" (Dennett 2006, 123). The tendency to make the safest bet is what got immortalized in our genes. You had some prehistoric ancestor, one of many, who was prone to hear rustling in the tall grass and run for her life. When it turned out not to be a false alarm one afternoon, she survived to conceive the next branch on the family tree that night. Her jaded, skeptical cousin did not.

So we look at ourselves, at a forest, in wonder, and we "see" the evidence of creation. Paul thought that, too, saying that the pagans were "without excuse" because they ought to have known the one true God from the "invisible things of him" that are "clearly seen" from creation (Rom. 1:19-20). The design argument is hard to relinquish, even when we have the alternative right in front of us. It all seems so clearly to be the work of a designer! We must "overcome the seductions of naïve intuition" to get beyond that illusion (Dawkins 2009, 371).

Ruse explains what he calls "a fundamental root metaphor in Darwinian evolutionary biology":

> Hands, eyes, teeth, noses, leaves, bark, roots, are all as if designed. They work to the ends of their possessors. Because I have hands and eyes and teeth and so forth, I can better succeed in life's struggles. I am more likely to survive and reproduce. I am more likely to be naturally selected. [2000, 24]

He is sympathetic to religion despite being an agnostic himself. It may be that organisms "were literally designed by God," Ruse allows, "but that is not the point in Darwinian biology. They are *as if* designed—truly not designed, but appearing so because of natural selection. Everything else is on top of—or, if you prefer, embedded within—this metaphor" (p. 24).

The classical formulation of the design argument is that of William Paley. He pointed out that no sane person, finding a ticking watch lying on the ground, would suppose it had simply grown naturally, like a flower. Such intricacy of structure and function surely demands an intelligent watchmaker. And so with the far more intricately functioning world around us. It, too, must have been designed by a Creator. Right? It seems so compelling—until one does a doubletake and then feels chagrin for having found it so persuasive.

> What's amusing about Paley's watch argument is that it defeats itself. Let's imagine his original situation. He's walking in a field and discovers the watch. It looks out of place, different from the plants and rocks. But if it looks different from nature because it looks designed, then nature must not look designed. You can't argue on the one hand that the watch looks remarkable and stands out from the natural background, and on the other that the watch looks similar to nature, so both must be designed. [Seidensticker 2012, loc. 2093]

Evolution is, as Peter Schuster lectured Pope Benedict XVI in 2006, "a process that goes on according to natural laws and needs no external intervention." There's no designer in sight:

> [T]he natural scientist at present is making not one single observation that could be explained compellingly only by the interference of a supernatural being, nor is one necessary for the extrapolation of our present knowledge to the interpretation of events in the past. [Horn and Wiedenhofer 2008, loc. 621]

Now, in addition to explaining away the design argument, modern science has added a twist: evidence *against* design. As we will see, that has generated a double standard big enough to drive a camel through. But first, let's look at why the designer is either absent or trying to trick us into thinking he is.

Dithering Design

If the evolutionary process were divinely guided, why didn't it take a lot less time? Why would a Creator "who desires all men to be saved and to come to the knowledge of the truth" (1 Tim. 2:4), who "has given us understanding so that we may know Him who is true" (1 John 5:20, both NASB), cover his tracks with evidence *against* design? It's not just fossils being planted in mountainsides to test our faith. This devious designer has constructed our entire genome to look exactly as naturalistic evolution would have produced it *without* his tinkering.

We have already discussed some of the genetic issues in "Apex or Ex-Ape": two chromosomes that are still separate in the great apes fusing together in humans, and the pattern of mutations in human DNA that matches up so remarkably with paleontological and archaeological evidence of our African origins. With refined sequencing technology, we can now examine those mutations closely enough to see that they have another amazing story to tell, of our common ancestry with other species and the slow, piecemeal changes that eventually separated us from them.

In "Let There Be Replicators," we discussed some engineering absurdities like the recurrent laryngeal nerve that are perfectly understandable in terms of gradual, naturalistic evolution but would be ridiculous for any conscious designer to implement. Dawkins mentions two other examples, the vas deferens and sinus drainage holes. Like the recurrent laryngeal nerve, the vas deferens "takes a ridiculous detour around the ureter, the pipe that carries urine from the kidney to the bladder." This is the unfortunate result of getting "hooked the wrong way around the ureter," something that gradual evolution cannot change (2009, 364-65). Our maxillary sinuses have their drainage holes *on top*, which is the worst possible location. But evolution can't make the drastic, sudden change of closing up some holes and drilling new ones, any more than it can cut and splice nerves or tubes to re-route them. The holes were in place, reasonably enough on the side of the sinuses, when our heads gradually started changing orientation to eventually wind up balanced atop upright-walking spines. And that's where they stayed when that became the top, against gravity and any sense of design (p. 370).

We also saw that there are vestigial features lingering in our bodies, which an *ex nihilo* Creator would have no more reason to put there than Gosse's navels on Adam and Eve. Those tiny muscles that give us goosebumps don't serve any purpose; our body hairs are now too sparse to give us warmth or convey threats, standing on end or not. But they remain part of the package, our *phenotype*, because natural selection hasn't bothered to disable the genes that produce them from our *genotype*. That's proof of evolution or at least a God unconcerned with details about dermatology, but the genotype itself contains more evidence of evolutionary benign neglect. We have inactive "pseudogenes" sitting on our chromosomes like old bums living out their days riding freight cars. They once served a purpose (in our distant ancestors), but no longer produce anything.

[A]s environments change over time and as organisms evolve to exploit new environments, a gene that at one time was essential may no longer be needed. At that point, mutations that disable the gene are no longer detrimental to the organism, and the mutated, nonfunctional gene can persist with no ill effects. It is now a pseudogene at the same position as the original gene. [Fairbanks 2007, 48-49]

To pursue this point, we need to review the process of going from a genotype to a phenotype. The fertilized egg is the first factory for producing proteins from the new genome, just formed from pairing up with the chromosomes supplied by one incredibly lucky sperm cell.[66] That initial cell is able to turn into anything from blood to brains, as are any of the first 32 cells of the zygote that forms from it over the next few days. But then the specialists start to appear, focusing on particular instructions from the chromosomes and, as an occasional part of the copying process, differentiating into ever more specific types. In almost every case (red blood cells being an exception), the parent cell supplies a copy of the entire DNA recipe, in all its incredible detail, to each of its copies, for them to consult and copy in turn. At last, the squalling new phenotype emerges from its mother with its own unique fingerprints and, occasionally, an atavistic tail extending from that vestigial tailbone.

Our goosebumps and tailbone correspond to leftovers that linger in our genotype. The pseudogenes don't even make it into the phenotype. In their case, it is only the vestigial DNA itself that remains, long since disabled by the mutations that, despite some amazing biological correction mechanisms, inevitably arise over the eons.[67] "Every animal and plant genome is subject to a constant bombardment of deleterious mutations: a hailstorm of attrition," Dawkins writes, describing it as "a bit like the moon's surface,

---

66. A man will produce many billions of sperm in his lifetime; a single ejaculation places about 250,000,000 of them into the vagina. Only about 200 of those will make the hour-long journey up the fallopian tube to meet the egg (Winston and Wilson 2006, 137), which is one of about 2,000,000 that were in the woman's ovaries when she was born (p. 132). Long odds, indeed!

67. "In prokaryotes, up to 10% of the coding capacity of the genomes may be dedicated to repair system components that act at all stages of DNA replication and also eliminate various mutational lesions that occur outside the replication process" (Koonin 2011, loc. 4513). Yet, in a fascinating twist, it actually may be *adaptive* for organisms sometimes to allow elevated mutation rates. "The evolutionary process cannot possibly 'know' what is about to come, but in stress-prone environments, organisms that evolve the capacity to transiently increase the mutation rate enjoy an increased chance of survival" (loc. 4684).

which becomes increasingly pitted with craters due to the steady bombardment of meteorites" (2009, 352).

Darwin used the metaphor of natural selection as what we might call an attentive engineering review board, "daily and hourly scrutinising, throughout the world, every variation, even the slightest; rejecting that which is bad, preserving and adding up all that is good; silently and insensibly working, whenever and wherever opportunity offers, at the improvement of each organic being in relation to its organic and inorganic conditions of life" (1859, 84). When a mutation disables an important function in an organism, natural selection limits its spread, in effect saying, "That change is a no-go, kill this new product line, you idiots!" Individuals possessing bad mutations "are more likely to die and less likely to reproduce, and this automatically removes the mutations from the gene pool" (Dawkins 2009, 352). But if the organism has a way to survive and replicate despite the mutation or, rarely, all the better because of it, the change will be immortalized in its DNA for generations to come. The pseudogenes are cases where the feature was flagged as removed ("leave this out") or deprecated ("you don't need to bother with this") and the engineers just disabled the code for it rather than make big changes to the design specification. It happens all the time in software and integrated circuit engineering where projects become huge and unwieldy; actively trimming out unused code can be more trouble than it's worth.[68]

A well-known example of a pseudogene is one called *GULO*, an operative version of which allows many animals (dogs and cats, for instance) to synthesize vitamin C. "By contrast, humans and all other primates lack this gene and must consume vitamin C to survive" (Fairbanks 2007, 53). Our degraded version is just a fragment of the original, and about 20% of the remaining DNA is mutated (p. 53). Like the fragmentary scraps of the New Testament found in the sands of Egypt, it is useless for anything but evidence of deep antiquity and subtle changes over time. In the case of *GULO*, the time frame is in the millions of years, not thousands; chimpanzees have a pseudogene that is 98% identical to ours, which means this mutation probably happened before our lineages split (p. 53).[69]

---

68. Please don't make too much of the engineering metaphor. We are not talking about actual, deliberate engineering any more than Darwin did when he frequently referred to Natural Selection as an active agent.

69. Chimps eat fruit, of course, and don't need to make their own Vitamin C any more than we do. Interestingly, "the same can be said for almost every other pseudogene in the human genome. With a few notable exceptions, chimpanzees and humans have the same pseudogenes in the same places, and they are, on average, about 98 percent similar" (Fairbanks 2007, 53-54). They *are* our cousins, like it or not.

There are many other fossils lurking in our genome; it "is littered throughout with millions of relics that tell an astonishing story of our evolutionary ancestry" (p. 29). Since we no longer have to smell food for ripeness or edibility, more than half of our olfactory genes are now inactive (Wade 2006, 270). We have a pseudogene called *GBAP1*, which is a mutated, non-functional duplicate of the *GBA* gene found a few thousand base pairs away on our longest chromosome. Two of our closest relatives, chimpanzees and gorillas, have the same gene and a pseudogene inactivated by the absence of the same 55 DNA base pairs (out of thousands). In Orangutans, however, the duplicate hasn't suffered that deletion of base pairs, and both of its copies of the genes are still working (Fairbanks 2007, 54-55). And guess what? Orangutans are from a split in the primate family tree that occurred millions of years before our split with the chimps and gorillas (Gibbons 2007, loc. 1306; Prothero 2007, 338-39). The mutation in the *GPAB1* pseudogene—deleting a chunk of 55 base pairs all at once—occurred in an ancestor which we share with the chimps and gorillas, well after the orangutan lineage had gone its own way.

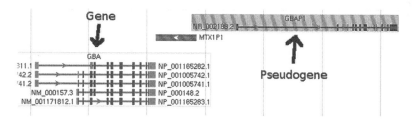

Figure 13: GBA and its duplication pseudogene on Human Chromosome 1

Thoughtful observers of the natural world have been puzzling about the anatomical oddities—in the phenotype—for a long time. Now those willing to look are seeing the same type of automatic engineering economy in the DNA itself. Sequencing technology has given us a glimpse deep inside the structure, right into the forbidden holy place where Moses' tablets supposedly bore the very writing of God. We've peeked behind the curtain looking for a relic written with "the language of God" as Francis Collins puts it. And what's there? Just more of the same damn naturalism. Despite the efforts of our theologians to put a positive spin on the story, the God of Israel remains a God who hides himself (Isa. 45:15).

The theist must contend not just with the appearance of suboptimal design, but evidence *against* design. Donald Prothero, the Christian paleontologist, points to the useless "tiny hips and thighbones buried deep in" the bodies of whales and snakes and says that "a God who would plant these vestiges of the past is a deceiver, tricking us by making it *look* like life evolved" (2007, 108). Yes, but that point goes well beyond Prothero's intent to criticize creationism. Even when the theist rejects the idea of a "cosmic trickster" (Collins 2006, 177) and embraces evolution as Prothero does, forcefully, he is still left with this blind emptiness that looks suspiciously like no God at all. Troubled Christians certainly might like "some indication of planning, purpose, design, at work in the history of life, a providential hand that reaches in and produces the truly important changes" (Kitcher 2007, 19). But what we see is quite to the contrary, to say nothing of the theological headaches caused by those chimps crashing the party at the family reunion in our genome.

A Two-Edged Standard

Despite the purported virtue of "childlike faith," fideists are happy to employ reason when it seems to be serving them. The design argument is one case of this. Another that Dennett notes is the eagerness of many religious organizations to scientifically confirm that religion promotes moral behavior: "They are quite impressed with the truth-finding power of science when it supports what they already believe" (2006, 280).

"There is no way around the fact that we crave justification for our core beliefs and believe them only because we think such justification is, at the very least, in the offing," Sam Harris observes. Christians would shout on the rooftops about a radiocarbon date of 29 CE for the Shroud of Turin: When there is evidence to be had, "the faithful prove as attentive to data as the damned" (Harris 2005, loc. 985). But this same faith refuses to "stoop to reason when it has no good reasons to believe" (loc. 990):

> People of faith naturally recognize the primacy of reasons and resort to reasoning whenever they possibly can. Faith is simply the license they give themselves to keep believing when reasons fail. When rational inquiry supports the creed it is championed; when it poses a threat, it is derided; sometimes in the same sentence. Faith is the mortar that fills the cracks in the evidence and the gaps in the logic, and thus it is faith that keeps the whole terrible edifice of religious certainty still looming dangerously over our world. [Harris 2005, loc. 3784]

Lamoureux is an example of the same author trying to have it both ways. On the one hand, he cites the "clear and plain knowledge" that nature supposedly reveals about God (2008, 68) and the "non-verbal revelation engraved in the cell" (p. 73) with its "elegance, intricacy, and efficiency" (p. 71). They are without excuse, he sniffs in Pauline fashion (Rom. 1:20) about the atheists who have not "all found the Lord who created proteins and DNA" (p. 69). Yet, despite that and God's supposedly leaving "many signposts pointing to Him," our cosmos "was made in such a way that humans are given true liberty to conceive His nonexistence, or to live as if He has nothing to do with them" (p. 99). Lamouroux gets splinters in awkward places from straddling the fence, even when writing a single sentence: "Sub-optimal design is perfect for a world intended to manifest both the noticeability and hiddenness of the Creator, and ultimately to provide an environment for us to develop a genuine relationship with Him" (pp. 100-101). Both noticeability *and* hiddenness? Oh, come on.

And all this piety is an attempt to avert our gaze from what he describes as the picture painted by evolutionary science—an accurate one, as we have seen, "of a world dominated exclusively by physical processes with no gaps in nature for purported divine interventions" (p. 381). God has given humans the freedom to focus on this apparent naturalism and the suffering involved with natural selection, Lamoureux says, even though it has in many cases "led to deism, agnosticism, and atheism" (p. 381). "The God of Love does not force Himself upon us, because that is not the nature of love" (p. 383). One wonders if Lamoureux's theology contemplates eternal torture as part of God's "love" for those whom he has thus allowed to be unconvinced of his existence.

A subtext of Irenaeus' appeal to the design argument shows how old the double standard is. As we saw earlier in this chapter, he praises the (one true) God's wondrous creation work. But then, a paragraph later, he makes an appeal to faith for those things that don't make sense. If we

> leave some questions in the hands of God, we shall both preserve our faith uninjured, and shall continue without danger; and all Scripture, which has been given to us by God, shall be found by us perfectly consistent; and the parables shall harmonize with those passages which are perfectly plain. [*Against Heresies*, Book 2, Ch. 28]

Keith Miller artfully turns the problem around by saying that natural processes provide us with evidence *for*, rather than against, "God's presence in creation, for the existence of a creator God." He claims support for this in the Bible: God is "providentially active in all natural processes, and all of creation declares the glory of God" (Miller 2003, loc. 137). These natural explanations pose no threat to our faith, he assures his fellow believers. "Rather, each new advance in our scientific understanding can be met with excitement and praise at the revelation of God's creative hand" (loc. 138).

Yet Miller is caught spinning the bottle once again when he writes later in his book, this time with David Campbell, about the apparent randomness of mutations. They cite "God's inclination to confound the wisdom of the wise and choose the weak and despised for honor," apparently a reference to 1 Corinthians 1:27, when explaining away the unpredictability of evolution (in Miller 2003, loc. 2340). "Only with a theological knowledge of God can we attribute to him the wisdom and creativity shown in creation" (loc. 2348). Wait a minute! What about nature plainly declaring God's glory for all to see, to the point where Paul said the pagans were "without excuse" (Rom. 1:20)? It's confounding, all right, when neither the Bible writers nor God's present-day apologists can keep their own stories straight.

Miller's original trick was transforming the water of naturalism into the wine of theism, thus placating the grumbling guests at his wedding of science and Christianity. It's a miracle that our Catholic contingent seeks to perform as well. Momentarily forgetting about the plagues, healing miracles, and Pentecostal tongues of flame that his Bible recounts as evidence for God's presence, Cardinal Christoph Schönborn claims, "God must be invisible to us, so that we do not mistake him for a bit of the world." You see, we *benefit* from all this hiddenness! "The less he manifests himself in a material way, the more we recognize that he stands opposite the world as its Creator" (Horn and Wiedenhofer 2008, loc. 1530). Kenneth Miller offers the same kind of cheery rationalization: Scientific findings that confirm our faith would just "undermine our independence." *Of course* God (conveniently) needs to keep himself hidden! "How could we fairly choose between God and man when the presence and the power of the divine so obviously and so literally controlled our every breath? Our freedom as His creatures requires a little space, some integrity, a consistency and self-sufficiency to the material world" (Miller 1999, 290).

Haught would direct us to the theater in search of God rather than the workshop, looking "not in the design but in the drama of life" (2010b, 58). Yes, that recurrent laryngeal nerve would be a lot shorter if the slightest forethought had been put into its layout. But "what looks to engineers like inefficiency looks to theology like dramatic suspense" (p. 60). Both "ID devotees" and "Darwinian materialists" miss the point of it all, he thinks, when they

> suppress any sense of life's dramatic depth, a dimension of nature that is much more remarkable and theologically interesting than design could ever be. They have failed to notice that the very features of evolution (contingency, regularity, and immense temporal depth) that seem separately to rule out the existence of an intelligent divine designer still point jointly to something much more interesting theologically. It is the drama underlying evolution more than the ephemeral designs that appear fleetingly on the surface of life that most awakens theological interest. [p. 58]

Drama, drama, drama. You know what would awaken our theological interest? Something other than the same tired excuse for seemingly every problem that evolution can throw at Christian theology. Sure, "if life in our universe is a drama, its present deficiencies in design may not be a sign of an overall senselessness but of narrative nuance instead" (p. 59). Whatever you say. And if life in our universe is actually the bungling effort of the Demiurge, its present deficiences may be a sign of that lesser deity's imperfections. That's what the Gnostics thought. Why is it any less plausible than this nonsense about the universe being a cosmic vaudeville show for an audience of one?

And still, despite his appeals to drama and nuance, design ultimately proves to be an irresistible temptation for Haught, too.

> Evolution by natural selection, for all of its jerry-rigged solutions, for all its failed experiments and blind alleys, is a wonderfully efficient way to populate a universe with diverse and interesting creatures. If I were an Intelligent Designer, and I had a hundred billion galaxies (at least) to fill with wonders, I can think of no way more efficient to do it than by genetic variations and natural selection of self-reproducing organisms. [p. 61]

Omphalos Again

Back in "Everybody's Working for the Weekend," we discussed the outrageous attempt by Philip Gosse "to untie the geological knot" (the subtitle of his 1857 book) of old earth evidence versus a young earth Bible. The book was published two years before Darwin's *Origin*, but Gosse also wound up wrestling with issues that relate more to the design argument than to the earth's age.[70]

One of those is the question of geographical distribution: Doesn't the dominance of marsupials in lonely Australia, together with their occupying the same ecological niches as their non-marsupial counterparts on other continents (e.g., marsupial versions of the rat, woodchuck, bear, and dog) count in favor of evolution? In isolation from competition with more efficient placental mammals, the Australian forms seem to have evolved in parallel fashion to their far-off counterparts. Now what does creationism have to say of this phenomenon? The creationist authors of Boardman et al. hastily disclaim:

> The general concept of world-wide dispersal of living things including ... limitation in migration by barriers and by diversification of isolated populations into related varieties or sometimes species is not disputed by creationists. [Nevertheless,] the

---

70. This section is adapted from Price 1980.

creationist believes that the basic forms of marsupials were created like the basic stocks of mammals and that they survived in Australia because of lack of competition due to isolation. [1973, 91]

It is not at all clear, however, that the authors are actually denying what they think they are denying. They almost seem to be espousing in the name of "creationism" what our theistic evolutionists are saying: God "created" the various species by evolving them. But since this would serve only to "refute" an opposing view by renaming it, we should look for an alternative meaning. In fact, their meaning seems to be that the processes that lead scientists to posit evolutionary speciation are quite real, but God *specially created* the various marsupials anyway, despite appearances. Why did God impose such patterns in nature which led naïve scientists to so faulty a conclusion? Well, God just wanted it that way.

One of the knots of absurdity in which Gosse found himself entangled was God's apparent deception in creating the world "with fossil skeletons in its crust, skeletons of animals that never really existed." He realized that it was a bit troublesome for a creator to form "objects whose sole purpose was to deceive us." His exasperating reply:

> Were the concentric timber-rings of a created tree formed merely to deceive? Were the growth lines of a created shell intended to deceive? Was the navel of the created Man intended to deceive him into the persuasion that he had had a parent? [Gosse 1857, 347-48]

It's a bit like the apologists citing the divinely directed massacres of Numbers and Joshua as a way of comforting those troubled about the suffering involved with evolution. (We'll see some of that in the next chapter, "Intelligent DeSade.") Gosse was so boxed in by the narrow confines of his biblicism that he appealed to one of its absurdities in order to address another. *Of course* God created Adam with a navel: "The Man would not have been a Man" without one (p. 349). Neither would God be a "he" without a penis, we suppose. (He's "omnipotent," not "impotent," after all.)

We can find the old belly button argument implicit in some forms of the creationist attack on comparative anatomy and physiology, too. One of the authors of (Boardman et al. 1973) wrote the following in his own book a few years earlier:

> On the assumption of creation, it is reasonable that there would be resemblances between creatures and that these resemblances would be stronger between those creatures living in similar environments and with similar physiological functions to fulfill. One could hardly imagine any more probable an arrangement than now prevails, if the

origin of all things actually were special creation. [from Morris 1969, 23]

What makes this or any other "arrangement" by a divine creator "probable"? Couldn't God theoretically have made humans that reproduce in some other way than urges his church would find sinful in all but the most restrictive circumstances, for example? Keep in mind that fundamentalists believe in plenty of zoological marvels, if they seriously envision the day when "the wolf and the lamb shall feed together, and the lion shall eat straw like the ox" (Isa. 65:25), that carnivores created in Eden originally did not eat meat (Gen. 1:30). Anything goes, or should, in that frame of reference. Nothing should be more "probable" than anything else since "with God nothing shall be impossible" (Luke 1:37).

This is not mere carping. The point is that by talking in terms of what is "probable" given the earth's environmental conditions, Morris is quietly admitting the evolutionist's criterion of environmental "fitness." In other words, he recognizes the validity of the processes of evolution but merely short-circuits the whole business at the last minute by appealing to the prescientific notion of teleology. Sure, it looks like creatures are fitted to survive in certain environments, and indeed they are. But this is because God arbitrarily wanted it that way.

As a result, God framed a riddle that would seem to cry out for the solution of evolutionary biology, Genesis or no Genesis: Those organisms that succeed in the race of differential replication are those that thrive in their environments. The creationist answer, however, is unrelated to the question, and not nearly so intellectually satisfying (or true). It is arbitrary fiat. God could have created grass-eating lions; he did in Eden, and he will again in the Millennium. But, in between, he put us on a false trail by creating the interlocking web of life that suggested the theory of evolution.

One more example of this argument crops up in the creationist repudiation of human evolution, the "descent of man" that we discussed in "Apex or Ex-Ape?" a few chapters back. The creationist, when he doesn't adopt the expedient of simply denying the existence of hominid fossils, finds himself (and hopes no one else will find him) in an odd position. He cannot deny the rather obvious chain of creatures (let's not prejudice the case by calling them "pre-human ancestors") that start out looking like lemurs and monkeys, and end up looking more and more like humans. But there must be no admission that these are "transitional forms." Instead, they must be declared extinct but independent life forms which *just happen* to look like they fall somewhere between monkey and man. The same goes for *Tiktaalik*, with its features of a fish that was beginning to find a better life on land. Transitional forms they may *seem*, but the creationist knows better.

Why do these fossils have the appearance of chains of development that never actually occurred?

Notice, please, that in none of these cases have the creationists explained the rationale of the omphalos argument as Gosse did. The creationists may not be aware of it themselves. But the implicit logic is the same: The evidence points in the direction of evolution, but that is because (for whatever reason) God simply wanted it that way. Kenneth Miller rightly scorns such thinking:

> In order to defend God from the challenge they see from evolution, they have had to make him into a schemer, a trickster, even a charlatan. Their version of God is one who intentionally plants misleading clues beneath our feet and in the heavens themselves. Their version of God is one who has filled the universe with so much bogus evidence that the tools of science can give us nothing more than a phony version of reality. [Miller 1999, 80]
>
> Intelligent design advocates have to account for patterns in the designer's work that clearly give the appearance of evolution. [p. 127]

It is a throwback not only to Gosse's esoteric argument, but also to the prescientific shrugging off of such questions by the catch-all appeal to teleology. Why do birds fly south? Because they were *made* to do this! As Jacques Monod observed, the notion of teleology is inimical to scientific inquiry, and has always served to nip it in the bud. How "scientific," then, can "scientific creationism" be? Can "Intelligent Design" ever get the academic respect it so fervently craves? Let us pursue this question along a slightly different avenue for a moment. Then we will be in a position to recognize the final irony of the omphalos argument as it reflects on creationism as "science."

Creationists often assume the pose of righteous prophets crying in the wilderness, ignored by pharisaical "establishment" scientists. If only their voice of truth were heeded! We would have a scientific revolution! Thomas Kuhn in his celebrated work *The Structure of Scientific Revolutions* has drawn a compelling picture of the history of science involving a series of turnabouts just like what the fundamentalists anticipate. Now, it is far from clear that the creationists are in reality the "scientific revolutionaries" in the scenario. But we will see that their polemical efforts are helpfully illuminated by Kuhn's schema.

Kuhn writes to correct the naïve notion that the progress of science is simply the accumulation of new discoveries. No, while new empirical discoveries do occur, real movement in science comes when scientists accept a new "paradigm," a conceptual model in the light of which the same old data may be better understood. A scientist will notice certain troublesome data that the current paradigm cannot accommodate. Such data sticks out like a sore thumb.

An example would be the retrograde motion of the planets in the geocentric Ptolemaic paradigm for astronomy. Everything else in the heavens moved like clockwork, and was tidily accounted for by Ptolemy, but a fantastic and elaborate series of "epicycles" (celestial wheels-within-wheels) was needed to make retrograde motion predictable. Copernicus was eventually to find this unsatisfactory. Could not some new paradigm be formulated that would deal more naturally, more economically, more inductively, with all the data, instead of dealing fairly with part of it and imposing contrivances on the rest? So Copernicus set to work and, going Archimedes one better, he moved the sun. He transferred it from the earth's periphery to the center of our orbit. Now everything seemed naturally explainable—no more epicycles. The lesson we are to learn from this brief history is that a scientific revolution occurs when somebody offers a new, more natural, way to construe the data. The new model must make economical sense of as much as possible of the data in its own right; it must make the most possible sense of it without reference to extraneous factors (e.g., invisible epicycles, dictated not by the evidence, but by the Ptolemaic model itself).

Though the model is imposed on the data by the theorist, he has derived the model from the suggestion of the evidence itself. It is like one of those puzzles where one must connect all the dots with the fewest possible lines.

On this basis, might the creationists be justified in expecting to usher in a new revolution in biology? How closely do their efforts match the pattern traced out by Kuhn? First we may observe that much (perhaps most) creationist literature concentrates on only half the job: pointing out epicycles. Creationists never tire of indicating (and exaggerating) data they find troublesome regarding the theory of evolution. Even if their claims were found accurate—and they never are—creationists could expect no "scientific revolution," according to Kuhn's scenario, until they had supplied an alternative model capable of doing a better job. But insofar as they restrict their efforts to demolition, they are committing one of the most blatant of logical fallacies. They assume that there are but two options, and that one of those must be true. And, as if we were all playing "Let's Make a Deal," the elimination of evolution automatically vindicates creationism. Not so fast. Lamarck, Lysenko, and a host of other contestants are waiting backstage.

Our second observation is that when creationists occasionally do try positively to defend the elusive "creation model," they violate the necessary criterion of inductiveness. That is, a paradigm must be derived as much as possible from the data itself, and as little as possible from outside considerations. But Duane T. Gish is forthright in his admission of where his model comes from; "a sound Biblical exegesis requires the acceptance of the catastrophist–recent creation interpretation of earth history. If this interpretation is accepted, the evolution model, of course, becomes inconceivable" (Gish 1973, 64). Morris is equally clear that "the general method of [Bishop] Ussher–that of relying on the Biblical data alone–is the only proper approach to determining the date of creation" (1969, 63).

So the hidden agenda is revealed. After all, "there is nothing hid except to be made manifest" (Mark 4:22). The "scientific" creationists, it would seem, are closer to the Inquisition than to Galileo in whose footsteps they claim to follow. They begin with a biblical dogma imposed heavily on the data.

It will put the efforts of creationists in proper perspective if we compare them to another famous school of pseudoscience, the offbeat astronomy of Immanuel Velikovsky. In fact the parallel is virtually exact. Velikovsky reads in Exodus that the Nile turned red ("to blood"), and in American Indian myths that the sky once turned red. First he concludes that Mars once must have nearly collided with the earth; then he reshuffles astronomy accordingly (Velikovsky 1950). In the same manner, Gish and Morris discover in Genesis that the earth is merely thousands of years old with a six-day period of creation; then they practice ventriloquism with the data of geology and biology. In both instances, the dusty pages of ancient legend dictate in advance the results of scientific "research."

Behold the methodological outworking of this *a priori* dogmatism. With their "paradigm" thus derived from an entirely different quarter, it would be the wildest stroke of luck if the data happened to conform spontaneously to the predetermined pattern. It doesn't, so it must be squeezed into place. Fundamentalists have well-developed skills in dealing with all the contradictions found in the Bible, and they put them to use trying to harmonize the data of science.

Let us return momentarily to the deliberations of Morris et al. on the question of starlight. Listing other options besides the unvarnished "omphalos" approach, they point out that there

> are several possible approaches to the solution of this problem, each of which is worthy of careful study by creationists. Some propose that the distance scale represented by the Hubble constant which relates distance to observed red shift is greatly in error and that the distance scale should be drastically reduced ... Another proposal made by creationist scientists is based upon the hypothesis made by Moon and Spencer in 1953, namely, that light travels not in Euclidean but in Riemannian curved space with a radius of curvature of five light

years, so that no transit time could exceed 15.71 years. And a third proposal ... is that further study of the meaning of the scriptural terms ... "[the heavens were] stretched out," etc., may give an understanding of how vast distances correlate with Biblical chronology. It is hoped that creationists may be able to gain a fuller understanding of this problem and attain a satisfactory solution in the near future. [Boardman et al. 1973, 26-27]

These aren't proposals for serious scientific study, but a roll call of available lifeboats being read out to those on the sinking ship. Pick whichever one seems most seaworthy to you as the waters close in. Despite the insistent claims that the "creation model" fits the data better than the evolution paradigm, the data has become a "problem" requiring a "solution." Notice how various hypotheses are being preferred on the basis, not of their inherent cogency, but rather of how much aid and comfort they provide for the creation model.

This case is symptomatic of the dilemma of creationism in general. The model is *prior to* the data, and the latter will be coerced and manipulated in any fashion in order to fit the Procrustean bed of the former. Alas, the creation paradigm is almost all epicycle! Obviously, this is the very opposite of what we would expect if the creationist model were the harbinger of a new "scientific revolution."

Now, what is the bearing of the unannounced rehabilitation of Gosse's omphalos argument on all this? Remember that the tendency of the navel argument is always to admit implicitly that the evidence actually does favor evolution, but that it is misleading. Fortuitously, God merely "did it that way."

In the original version—Gosse's—there were two possible explanations for this. One is that God made it all *look like* it had been done by mere naturalistic evolution in order to test our faith. This is still suggested by some fundamentalists in order to explain away dinosaur bones. Gosse's own preference was that God created the world as if the very real processes now observed in nature had always been in operation, just so that the curtains could open on a fully set stage. In either case, every time the omphalos argument is invoked, even anonymously, creationists are admitting that they hold to their "new" paradigm despite the fact that the old paradigm (evolution) fits the data better.

Creationist arguments evolve as everything else does, reluctant though some are to admit it. And just as in biological evolution we occasionally run across cases of atavism, such a throwback reveals the origins of fundamentalist pseudoscience. No matter how much "scientific" creationists or their more evolved "Intelligent Design" counterparts would like to forget that "black sheep of the family," the omphalos argument of Philip Gosse, it stubbornly reappears in the population every now and then. And when it does, we see what sort of animal we have been dealing with all along: not scientific theory, but religious propaganda.

Quantum Apologetics

> *To have full control over all events God would have to manage the motion of every fundamental particle in the universe in a nanosecond-by-nanosecond basis. I suppose, being omnipotent, he could do that. But I get the impression in my reading that most theologians would not be happy with that solution. Furthermore, such micromanagement would still not guarantee a predetermined outcome on the macroscale.*
>
> —Victor J. Stenger, *Quantum Gods*

The God of Christianity is an active deity, apparently causing barley loaves and dead fish to divide by mitosis to feed crowds of thousands, nullifying gravity in the case of an inconveniently sunken ax-head, destroying recalcitrant fig trees, reintegrating severed ears with their former heads, raising certain persons from their graves. Nowadays God is much more subtle, working in the background to guide surgeons' hands, favor football teams, and facilitate the transfer of funds from widows to televangelists. But a universe governed by uniform laws doesn't leave much wiggle room for active intervention, overtly or hidden. One of the few openings through which God might still poke his fingers, say some of our science-savvy theologians, is the *indeterminism* of quantum mechanics, a fundamental uncertainty about the position and state of things on the smallest scale of the universe.

Deconstructing Divine Action

Envisioning God as performing miraculous feats (or indeed any feats at all!) has been theologically problematic ever since Augustine and Aquinas reinterpreted biblical belief according to the abstractions of Plato and Aristotle. Christian theology came to hold that God existed outside of the time stream. This made it difficult, really absurd, to think of "him" as "doing" anything at all. If you suggest he was *always* "doing" everything at once, but that it *appeared* in our world, inside the time stream, as a chain of serial actions, you have essentially switched over to the Hindu notion of *maya*, whereby the world we perceive is an illusory shadow play. The true world is distorted, refracted by the conditions of temporal existence, into something completely misleading. Only the illuminated mystic can awaken from it to perceive the unchanging Reality. Then God "acts" only in our dream of the world. Oops! Theology finds itself stuck in another bog of unintended consequences.

Divine action encountered another obstacle with the advent of experimental and observational science: The laws of nature finally achieved primacy over the lawgiver. The theologians resisted this development for a thousand years. In the Middle Ages, it was "the accepted idea that, as soon as a man conceived a wish to study the works of God, his first step must be a league with the devil" (White 1895, Ch. 12).

The 13th century monk Roger Bacon came to lament how he "had given myself so much trouble for the love of science!" His church harassed and suppressed him for daring "to attempt scientific explanations of natural phenomena, which under the mystic theology of the Middle Ages had been referred simply to supernatural causes" (Ch. 12). The obstructionism would continue with Bruno, Galileo, Copernicus, and a 17th century scientist who confirmed the ideas of several earlier thinkers that comets were just objects in highly elliptical orbits around the sun, not fireballs "flung from the hand of an avenging God at a guilty world" as had been preached by theologians from Origen to Luther (Ch. 4, §2). For that, the pious Isaac Newton was denounced as "atheistic," his discoveries "built on fallible phenomena and advanced by many arbitrary presumptions against evident testimonies of Scripture" (Ch. 3, §4).

Despite Newton's own belief in God "as an unmoved mover who sets things in motion and then stands back and lets things happen" (Ruse 2010, 49), the shrieking churchmen were right to be concerned no matter how wrong they were about the facts, just as we see with evolution now. There was a big theological problem with the clockwork universe he postulated. It was so orderly, so mechanistic, that it "held the seeds of the expulsion of God from the cosmos—with the possibility of denying His existence, certainly His pertinence, altogether" (p. 33).

Ways Past Finding Out

Newton allowed for some divine intervention here and there to keep things on track (pp. 49, 53). But how? One possibility finally opened up, ironically, from science itself.

As the experimenters of the early 20th century refined their instruments and peered ever deeper into the details of physics, they started seeing some fuzziness in the clockwork determinism. This we now recognize as having nothing to do with limitations of our observational abilities, but a *fundamental imprecision* in the fine structure of the universe: quantum indeterminacy. This is "a very serious kind of unpredictability" we find on the smallest scale of things "that cannot be blamed on our ignorance of the details of the system being studied, or a practical inability to set initial conditions. Instead, it turns out to be a fundamental feature of nature itself at this level" (Al-Khalili 2004, 59). It is hard to accept, like looking at an old photograph under a magnifying glass and being told that the graininess you see is not just in the film but was inherent in the objects being filmed. But when you look at things on the scale of an atom, that turns out to be the truth.

From high school chemistry class, you may remember the electron as a little ball whizzing around a slightly bigger ball of the atomic nucleus. But it's more like a spread-out smudge of negative charge. There is no actual hard-edged object that you could take a snapshot of and say, "See, that's the electron!" It just isn't something whose position and movement you can determine so precisely. Its "orbit" around the nucleus is really just a cloud of simultaneous possibility defined by an odd equation called the *wave function*.

The name comes from the fact that objects at the quantum level act a bit like waves, with their influence spread out in space (p. 67). This is sort of like the oscillating type of wave that defines the time-varying water level on the ocean surface or voltage at the antenna connection of a radio receiver. But not exactly. A wave function "does not describe a vibration of any kind of medium and need not even be mathematically a wave. A much better designation is *state* vector. This is a vector in an abstract space that specifies the state of a system" (Stenger 2009, 186).

One state of an electron is its location in space. Remember, this cannot be determined exactly; it is described only as a *probability amplitude* of a wave function. The wave function of an electron's position relative to its atomic nucleus is like the standing wave you observe along the length of a jump rope that you are wiggling at just the right frequency. Your eyes don't track the up-and-down motion of the rope, only the pattern it makes in the air over time. Likewise, the squared magnitude of a position-space wave function Ψ defines a particle's *probabilistic* pattern of existence.

Figure 14 illustrates this with a plot of the probability density function of a simple wave function for the spatial position of a particle, viewed in just one dimension. Below the plot are some circles representing potential locations for the particle, with darker shading indicating greater probability.

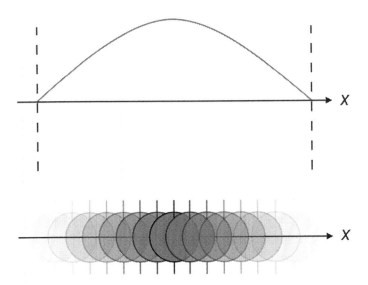

FIGURE 14: Particle position along X as a probability density function.

The value of $|\Psi|^2$ for a given position in space is the relative probability that the particle could be considered "at" that point. But there really is no such point, any more than there is an average American family having 1.8 kids. Nobody can track the position of an "orbiting" electron precisely, because such precision simply does not exist. We must regard it as being spread, unevenly, "throughout the volume of its atom. This is the only valid picture of the atom that quantum mechanics allows us to have" (p. 162). Grandma's house really was blurry; it's not the camera's fault for making it look that way.

The "wave nature of the electron" is what "gives bulk to every atom and prevents people, baseballs, and bacteria from collapsing" (Ford 2005, 190). This, at least, makes some sense in an otherwise very counterintuitive field of science. Little ball-electrons circling around a ball-nucleus, no matter how fast, would not define the three-dimensional structure that atoms have. But standing waves of electron probability density do just that, if in a fuzzy, somewhat intangible manner.

The wave function also explains two very odd behaviors of subatomic particles that contribute, as we shall see, to evolution's randomness: quantum tunneling through what might seem like impossible barriers, and radioactive decay. The latter is actually an instance of the former: Very rarely, in certain types of atoms, a particle will tunnel away from the rest of the nucleus. The decay particle is a piece of the nucleus, e.g., an α particle made up of two protons and two neutrons, that finally escaped the strong nuclear force holding it together with the rest of the nucleus's protons and neutrons.

How often this happens, if at all, depends on the kind of nucleus you are dealing with. With a half-life of just 138 days, atoms of polonium-210 unleash α particles at a furious pace. That makes it extremely toxic; ingesting *one microgram* will irradiate your spleen and liver with over a hundred million α particles per second and kill you in a matter of weeks. The most common (by far) isotope of uranium, on the other hand, is pretty tame stuff. Half of the U-238 atoms in a lump of uranium ore will still be waiting for their α decay event after four *billion* years.[71]

The probability of this "tunneling out of the nucleus" decay event is determined by the forces holding the nucleus together—the barrier to the decay—and the wave function. Almost always, the strong nuclear force overcomes "the electric repulsion between positive protons sharing the nucleus, which tries to break it up. The nuclear force wins out at short distances, which is why nuclei exist at all, but since it falls off rapidly with distance, far away the electric repulsion dominates" (Stern 2005). In a Newtonian world, that's the way things would remain forever and ever, amen. The forces don't change, and the situation in the nucleus wouldn't either.

But the particle is defined as the smear of a quantum wave function (slightly) spread out in space, not as some little ball (or clump of balls) stuck rigidly in place. That wave function exists "as a superposition of being on both sides of the barrier at once. And it is the wavefunction that is penetrating the barrier" (Al-Khalili 2004, 176).[72] Why? Because, more than anything (including a vacuum), nature abhors discontinuity. The wave function must remain continuous in both amplitude and slope. No abrupt

---

71. Wikipedia has some highly informative articles on radioactive decay, including alpha decay, and the half-lives of particular elements. It also discusses the creepy poisoning of Alexander Litvinenko, most likely with a tiny dose of polonium-210.

72. Remember, the wave function defines something a bit different than the waves we're used to—ocean swells, sound waves, radio waves, visible light. Those all convey energy, whereas the wave function has a *probability amplitude*. There are many commonalities, though. And for the massless photon, the wave function can be understood as equivalent to the electromagnetic wave (Stenger 2009, 124). See also Iwo Bialynicki-Birula, "On the Wave Function of the Photon," *Acta Physica Polonica A* 1-2, vol. 86, pp. 97-116 (1994). Your retinas are doing countless acts of wave function decoherence right now, with each photon that hits them.

turns or jagged edges allowed; that's just the way things are.[73] The wave function amplitude has to fade out gracefully after encountering the barrier.

The lighter a particle is, the bigger (more spread out) is its wave function. That's why quantum mechanics is really only of significance at the atomic and subatomic level. You won't find people doing quantum tunneling through walls, except perhaps in this familiar story: "Then the same day at evening, being the first day of the week, when the doors were shut where the disciples were assembled for fear of the Jews, came Jesus and stood in the midst, and saith unto them, Peace be unto you," John 20:19. (To tunnel that far, the resurrected Jesus must have weighed very little indeed.)

The weirdness of quantum mechanics quickly disappears from view as things scale up, as all those wave functions interact with each other gazillions of times and *decoherence* sets in. Then the oddities of the individual wave function are lost, as is the intricate structure of a snowflake when it hits the ground and joins the others piling up on your driveway. A single snowflake, carefully isolated in a cold room under a microscope, is a wonder to behold. Millions of snowflakes are just something to shovel.

But on the scale of an atomic nucleus, the wave function of, say, an α particle, may be sizable, extending to "very large distances outside the nucleus" (Al-Khalili 2004, 76). So the barrier of the balanced forces keeping it there is not insurmountable after all. Every once in a while, the little bit of the wave function that gets past the barrier will result in the particle "being" past it, too.

> That wave is highest inside the "crater" of the nucleus, and if the proton [or other decay particle] materializes there, it stays trapped ... The fringes of the wave, however, extend further out, and

---

73. Saying this is not the same shrugging of the shoulders as the religious do about incomprehensible doctrines like substitutionary atonement or the Trinity. It is true that we "are not meant to be comfortable with the conclusions of quantum mechanics" (Al-Khalili 2004, 85), and that "there is no simple answer or straightforward and intuitive explanation" for its mysteries (p. 81). But that doesn't mean it is unproven or imprecise science.

The probabilities of the things that quantum mechanics makes indeterminate—an electron's position in its orbital around an atomic nucleus, when a nucleus will undergo radioactive decay—are fully accounted for by long equations with very precise coefficients. Perhaps the main one is Planck's constant, which defines the level of quantum weirdness; if it were zero, there would be no quantum effects. That constant, which *defines uncertainty*, is known to a precision of better than one part in a hundred million! Like classical physics, quantum mechanics is "unambiguous and exact, but in a different sense: in the sense that *probabilities* can be calculated precisely" (Ford 2005, 114-15).

it always has a finite (though very small) strength beyond the crater, giving a finite chance for the proton to materialize on the outside and escape. It is as if quantum laws gave it a tiny chance to "tunnel" through the barrier to the slope outside. [Stern 2005]

Fans of Godzilla and Spider-Man know that radiation can cause mutations. Actually, it's not just science fiction; Forster et al. point out that natural radiation from terrestrial, dietary, and cosmic sources "has been irradiating all forms of life since the beginning of evolution" (2002, 13950). They studied the mitochondrial DNA of 730 healthy people living in areas with high levels of natural radioactivity along with that of 258 controls in nearby low-radiation areas, and found their results to "strongly support an acceleration of the evolutionary DNA mutation mechanism through radiation" (p. 13954).

Quantum tunneling can also cause mutation in a more direct way: protons migrating through one of the hydrogen bonds connecting DNA base pairs (Stamos 2001, 176-78).

> Ever since Francis Crick and James Watson discovered the double-helix structure of DNA, it has been known that certain natural mutations can occur due to the random quantum tunneling of protons from one site on the DNA [molecule] to a nearby one, forming a different chemical bond. This sort of accidental error in the coding of the DNA will occur in one out of a billion sites, but when it does we get a quantum mutation. [Al-Khalili 2004, 232]

So, Al-Khalili concludes, "quantum mechanics certainly has some part to play in evolution" (p. 232). Really, though, it seems that *all* mutations—and perhaps all the chaotic events that we view as "random"—have their origins in quantum indeterminacy.[74]

Consider ultraviolet light, another significant cause of DNA damage and hence mutation. The UV-B photons that have bombarded the earth for billions of years originated in the fusion furnace of the sun, from a process

---

74. Bishop provides an interesting discussion of "sensitive dependence on initial conditions" of chaotic systems, or SDIC. "One of the exciting features of SDIC is that there is no lower limit on just how small some change or perturbation can be—the smallest of effects will eventually be amplified up affecting the behavior of any system exhibiting SDIC." He and other authors argue "that chaos through SDIC opens a door for quantum mechanics to 'infect' chaotic classical mechanics," but he candidly notes some reasons why the "argument does not go through as smoothly as some of its advocates have thought." Just "because quantum effects might influence macroscopic chaotic systems doesn't guarantee that determinism fails for such systems" (Bishop 2009, §4).

that also involves quantum tunneling.[75] Adding to the indeterminacy from that, there is a lot of quantum-level interaction between those photons and gas molecules of the earth's atmosphere.[76]

The philosopher of biology David N. Stamos is an advocate for the idea of "quantum indeterminism 'percolating up' to the level of DNA mutations, and therefore phenotypic traits" (2001, 165). He points out how even mutations caused by thermal motion of DNA polymerase (the enzyme involved with DNA replication) can be influenced by quantum behavior of the surrounding water molecules wiggling it around. "Not only must their motion be affected by the quantum statistical nature of their constituent electron orbitals, but recent experimental evidence on water polymers has revealed significant proton tunneling through their constituent hydrogen bonds" (p. 179).

The "percolation argument" is part of a debate about *evolutionary indeterminism*. Timothy Shanahan frames the argument as follows: "Indeterminism in the most fundamental physical processes, as described by quantum mechanics, can 'percolate up' into (i.e., have effects in) macrolevel physical systems. Biological systems are macrolevel physical systems. Therefore, indeterminism in the most fundamental physical processes can have effects in biological systems" (2003, 165).

---

75. "The fusing of two protons which is the first step of the proton-proton cycle created great problems for early theorists because they recognized that the interior temperature of the sun (some 14 million Kelvins) would not provide nearly enough energy to overcome the coulomb barrier of electric repulsion between two protons. With the development of quantum mechanics, it was realized that on this scale the protons must be considered to have wave properties and that there was the possibility of tunneling through the coulomb barrier." R. Nave, "Proton-Proton Cycle." *HyperPhysics*. Georgia State University, Department of Physics and Astronomy (hyperphysics.phy-astr.gsu.edu/hbase/astro/procyc.html, accessed December 2012). "One example of tunneling phenomena in nuclear physics is the fusion of two nuclei at very low energies," which is "of central importance for stellar energy production" (Balantekin and Takigawa 1998).

76. This photon-molecule interaction is called *Rayleigh scattering*. It is what causes the white light from the sun to light up the atmosphere with that vivid blue sky of a sunny day. Blue light photons, with their shorter wavelength, scatter more efficiently off the air molecules than do red photons, ultraviolet photons even more so. The quantum-mechanical analysis of this scattering is part of *quantum electrodynamics*, which D.P. Craig and T. Thirunamachandran introduce as "the most precise and widely applicable theory so far found for calculating the interaction of electromagnetic radiation with atomic and molecular matter, and the interactions between molecules" (1998, §1.1). Rayleigh scattering is a "two-photon scattering processes [involving] one-photon absorption and one-photon emission," §6.1. The quantum indeterminacy of the original photon's emission from the sun is compounded by the indeterminacy of its re-emission as a scattered photon.

This, says Shanahan, is actually a point of agreement in the debate, which is about whether at least some evolutionary processes are *fundamentally indeterministic* apart from the influence of quantum effects (pp. 163-65). Although the debate "appears to be about the nature of evolutionary processes," he agrees with David Stamos (2001) that it really "concerns the more fundamental issue of the relationship between biology and the physical sciences." The evolutionary indeterminist would like us to consider the probabilities involved with evolutionary theory "objective features of the processes it describes, even if this requires positing indeterminacies that have no explanation in chemistry or physics," i.e., other than the contributions of quantum indeterminacy (Shanahan 2003, 169).[77]

The key for our discussion is this: There is no way, not even theoretically with infinitely precise instruments, to look back on the countless mutations that have driven the engine of evolution and see some original winding of the clock. Those mutations occurred, and continue to occur, as *utterly random* acts of nature whose unpredictability is guaranteed by a firmly established science. And "the fact that mutation and variation are inherently unpredictable means that the course of evolution is, too" (Miller 1999, 207). There may even be indeterminacy at the higher levels of evolution, genetic drift and natural selection (Shanahan 2003; Stamos 2001; but cf. Graves et al. 1999).

For My Mutation They Cast Lots

Christian scientists have seized on intederminacy as a means for God to influence things without doing violence to the laws he set up. As the philosopher Robert Bishop observes, the theistic proposal (from John Polkinghorne and others) is to interpret "the randomness in macroscopic chaotic models and systems as representing a genuine indeterminism rather than merely a measure of our ignorance," which leaves an opening for divine action. "In essence, the sensitivity to small changes exhibited by the systems and models studied in chaotic dynamics, complexity theory and nonequilibrium statistical mechanics is taken to represent an ontological opening in the physical order for divine activity" (Bishop 2009, §6.2).

---

77. Graves et al. sharply disagree, saying that evolutionary theory "can have no truck with such probabilities or the indeterminism they engender. Unlike quantum mechanics, evolutionary biology is not a science at the basement level, and its variables are realized presumably by complex combinations of more fundamental variables. To posit the existence of ungrounded dispositions, i.e., pure probabilistic propensities, in evolutionary theory does not aid in the development of the theory, and the evidence across the sciences weighs against this posit" (1999, 154). "At the level of the macromolecule and above," they say, "biological processes asymptotically approach determinism" (p. 153), a conclusion that Stamos disputes (2001, 80-81).

Bishop doesn't include biological evolution in his discussion of chaotic systems, but it certainly seems to fit. Mendelian genetics preserves the effect of single DNA mutations rather than blurring them into an average, allowing genetic drift and natural selection (sometimes dramatic "runaway" selection) to amplify those effects into entirely new species. This is the *exponential divergence* of a chaotic system (Bishop 2009, §1.2.5). Evolving organisms do not regress to their more primitive states, which fulfills the further criterion of *aperiodicity* (§1.2.6).

So predestination, the ultimate outcome of everything determined from the beginning of time in God's initial plan, has become passé. "A clockwork world of mere mechanism could only be the endlessly spinning system kept in place by the God of deism," Polkinghorne claims. He finds the world of modern science to have "an openness in its becoming which is consonant with ... its being a world which is the creation of the true and living God, continually at work within its process" (2005, 112).

Now, with this view ascendant, we have Proverbs 16:33 ("The lot is cast into the lap, but its every decision is from the LORD") being cited for the proposition that God is "sovereign over random events like casting lots" (Haarsma 2003, 76). Kenneth Miller is writing about a God who exploits "the indeterminate nature of quantum events" to "influence events in ways that are profound, but scientifically undetectable to us" (1999, 241). And Francis Collins, writing with co-author Karl Giberson, is assuring the faithful that it is

> perfectly possible that God might influence the creation in subtle ways that are unrecognizable to scientific observation. In this way modern science opens the door to divine action without the need for law-breaking miracles. Given the impossibility of absolute prediction or explanation, the laws of nature no longer [seem to] preclude God's action in the world. [Giberson and Collins 2011, 119]

Miller finds indeterminacy "a key feature of the mind of God" (1999, 213). But Stamos, a champion of evolutionary indeterminacy as much as Miller, doesn't see it that way. If "it is indeed true that at least many of the mutations that feed natural selection are the product of quantum chance, then it does indeed seem quite difficult if not impossible to reconcile this in any rational way with the hope of God-directed evolution" (2001, 182). In the gaps where Miller finds God lurking (conveniently hidden!), Stamos sees nothing but chance.

And there are an awful lot of these little God-gaps. A ridiculously large number of them, in fact. The visible universe alone contains 10^{79} electrons, protons, and neutrons (Stenger 2009, 221-22), all with their own wave functions doing their little dances of indeterminacy and entanglement. That number is simply incomprehensible to human minds. Assume 10^{19} as a very rough estimate of the number of grains of sand in the earth's beaches and deserts.[78] You would need to stuff a whole earth full of tinier sub-grains into each sand grain to get 10^{38}, and repeat the process again to get 10^{57} sub-sub-grains. But you'd still be a *long* ways from 10 to the 79th power. And, Stenger informs us, there are a billion times more massless photons and neutrinos for God to keep track of, too.

This is a very busy God, having "to somehow maintain control over countless events taking place at the submicroscopic level over extended periods of time" (p. 222). It's amazing that he still has time to get worked up about contraception and gay marriage! The absurd idea of God as a frantic cosmic juggler

> micromanaging all these particles throughout the universe (and perhaps many other universes) does not appeal to many of today's theologians. They are looking for ways for God to act on the everyday scale of human experience where that action is meaningful to humanity. [p. 222]

As if that weren't bad enough, consider how much a quantum-diddler God is constrained in what he can accomplish. Yes, the wave function of an electron bound to a nucleus defines a probability density that is smeared around a little bit of space. But only a little bit. The "electron may be here" probability drops off dramatically as you look further away from the most likely region, and the dimensions we are talking about are tiny indeed. For example, the single electron of a ground-state hydrogen atom is most likely found at a distance of about 50 *trillionths* of a meter from the proton to which it is bound, a constant known as the *Bohr radius*. It will never be exactly in the middle of the atom, and it has only about a 1% probability of being more than four times the Bohr radius from the proton. If you're a God who wants to intervene in an apparently naturalistic universe and needs to push the electron out to, say, 10 times the Bohr radius, you'd better not let it happen in more than one out of a million quantum interactions, or things will start looking miraculous.[79]

The God of the quantum gaps is a cramped, constrained deity, who "by

---

78. See hawaii.edu/suremath/jsand.html for one calculation.

79. The Bohr radius is known to a pretty exact value: $5.2917721092 \times 10^{-11}$ meters. Again, there is nothing uncertain about the equations and probability calculations of quantum mechanics; the indeterminacy is in the individual quantum-level events. The radial probability density of an electron in a ground-state hydrogen atom is given by the equation $P(r) = 4/R^3 \times r^2 \exp(-2r/R)$, where r is the electron-proton distance in meters and R is the Bohr radius. To compute the probability of finding the electron in a given range of distances from the proton nucleus (never at a single exact point, remember), you integrate the probability density function over that range.

limiting himself to placing the electron within a finite region of space, . . . is surrendering some of his omnipotence" (Stenger 2009, 221). While "quantum mechanics with chaotic amplification may provide a place for God to act to change a natural event, it will not always prove possible, rendering God as somewhat less than omnipotent" (pp. 223-24).

You cannot cure fungal meningitis or make a morbidly premature infant whole in response to a prayer uttered a day or week earlier by fiddling with quantum events. There just isn't enough wiggle room, not even with trillions of individual indeterminacies. Not even if you're God.

Unintended consequences again. What a mess Miller and friends have made by trying to put new science into stretched old theological wineskins! It is telling that Haarsma, despite having God rolling quantum dice to influence things, allows that "God might use random processes" to "give the created world a bit of freedom. Through the laws of nature, God has given the material creation a range of possibilities to explore, and he gives his creation the freedom to explore that range" (2003, 77). Apparently, God's *presence* is indeterminate, too. He exists and acts, except when he doesn't.

Elohist Evolution

Well, they might protest, we are mostly talking about evolution. That operates on a much longer time scale and provides genuine opportunities for chaotic amplification. Fine. Let's step back and consider the absurdity of what is really being advocated for that, too.

Let's say you are the "clever and subtle God" that Kenneth Miller offers by way of example, who wants "to influence events in ways that are profound, but scientifically undetectable to us" (1999, 241). We certainly wonder why you have gotten so shy of late, considering that you killed innocent Egyptian children to show your power and have your name known throughout all the earth (Exod. 9:16). You must have your reasons.

So how do you go about doing a bit of profound but hidden influencing? Let's say you have your eye on one of the point mutations in the FOXP2 gene that seems to have a lot to do with our unique language abilities.[80] You

---

80. Sean Carroll notes that "there are hundreds more differences" between humans and chimps "in noncoding DNA around FOXP2, in switches and regions that affect the place and amount of FOXP2 expression." He believes that changes in *those* regions, which don't directly code for the FOXP2 protein, are what "enabled the evolution of fine-scale differences in individual brain regions" resulting in human speech (2005, 276). We use a mutation in the gene itself as our example, but it doesn't matter. Any divine quantum dithering would have to happen the same way for this "dark DNA," as Carroll calls it (p. 111).

are intrigued by how these two-legged apes are making stone tools and building campfires, but you want someone to preach the Word eventually, and they aren't saying much. So a divinely directed mutation is in order.

We have already reached crazy town, have we not? This God may be subtle and hidden, and respectful of all the constraints of physics, but he can still sense what is happening to a band of primates squatting on an insignificant speck of a planet orbiting one of more than a hundred billion stars in one of more than a hundred billion galaxies. While doing their regular business of scattering off noses and cave walls, those photons whose wave functions are so fully defined by quantum mechanics are also somehow reaching God's omniscient eyes (located where?) to let him see the *Homo erectus* hunters grunting at each other.

Or maybe God has the coordinates of every atom being transmitted back to him via a back-channel connection—probably with the quantum indeterminacy turned off, like GPS after the military finally relinquished its privileged accuracy. His sentient-being alert system reconstructs the shapes and movements so he can tell when a candidate for *Homo divinus* is in the offing. Sarcasm aside, is there any way to explain how any deity could possibly know all this stuff? We dare Miller et al. to say, "Because he's God, that's why." Then they are no better off, science-wise, than the six-day creationists from whom they try so hard to distinguish themselves.

Let's grant the point, though, and move on. God knows, somehow, what is needed and decides to produce this key mutation. The target is precise: You want to effect language, and you must hit the FOXP2 gene, and probably not just anywhere in its six hundred thousand or so base pairs, either. Presumably, you know how to aim the bow to hit the apple off the kid's head without hurting anybody.

Having done the necessary planning for the target of your mutation, you now must figure out how to get something there to dislodge the offending DNA nucleotide. Let's see . . . how about a bit of natural radiation? There is a suitable isotope in the rocks near the testicles of the guy you have in mind as father of the first speaking ape. Quantum tunnel that particle out, now! Off it goes, in exactly in the right direction to enter the seminiferous tubules coiled up inside a testicle, enter a sperm cell or one of its precursors, and hit a single tiny spot among three billion nucleotides of its coiled-up DNA. *Voilà!* The gift of language (or what you hope will be progress in that direction) will be carried to all the generations of humanity the next time your target male and his mate get a gleam in their eyes.

Of course, you must account for any interactions the particle might make with the countless others that lie along the path between the radioactive bit of rock and the nucleotide target of your efforts. You must pull the particle-decay trigger at just the right moment and locate its wave function decoherence at exactly the right location. You must know that the mutation will be propagated into viable offspring, or take a shotgun approach where you are mutating sperm left and right in hopes of getting the new innovation established by one of these slouching Romeos.

Now, *this* is supposed to be the more reasonable, scientifically palatable story about divinely directed evolution? It is clear that God has long since lost his explanatory powers, and now is relegated to being a concept that *requires* explanation. The sophisticated theologians who are grasping at straws like quantum apologetics must realize, somewhere deep down, that their efforts are really borne of desperation—their own, or perhaps that of their sponsors or readers.

Intelligent DeSade

> *Biological truths are simply not commensurate with a designer God, or even a good one. The perverse wonder of evolution is this: the very mechanisms that create the incredible beauty and diversity of the living world guarantee monstrosity and death. The child born without limbs, the sightless fly, the vanished species—these are nothing less than Mother Nature caught in the act of throwing her clay. No perfect God could maintain such incongruities. It is worth remembering that if God created the world and all things in it, he created smallpox, plague, and filariasis. Any person who intentionally loosed such horrors upon the earth would be ground to dust for his crimes.*
>
> —Sam Harris, *The End of Faith*

Have you ever heard a report of abuse inflicted on an innocent animal by some cruel human and, as a gut reaction, wished death for him? Even if you don't believe in capital punishment for the murderers of fellow humans? Suppose you discovered your boss was an animal abuser, and you had to flatter him, tell him he was right to do what he did, lest you lose your job. Would you do it? Would you be that yellow? That spineless? That obsequious?

We must be candid here: Evolution-accepting theists are doing the same thing when it comes to their attitude toward a God who not only permits animal cruelty, but has erected a whole system of natural selection based *centrally and entirely* on undeserved animal suffering. Evolution is a grim struggle to survive and reproduce in a world of parasites, predators, and rivals. If there is a divine guiding hand behind evolution with its awful amount of dog-eat-dog destruction, it would seem to be the same hand that wields the club against the baby seal.

A Creator Red in Tooth and Claw

In Dostoyevsky's *The Brothers Karamazov*, Alyosha tries to persuade his brother to return to faith. Ivan explains why he cannot oblige him. If he were to accept the sovereignty of a God who allows the suffering of innocent children, he would have to swallow his scruples and parrot the official party line that the Almighty must have some reason for what he permits. And in endorsing such excuses, he would be signaling that the suffering of children was acceptable after all. If Ivan were to do that, he would have lost his soul in the very attempt to save it. Well, you must ask yourself if you are willing to make the same sacrifice by accepting some sophistry to excuse the innocent suffering of animals in the process of natural selection—if it is guided (or even permitted) by a Creator God.

Kitcher notes that "a history of life dominated by natural selection is extremely hard to understand in providentialist terms." He explains why in a few concise sentences:

> Mutations arise without any direction toward the needs of organisms —and the vast majority of them turn out to be highly damaging. The environments that set new challenges for organic adaptation succeed one another by processes largely independent of the activities and requirements of the living things that inhabit them. Even if the succession of environments on earth has some hidden plan, Darwinism denies that the variations that enable organisms to adapt and to cope are directed by those environments. Evolutionary arms races abound. If prey animals are lucky enough to acquire a favorable variation, then some predators will starve. If the predators are the fortunate ones, then more of the prey will die messy and agonizing deaths. [2007, 124]

His conclusion is equally concise, and theologically devastating: "There is nothing kindly or providential about any of this, and it seems breathtakingly wasteful and inefficient." If we were to "imagine a human observer presiding over a miniaturized version of the whole show," Kitcher says, we would be hard pressed "to equip the face with a kindly expression" (p. 124). Ever read the Superman comics about the evil genius Brainiac shrinking the city Kandor, which he kept in a bottle along with its mite-sized inhabitants? He peered into the glass with malevolent glee, but at least he didn't stage a trillion death-duels of disease, starvation, and rivalry to get things set up there.

Even when evolution finally gets around to fixing some of the problems it has created, its blunt instrument of natural selection often winds up inflicting suffering in the process. Malaria became a threat to humans around 5,000 years ago, and what sustained the human population still remaining in Africa was an evolutionary "quick fix," a mutation in the hemoglobin beta gene. If just one of your parents has the mutated gene, you get substantial protection against malaria. If both of them have it, you get sickle-cell anemia (Wade 2006, 252). The cold calculus of evolution doesn't care about that. It stumbled upon a way to keep human genes propagating in the face of a disease that was killing a lot of people before reproductive age, and that was good enough. People whose ancestors sailed away from Africa centuries ago still occasionally suffer the devastating side effect of its crude genetic surgery.

Pilgrim's Progress

Annie Dillard's masterpiece *Pilgrim at Tinker Creek* is the result of a year spent contemplating nature in both its beauty and its horror. The work is suffused with a deep though unorthodox spirituality, but it is clear-headed about how evolution gets its dirty work done: "Evolution loves death more than it loves you or me." It is a difficult thing for any of us to accept, being products of evolution ourselves. "The words are simple, the concept clear— but you don't believe it, do you? Nor do I. How could I, when we're both so lovable? Are my values then so diametrically opposed to those that nature preserves?" (Dillard 1974, 178). Yes, they are, because nature doesn't care. It just *is*, and we are here—a part of it—because of its ruthlessness.

The six-foot long wooden spears recovered, astoundingly, from a *Homo heidelbergensis* hunting site weren't just for show. They, along with the stone tools and thousands of bone fragments found at the site, show what a brutal, dangerous world it was for our predecessors hundreds of thousands of years ago, and the determination they had to survive in it.[81] Fagan describes the butchering of elephants with hand axes, their massive skulls broken open for the brains inside,

> as gore-covered men and women clamber over the massive beasts, peeling back the thick hides with sharp-edged stone cleavers, hacking through joints and along ribs to dismember the carcasses, cutting strips from choice cuts to be carried away and dried in the sun and wind. Ever watchful, men with spears keep a close lookout for lions, while hyenas hover close at hand. Come twilight, the band moves away to the protection of hearth and camp, and the waiting predators move in for their turn. [2010, loc. 810]

Along with Dillard, we "have to acknowledge that the sea is a cup of death and the land is a stained altar stone. We the living are survivors huddled on flotsam, living on jetsam. We are escapees. We wake in terror, eat in hunger, sleep with a mouthful of blood" (Dillard 1974, 177). In the face of such

---

81. Incredibly, these spears are at least 240,000 years old, perhaps much older. "The site lay in an organic mud deposit, which offered preservation conditions so perfect that [Hartmut] Thieme recovered not only the bones of the butchered horses and the stone tools used to dismember them but also the long wooden spears used in the chase" (Fagan 2010, locs. 793 & 1338). An age of 400,000 years was established by radiocarbon testing (Kouwenhoven 1997). But that technique is not really suitable for dates so far beyond the 5,700 year half-life of Carbon-14. Klaus-Dieter Meyer proposed two possible figures, either 270,000 or 240,000 years ago, based on a study of the geological layers of the site and known dates of interglacial periods (2005). In any event, it is remarkable that a finely crafted wooden artifact would be preserved for our study well after the extinction of the species whose members created it.

wanton and widespread death, populations maintain themselves through a "mindless stutter" of repetition in new life:

> Creatures extrude or vent eggs; larvae fatten, split their shells, and eat them; spores dissolve or explode; root hairs multiply, corn puffs on the stalk, grass yields seed, shoots erupt from the earth turgid and sheathed; wet muskrats, rabbits, and squirrels slide into the sunlight, mewling and blind; and everywhere watery cells divide and swell, swell and divide. [p. 163]

She does not portray the "driving force behind all this fecundity" as either the simple beneficent Creator of Sunday School or the sophisticated mumbo-jumbo of some "wondrous ground consisting of the cosmic narrativity that makes the drama of evolution possible" (Haught 2010b, 92). No, it is simply

> the pressure of birth and growth, the pressure that splits the bark of trees and shoots out seeds, that squeezes out the egg and bursts the pupa, that hungers and lusts and drives the creature relentlessly towards its own death. [Dillard 1974, 163]

"The faster death goes, the faster evolution goes," she observes. "It's a wretched system" (p. 177). This "whole business of reproducing and dying by the billions" is as "unsatisfactory and clumsy" to us, and to the theologians, as it was to her. "We have not yet encountered any god who is as merciful as a man who flicks a beetle over on its feet" (p. 179).

Justification by Faith

We have seen a couple of attempts at God-justifying explanations so far. There's more to come. In "Passive-Aggressive Creationism," we will see how Process theologians use essentially the old Zoroastrian conception that God would eradicate evil and suffering if he could, but his power is limited to persuasion, and some creatures are just never going to listen to him. Haught and others add the disturbing note that all the senseless bloodshed is part of making the cosmic spectacle interesting for its main audience, God.

Haught celebrates "the mystery of a God who pours the divine selfhood into the world in an act of unreserved self-abandonment" (2000, 48). This isn't a denial of God's power, but an acknowledgment of "the divine participation in the world's suffering." Somehow, this God is not weak or powerless, but has "a kind of 'defenselessness' or 'vulnerability,'" a "divine humility" through which "the power of God becomes most effective." Yeah, right. Tell that to the forty million people disfigured and incapacitated by lymphatic filariasis (World Health Organization 2012) because Haught's optimally effective God humbly watched as filarial parasites evolved to thrive and grow in human flesh.

Parasites are the most successful of evolution's product lines. In all their forms—viral, bacterial, filarial (worms), even psychopathic humans—they have inflicted untold death and suffering. Parasitic insects comprise *one tenth* of the species in this evolved world, as Annie Dillard reminds us. "What if you were an inventor, and you made ten percent of your inventions in such a way that they could only work by harassing, disfiguring, or totally destroying the other ninety percent?" (Dillard 1974, 23). Haught recognizes this problem as part of the "dithering design" issue we just discussed in "Peekaboo Deity":

> Nature's selection of the lucky organisms occurs in complete blindness and with no sense of fairness, compassion, or justice, as far as individuals are concerned. Natural selection entails a "struggle for existence" among unequal organisms and species, and this contest causes considerable loss and pain. The unfairness and impersonality associated with natural selection seems to many evolutionists and creationists alike to contrast sharply with the infinite intelligence and beneficence that the Bible attributes to the Creator. [Haught 2010b, 33]

Parasitism is no puzzle for a naturalistic world. Indeed, Koonin considers it "an inevitable consequence of any evolutionary process in which there is a distinction between the genome (genotype) and phenotype" (2011, loc. 5198). It is only a dilemma when you must ask why "an infinite wisdom if it exists," wouldn't "do a more efficient job of engineering life's diversity" (Haught 2010b, 33). Haught's excuse is to make it part of God's "self-effacing persuasive love," his "self-emptying humility" in which "the fullest effectiveness resides" (2000, 97). Such drivel betrays an utter surrender of theism to science and the stark realities of our world. God bowing out and "letting evolution happen" is just a mythic narratizing of theology throwing in the towel.

For Haught, it all seems to be about the stagecraft; "theology may plausibly claim that biodiversity exists ultimately because of an extravagant divine generosity that provides the enabling conditions that invite the universe to become as interesting, various, and hence beautiful as possible" (2010b, 35-36). You have grossly deformed legs (elephantitis) because some *Wuchereria bancrofti* roundworms have taken up residence in your lymphatic system? Too bad. The show must go on! Look at how interesting, how varied it all is! Your awful disease is the "beautifully" successful result of *W. bancrofti* genes evolving to form tiny larvae that hitch a ride into your bloodstream via mosquito, then develop into worms. God's "unrestrained generosity" extends to parasites who find "a surfeit of opportunities for variety as the world comes into being" (p. 40), including the world of lymph and blood within your poor ravaged body. Let not your heart be troubled, though: "The divine maker of such a self-creative world is arguably much more impressive—hence worthier of human reverence and gratitude—than is a 'designer' who molds and micromanages everything directly" (p. 42).

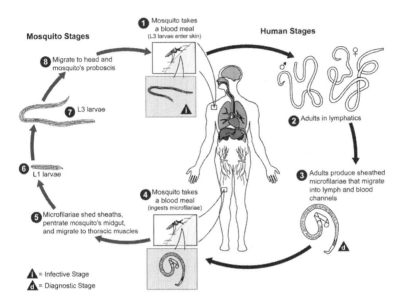

FIGURE 15: Life cycle of *Wuchereria bancrofti*.

It should be obvious to all but the most devotion-clouded minds that a God who creates in this way is no more worthy of genuine gratitude than was Nero as he watched his gladiators fight to the death for the spectacle of it all. In this respect, Haught's sophisticated-theology God doesn't seem much different from the divine slaveholder of fundamentalist Christianity. The loser lying on the arena sand is just as dead, regardless of whether the fatal sword thrust was done at the tyrant's direct "thumbs-down" command or because he merely declined to stop the show before its inevitably violent ending.

Others, however, pass the buck, observing that the eons-long bloodbath is indeed a problem, but at least it is no *new* problem! Isn't it exactly the same challenge to faith when you look at the chaos of the world around you every day? If you say you believe God is in control, you have a lot of explaining to do. And yet believers have come to feel they can live with that bafflement. The red randomness of evolution is simply more of the same. Just put it on the tab. So say Karl Giberson and Francis Collins:

> When God, as a loving Creator, *withdraws* from complete sovereign control over his creatures and grants them freedom, this means—in ways often difficult to understand—that those creatures can now act independently of God.... Such reflections have long characterized Christian thinking about the problem of evil. All we need to do now is enlarge this general concept to include the sorts of things nature is doing on its own. [Giberson and Collins 2011, 138]

Lamoureux acknowledges God's responsibility for suffering and death, claiming that those unpleasantries "are part of God's creative method. Evolutionary creation argues that as Christians develop a theodicy to justify this divine activity, they should then apply it to the evolution of life" (2008, 302). He notes, "Most Christians are repulsed by the evolutionary concepts of natural selection and the survival of the fittest, and quickly dismiss the thought that God would use such cruel mechanisms in creating life and humans" (p. 294). But Lamoureux confronts those believers with the harsh reality of their soteriology:

> Does the gospel message not include notions that could be termed "spiritual selection" and "the survival of the spiritually fittest"? At the final judgment ... people whose names are not in the "book of life," are they not going to be thrown into the lake of fire (Rev 20:15)? ... As believers come to terms with the reality that many will be lost forever, their justification for this harsh future reality can be applied to the evolutionary process, putting some divine perspective on natural selection and the survival of the fittest. [p. 294]

It's no new problem, then, but Lamoureux's cure seems worse than the disease! He adds (p. 296) the merciless slaughter of Canaanite babies as well as the infants of Bethlehem whose little lives were sacrificed to fulfilled prophecy. "Christians will come to realize that concerns about biological evolution simply pale and gradually dissipate" (p. 297). Why strain out a gnat about evolution now when you've been willing to swallow a whole herd of camels for all this time? In for a penny, in for a pound.

Kenneth Miller agrees: We've gotten used to being baffled by the larger problem of justifying God's ways, and we'll get numb to this version soon enough, if we just try to think of happier things.

> Ironically, anyone who believes that evolution introduced the idea of systematic cruelty to human nature is likely overlooking many acts of biblical cruelty, especially God's killing of the Egyptian firstborns, Herod's murder of the innocents, or the complete and intentional destruction of the cities of Sodom and Gomorrah. [Miller 1999, 246]

> Since chance—and for that matter, free will—played such an important role in bringing me to life, does this mean I cannot view my own existence as part of God's plan? Of course not. Any clergyman, very much in the Christian tradition, would caution me that God's purpose does not always submit to human analysis. God's means are beyond our ability to fathom, and just because events seem to have ordinary causes, or seem to be the result of chance, does not mean that they are not part of that divine plan ... God, if He exists, surpasses our ordinary understanding of chance and causality. Christians know that chance plays an undeniable role in history, and nonetheless accept the events that affect them in their daily lives as part of God's plan for each of them. This means that Christians already agree that the details of a historical process can be driven by chance ... and that the final result of the process may nonetheless be seen as part of God's will. [p. 236]

But does this make any more sense than the problem you're trying to explain? This is just to say, hey, we are already happy to go without a solution! You're just deferring any solution and claiming your inaction as the solution.

Polkinghorne sees God not so much as the Creator of the universe as the Master of disaster. God is, after all, "the one who allowed the wastefulness of evolution, with its blind alleys and competition for limited resources." Polkinghorne finds his solution "in a variation of the free-will defense, applied to the whole created world." In God's "great act of creation," he "allowed the world to be itself,"

> endowed in its fundamental constitution with an anthropic potentiality which makes it capable of fruitful evolution. The exploration and realization of that potentiality is achieved by the universe though the continual interplay of chance and necessity within its unfolding process. [Polkinghorne 2005, 77]

And this despite Polkinghorne's assurance that "we have been at pains to exclude any appeal to God as just a physical agent among other agencies ... He is not an alternative source of energetic causation, competing with the effects of physical principles from time to time and overriding them" (p. 42). But then how did he "endow" it? At some point, mustn't God have reached his finger down through the levels of reality to press a button? How else can he "influence," "lure," etc.? We have a kind of Cartesian dualism here.

Rene Descartes puzzled over the conundrum of how an immaterial substance (Mind) could cause a material substance (Body) to move. We aren't surprised that we can't move objects through the air without touching them, just by thinking about it. Why should it be any different for the mind to move the body? How can it happen? In the case of evolution as these writers envision it, how can God, who ostensibly occupies a higher position in a transcendent plane of reality, become a cause among causes in the material world? Again, these Christian evolutionists pretend to embrace the less "crass" supernaturalism that William James calls the "refined," but then they set it aside when they need God to have stirred the pot.

Actually, the whole "lead from behind" strategy that Process Theology ascribes to the "Creator" is itself a large-scale theodicy. For Russell, "the way forward will be a version of the 'crucified God,' developed originally in the context of the human atrocities of the 20th century, and now reformulated in light of the cruciform character of the history of life on Earth" (2003, 368). For Haught, too, reflection on Darwinian suffering leads to the cross, in the symbol of which "Christian belief discovers a God who participates in the world's struggle and pain. The cruciform visage of nature reflected in Darwinian science invites us to depart, perhaps more decisively than ever before, from all notions of a deity untouched by the world's suffering" (2000, 46). Haught's esoteric Christianity "has clearly made the crucifixion of Jesus an inner dimension of God's experience rather than something external to the deity" (p. 47).

This theology makes God into a victim of survivor guilt: As the impassive deity, he does not suffer. So he gratuitously subjects himself to it like Father Paneloux in Camus's *The Plague*.

Teilhard de Chardin "always found it impossible to be sincerely moved to pity by a crucifix so long as this suffering was presented to me as the expiation of a transgression which God could have averted—either because he had no need of man, or because he could have made him in some other way" (1933, loc. 1075). So how did Teilhard deal with this blatant problem he identified, accurately, with the doctrine of Original Sin? By projecting the cross onto the background of theodicy! Two problems solved for the price of one.

> Seen, however, on the panoramic screen of an evolutive world which we have just erected, the whole picture undergoes a most impressive change. When the Cross is projected upon such a universe, in which struggle against evil is the *sine qua non* of existence, it takes on new importance and beauty—such, moreover, as are just the most capable of appealing to us. Christ, it is true, is still he who bears the sins of the world; moral evil is in some mysterious way paid for by suffering. But, even more essentially, Christ is he who structurally in himself, and for all of us, overcomes the resistance to unification offered by the multiple, resistance to the rise of spirit inherent in matter. Christ

> is he who bears the burden, constructionally inevitable, of every sort of creation. [loc. 1078]

But how does this benefit anyone? What good does God's vicarious and gratuitous survivor-guilt suffering do those who are actually doing the suffering? Haught employs metaphors to create a web of vague affect. His is a prime case of what fundamentalist apologist Francis A. Schaeffer used to call the use of "connotation words" in lieu of real theological content. "The cross reminds me of the suffering of nature." *So what?*

> A vulnerable God, as the Trinitarian nature of Christian theism requires, could not fail to feel intimately and to "remember" everlastingly all the sufferings, struggles, and achievements in the entire story of cosmic and biological evolution. By holding these and all cosmic occurrences in the heart of divine compassion, God redeems them from all loss and gives eternal meaning to everything. [Haught 2000, 56]

Actually, Trinitarian theology said that God suffers on the cross only by virtue of the "communication of properties" in Christ. God cannot suffer, being impassible and invulnerable. So Jesus, who is both God and man, suffers as a man. Since this man also has the very nature of God and thus can be called either God or man, it is permissible to say that "God suffered on the cross." Suffering is most definitely not an inner experience of the omnipotent Creator of the universe. Haught has no business invoking Trinitarianism to support his science-inspired schema.

Molecular biologist Jacques Monod was "shocked by the intellectual spinelessness" of Teilhard de Chardin's efforts to attribute teleology to evolutionary biology. It appeared to Monod, "above all to show a systematic complacency, a willingness to conciliate at any price, to come to any compromise" (1974, 39). Is Haught's Trinitarianism-of-the-gaps any better?

Damage Control

Any weakening, enervating, softening, hedging or compromising of the creationism of the Bible is not true to the Bible, and already is a crack in the wall which unbelief will smash open into a huge crevice.

—Bernard Ramm,
The Christian View of Science and Scripture

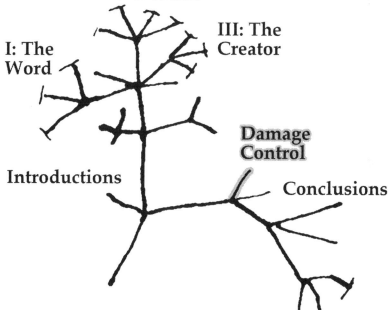

Let Not Your Left Brain Know What Your Right Brain Is Doing

> *The more committed we are to a belief, the harder it is to relinquish, even in the face of overwhelming contradictory evidence. Instead of acknowledging an error in judgment and abandoning the opinion, we tend to develop a new attitude or belief that will justify retaining it.*
>
> —Robert Burton, *On Being Certain*

Christian evolutionists seem to portray their intellectual inquiries as being first to the Bible and then to the scientific evidence with grateful awe that the two fit seamlessly together. Do their writings make it sound as if they are really accomplishing that? Or is it not obvious they find themselves in a bind and are desperately looking for a way out of it, like James Bond or Batman rapidly scanning a deathtrap to find some hidden way of escaping? They owe a debt, they think, to two masters: Darwin on the one hand and God on the other. But it turns out to be the same sort of dilemma we face between God and Mammon. As Jesus warned, you just can't serve two masters, yet Christian evolutionists would rather split the difference. One could almost see the need for this if they were caught in the middle between two seemingly conflicting bodies of research data. But that is not the case: Evolution is true, and they know it. So they insist on trying to mix oil and water, subjective feelings of religious sentiment on the one hand with hard scientific data on the other.

Desperate Dissonance

Secular scientists seem to have little interest in reconciling evolution and the Bible, or indeed science and religion in general. They've got real work to do, and Christianity's theological dilemmas are just distractions. Even Stephen Jay Gould's "non-overlapping magisteria" proposal (which seems to have gone over like a lead balloon) was mainly an attempt to get religious people to stop attacking evolution, to get creationists off the backs of science educators.[82] He was just playing the role of a ring referee trying to

---

82. The criticism of NOMA wasn't just from secular scientists. Cardinal Christoph Schönborn recognized the problem from the churchman's perspective, too: "There must be 'intersections' between theology and the natural sciences, between believing, thinking, and investigating. Belief in a Creator, in his plan, his 'rule over the earth,' his guidance of the world to a destination set by him, *cannot* remain without points of contact with the concrete exploration of the world" (Horn and Wiedenhofer 2008, loc. 1080).

pry apart two weary boxers, locked into a stalemate embrace. No, it is the religious folks, at least the ones who do not simply ignore science, who seek some peace proposal, and this for apologetical reasons. They know how damaging science's impacts on religion and scripture have been, discrediting Christian mythology, and they are desperate to engage in some damage control.

Is there a way to disarm the weapons of science and evolution? It is far from clear that the two sides need each other or that both have any interest in reconciliation. And among the religious people who have taken an interest in science, we might distinguish two nuances of motivation. One has to do with personal faith. Whether you began as a Bible-believer and then were overwhelmed by the evidence for evolution (like Denis Lamoureux), or you began as a scientist but found yourself converting to religion for existential reasons (like Francis Collins), you soon realize that "an accord between beliefs and reality is necessary for psychological well-being" (Lamoureux 2008, 14). Thus the need "to reflect upon the ways that a modern understanding of science can be harmonized with a belief in God" (Collins 2006, 124). It is just the latest example of the posturing that White already in 1895 was calling an "effort at a quasi-scientific explanation which should satisfy the theological spirit," an "inevitable effort at compromise which we see in the history of every science when it begins to appear triumphant" (Ch. 10).

Lamoureux had been raised a fundamentalist-literalist-creationist, but exposure to science education in college caused him to reject biblical religion. He embarked on the path of secularism and atheism but eventually returned to fundamentalism and creationism by way of a crisis conversion experience. But he found he was unable to switch off his brain completely, and the more he studied evolutionary theory to prepare himself to refute it, the less merit he saw in creationism. So he began to work up a compromise position.

His newly regained faith meant too much to him to surrender it again. He found what appeared to him a viable compromise solution: theistic evolution (or as he seeks to rebrand it, "evolutionary creationism" so as to make the pill less bitter for his readers). Not surprisingly, he remains essentially captive to the working assumption of young-earth creationists like his erstwhile heroes Duane Gish and Henry Morris: He cannot proceed simply on the basis of the evidence but has to navigate between the Bible and the scientific facts. As we will soon see, it is his personal agenda of protecting his narrowly regained piety that compels him to harmonize faith and science.

It is not that any gap in science or scientific evidence is in need of some data that religion can supply. Rather, it seems clear that most of these theistic evolutionists are just trying to reduce their personal cognitive dissonance and minister to those suffering the same affliction. It is laughable to see Lamoureux try to put the shoe on the other foot, claiming that "religious skeptics" are refusing to "acknowledge God" and need "to concoct 'reasons' to explain away the powerful impact of intelligent design in order to maintain their own psychological stability and comfort" (2008, 74-75). Dr. Lamoureux, by all means enjoy the comforts of your faith, your teaching position at a Catholic college, and—last but not least—the support of the Templeton Foundation (p. xx). But please put that shoe back on the foot where it fits. It's starting to smell in here.

Rammifications of Science

Neo-evangelical elder statesman Bernard Ramm seems not to have cared much for science for its own sake. He was simply dismayed at the widespread discredit in which fundamentalism stood largely thanks to the kind of obscurantism on display in the Scopes Trial. And though much more open to scientific data than someone like William Jennings Bryan, Ramm did not even feel he needed to go as far as theistic evolution. He stuck with progressive creationism.

We have quoted extensively from Ramm's classic treatment *The Christian View of Science and Scripture*. There Ramm raised the question of why the academy rejected scripture and exalted science in the nineteenth century, leading to the reproach he wants, as a neo-evangelical, to deflect. He lists several factors (1954, 16-18), which we will summarize only briefly.

First, there was a revolt against religious authority, begun in the Renaissance and Enlightenment, leading to secularism. Then the establishment of what Tillich called methodological skepticism left little room for faith and giving the benefit of the doubt to biblical assertions about nature. Meanwhile, science and invention were racking up amazing accomplishments that made Christianity seem old fashioned. Theology just produced theology books. The mismatch in results was embarassing, though we might be charitable about that point: Would anyone expect art or literary criticism to match the results of scientific inventions? It would seem to be a case of apples and oranges.

Old (Protestant versus Catholic) and new (Fundamentalist versus Modernist) divisions among Christians gave the impression that theology had no objective or reliable method and thus was epistemologically bankrupt, unlike science, where agreement was easier given shared standards of empirical verification. Evangelicals lacked a philosophy of science and were content to take the lazy path of bifurcation: It's either fiat creation or atheistic evolutionism!

These are all legitimate points. And Ramm anticipated the glee that the New Atheists now seem to take in smashing the whole edifice of religion into bits with the hammer of science: He remarks on how "anti-Christian man takes pleasure in making the gap between science and Christianity as wide as he can make it, and will heartlessly ridicule any efforts at reconciliation" (1954, 38). Such efforts are now referred to disparagingly as "accommodationism," and sympathetic non-believers like Michael Ruse who try to give the pious some philosophical cover find themselves subject to withering criticism. In some quarters, it seems that the credal purity demands of fundamentalist religion have come full circle: If you aren't fully on board with us as a *true* atheist, then you're against us.

But Ramm was not even on the front lines of that battlefield. Yes, it was bold of him in 1954—when Senator McCarthy's witch hunt was just winding down—to remind "the most hyper-orthodox among us," that "most of the views he now holds about the Bible, medicine, science, and progress which he thinks are now so orthodox, safe, sane, and Biblical, would, a few centuries ago, have cost him his life" (p. 22). Yes, he waved the white flag by saying, "Truth is truth and facts are facts no matter who develops them" (p. 38), and "evolution is not metaphysically incompatible with Christianity" (p. 204). Despite all that, however, he remained firmly in the evangelical Christian camp, keeping his powder dry. He revealed his true colors with statements like this observation about the state of mind adopted by most scientists: A "scientist's lack of faith incapacitates him from truly harmonizing science and Scripture" (pp. 38-39).

Ramm didn't like the professional neutrality that made unbelief, e.g., toward biblical supernaturalism, a default state. "Only the man of faith has the correct perspective and motivation to harmonize Scripture and science" (p. 39). In this, Ramm anticipated most of today's Christian evolutionists who whine about the bigoted "scientism" of those who "arbitrarily" eliminate the spiritual or supernatural dimension—as if there were some evident reason to factor such an elusive element into the equation. Keep in mind, it is the religious who would find something missing (from their personal lives and habits) if they capitulated to the consistent methods of empirical science.

His citation of *motivation* "to harmonize Scripture and science" is telling, too. It is indeed "the man of faith" who wants to make it all work, evidence or no evidence, because "faith is the substance of things hoped for, the evidence of things not seen" (Heb. 11:1). As evidence of a natural world piled up and Christian theology remained fixed on "Jesus Christ the same yesterday, and to day, and forever" (Heb. 13:8), it was inevitable that science would wind up pulling ahead. This Ramm lamented, even as he acknowledged it:

> The result of losing the battle of the Bible and science in the nineteenth century is simply and tragically this: Physics, astronomy, chemistry, botany, geology, psychology, medicine and the rest of the sciences are taught in disregard of Biblical statements and Christian perspectives, and with no interest in the Biblical data on the sciences, and no confidence in what the Bible might even say about the same. [p. 19]

Uh, just what "Biblical data on the sciences" could Ramm possibly have had in mind? Are scientists missing out by leaving out the "data" from the *Gilgamesh Epic* concerning a physical netherworld and a subterranean ocean, too? Ramm looked for political and public relations reasons for the discrediting of Christianity and the Bible. It seemed never to occur to him that the critics and scientists might be *right*. He didn't even weigh their criticisms. He saw it like a labor vs. management dispute which his side was losing because they had little clout and no sense of strategy. His goal was not to arrive at scientific truth for its own sake, but rather "making clear the scientific respectability of Scripture" (p. 20), "the *rapprochement* of science and evangelicalism" (p. 25) so that people would again take its *religious* teaching seriously:

> If we believe that the God of creation is the God of redemption and that the God of redemption is the God of creation, then we are committed to some very positive theory of harmonization between science and evangelicalism. God cannot contradict His speech in Nature by His speech in Scripture. If the Author of Nature and Scripture are the same God, then the two books of God must eventually recite the same story. [p. 25]

By hook or by crook! Again, Ramm's goal was not to vindicate the Bible as a source of scientific knowledge with a view toward increasing scientific knowledge, but rather to restore Bible credibility on *theological* matters by making it look like it is not primitive and anti-scientific.

Reductionism or Mystification?

Let's take a closer look at the "scientism" charge, which Korsmeyer calls "the belief that scientific knowledge is the only kind of knowledge." Those fanatical anti-religious scientists "consider that if something cannot be verified scientifically, it is just subjective opinion, and so without any objective value" (Korsmeyer 1998, 73). Daniel Dennett is a philosopher who is quick to defend science against those who try to fuzz it up, both philosophers and theologians being guilty of this. "It is not 'scientism,'" he says, "to concede the objectivity and precision of good science, any more than it is history worship to concede that Napoleon did once rule in France and the Holocaust actually happened." Then he points out what is really going on here: "Those who fear the facts will forever try to discredit the fact-finders" (Dennett 1995, 495).

This may remind some readers of an old theological debate from some fifty to sixty years ago. It recalls the objection of evangelicals like Carl F.H. Henry and J.I. Packer to the belief of Neo-Orthodox theologians (e.g., Donald M. Baillie and Emil Brunner) that in the Bible and biblical history God reveals not information but rather *himself*. But surely the two are by no means inconsistent! Why not both? says Packer. But all agree that both are *theoretically* possible; it's just that biblical criticism has destroyed every reason to posit a biblical revelation of information. (Lamoureux and Enns largely admit this today, do they not?)

So what's left? Well, people still claim to experience a spiritual encounter as they read scripture devotionally. So maybe that's the most important thing, even if the Bible doesn't really have any information about unseen worlds and the long-dead past. It is the same situation with regard to science. Why posit anything beyond the reach of empirical verification? Isn't it merely the will to believe? Freud made that observation long ago vis-a-vis the notion of life after death in heaven: Sure, it's possible, it's conceivable, but what evidence does anybody have to show for it? (Freud 1964, 49 & 55).

And note Korsmeyer's question-begging: Scientists who are already anti-religious hold a "belief" (as if by faith) that forbids them from acknowledging a reality ("something") that can't be fed through their arbitrary meat grinder. It's as bad, in its own way, as an anti-Semite saying, "I'm not going to consider any theory proposed by a Jew."

Kenneth Miller sounds the same note about the "veiled condescension" that he thinks Stephen Jay Gould displays in questioning God's apparent wastefulness in creating via the evolutionary route, the meaninglessness that Richard Dawkins supposedly finds in life, and E.O. Wilson's finding, as Miller puts it, that "evolution can explain God away as an artifact of sociobiology." Daniel Dennett, too, annoys him with his readiness "to dig quarantine fences around zoos in which religion, held safely in check, can be appreciated from a distance" (1999, 185). All of those writers, without "saying so directly, have embraced a brand of materialism that excludes from serious consideration any source of knowledge other than science" (p. 185). They just don't get it, Miller protests:

For those who lack a sense of the spiritual, the reality of that [biological] heritage may be all there is to mankind, but for people of faith there is something more. There is no scientific way to describe the spiritual concept of grace, which makes it less than real to an absolute materialist. To a believer, grace is as real as the presence of God himself. [p. 280]

But who really has the burden of proof here? Is it not the people who are urging belief in *an invisible empire of pure surmise?*

The Experiential "Evidence"

In their own way, Christians seeking to add biblical faith to science do try to bear the burden of proof. But we will have to ask whether the subjective religious experience of believers provides any adequate reason to harmonize science and evolution with the Bible and Christianity. Listen to Bernard Ramm: "Through revelation we know that this great system we call the universe ... is from God" (1954, 27). "Truth established through the reliability of the scientific method [needs to be] guided in its use [e.g., morally] through the reliability of revelation" (p. 27).

And how does one know one *has* a true revelation with which to supplement empirical research? You guessed it: "Religious thinkers can deal with evolution in a meaningful way only if they do so on the basis of their own experience of the sacred as mediated through the faith communities to which they belong" (Haught 2000, x). To his credit, Miller admits that the basis on which to add religion to science might not be so sound. He speaks of "the God we think we know through prayer and worship" (Miller 1999, 277). *Think* we know?

How can these writers blithely assume the "reliability" of *revelation*, really of faith and subjectivity, based on the sheer will to believe? Ramm does not shrink from the claim of divine guidance: "Although in theology we believe we have objective and real knowledge, we believe that in large part it appears credible only to those who have had an inward experience of the grace of the Holy Spirit" (1954, 35). "In the Christian community, belief in God rests primarily on the historical witness to redemption in the covenant with Israel and the person of Christ and on the personal experience of forgiveness and renewal" (Barbour 2000, 59). *This is data?* In the final analysis it is just subjectivity based on subjectivity, like those cosmologies where the world is balanced on the back of a turtle, which is balanced on the back of an elephant, and so on. How can one possibly think scientists are going to take this stuff seriously?

Francis Collins declaims, "Science is not the only way of knowing. The spiritual worldview provides another way of finding truth" (Collins 2006, 229). Mike Aus, having done his time in Protestant Christianity as a minister and then walking away, says it like it is: "There are not different ways of knowing. There is knowing and not knowing, and those are the only two options in this world" (2012).

And when did Collins the geneticist become an expert in epistemology? "The hour I first believed," that's when. Lamoureux waxes misty-eyed about his own campfire piety and forgets all about science. He is really looking for something else: "Evolutionary creationists experience God's presence and love in their lives. In particular, these Christians have a personal relationship with the Lord that includes miraculous signs and wonders" (Lamoureux 2008, 30). What happened to natural law and the rejection of gaps?

On the Island of Dr. Lamoureux, it is not only science but theology that is tested by emotion and emotional preference. "The greatest problem with deistic evolution is that a god who winds the clock of the universe and then leaves it to run down on its own rarely meets the spiritual needs of people" (p. 36). "The gospel of deism rarely if ever transforms lives in the way that the gospel of Jesus Christ throughout history has led men and women to be born-again" (p. 37).

He rattles off a list of cherished "foundational" beliefs—"the Holy Trinity, creation of the world, personal divine action, inspiration of Scripture, intelligent design, Image of God, human sin, and biblical ethics"—and says, "Anyone who knows the Lord personally will agree that these are basic tenets of the faith" (p. 47). Is this naïveté any better than that of teenage Mormon missionaries who tell us they know the Book of Mormon is historically accurate because they have a warm, swelling feeling in their stomachs when they read it?

Behold the sort of data that eventually made Lamoureux into an "evolutionary creationist":

> I walked into a dingy little bookstore in Tel Aviv, and near the back in a dark and dusty corner was Duane Gish's famed *Evolution: The Fossils Say NO!* (1972). I remembered exactly who Gish was, and how could I interpret finding his book in such an unusual place other than God putting it in my hands? [Lamoureux 2008, 341]

Despite his claim that "now creation science undergirded my religious worldview" and "despite an increasing focus on the issue of origins," Lamoureux confesses, "I knew that Jesus was the center of my belief" (p. 344). This fideism is the starting point for each and every one of these writers, not with the arguments that they later concoct to reconcile it with the scientific realities they are too intelligent to dismiss.

Thus convicted, Lamoureux "began to sense a calling to become a creation scientist" (p. 344). But to take that career path meant leaving the military. He struggled but finally yielded to God's leading and quit the Canadian army (pp. 345-46). Then, having enrolled at evangelical Regent College, he took a course with Loren Wilkinson, a theistic evolutionist who rejected young earth creationism. Despite Lamoureux's militant opposition, he had to admit,

> The Holy Spirit was flowing through [the professor's] words and casting a light on the foundations of my Christianity. I stammered, stumbled, and didn't really answer. Deep in my heart I knew that my relationship with Jesus was more important than any position on origins. [pp. 347-48]

One day Lamoureux noticed how Genesis 1 said nothing definitive about the world being created *ex nihilo* or how long ago it might have happened. "Genesis 1:2 exploded and shattered my calling to be a creation scientist" (p. 349). Wait a minute! Wasn't he being led by the Holy Ghost before this, when the Spirit dropped Gish's book into his hands? That should have shown Lamoureux how utterly vacuous such subjectivity is.

John Polkinghorne, too, makes a virtue of subjectivity: "Involved in all our discussion is an inescapable dialectic tension present in all Christian talk about God: the ground of being, encountered in mystical experience ... [and] the sovereign Lord, encountered in personal confrontation" (2005, 22). Polkinghorne is a credentialed scientist, not some diploma mill fake like many fundamentalist creationists. But he is also an ordained Church of England priest. It is safe to say that it is his ecclesiastical Hyde persona, and not his scientific Jekyll side, that drives him to cobble together a religion-and-science Frankenstein.

Ian Barbour, in a comprehensive survey of the contemporary discussion, observes, "Some of the most creative work today involves collaboration between scientists and theologians in drawing from the ongoing experience of a religious community while taking seriously the discoveries of modern science" (Barbour 2000, xiv). *Listen to yourself!* Isn't it obvious that, as even your own description suggests, this is all negotiation toward a compromise *on paper*, having no likely connection with reality at all?

Demonic Delusion and Noetic Dysfunction

We occasionally catch our sophisticated evangelical pro-evolutionists in what looks like an unguarded moment, when they think, so to speak, that the microphone is off. And then we hear something that makes us wonder if all the high-sounding civility is a mere cloak for a never-repented fundamentalism, even fanaticism. Their "we're all scientists" bluster is exposed, if we're quick enough to catch a glimpse.

Ramm lets the mask drop: "Even though the non-Christian scientist is unsaved and even though he is spiritually ruled by the Evil One, yet that is no basis for writing scientists and science off the record" (1954, 24). Is this an example of the "reliable knowledge" the Bible gives us to supply what is missing via mere scientific method? *Ruled by the Evil One*? Is he kidding? Does he think he is a character in an *Omen* movie or something? The impartial evidentiary rigor of science is clearly not what is driving this man's supposed investigations. It is just like when, in the early 1970s, the marketing department of Transcendental Meditation decided they could interest more American customers by soft-selling the Hinduism and pretending TM was the key to a better golf game.

Lamoureux's pious autobiographical musings betray the same internal agenda. He fares no better than Ramm, whose position Lamoureux bemoans when it is directed against revisionists like himself by those further to the right theologically. "In order to explain the unwavering acceptance of evolution by modern science, six-day creationists claim that there is a demonic delusion of incredible proportions that blinds the minds of hundreds of thousands of scientists." But that can't be, because the claim "that Satan blinds evolutionists also includes many Christians who accept this scientific theory as God's method of creation" (Lamoureux 2008, 24).

And yet Lamoureux himself thinks virtually the same thing when it comes to the rejection of Intelligent Design, as we subsequently find out. He thinks "religious skeptics need to concoct 'reasons' to explain away the powerful impact of intelligent design in order to maintain their own psychological stability and comfort" as they stubbornly refuse to acknowledge God. In his final analysis about the strong anthropic principle (there must have been a fine tuner), "sin is at the root of arguments against" it; the godless skeptics have "a problem with the First Commandment. They simply do not want to acknowledge their Creator" (p. 93). This is the self-righteous and deductive mind-reading we might have thought we had left behind with uneducated fundamentalist zealots. They are so cock-sure of their doctrines that they cannot imagine any honest person failing to see their self-evident truth. There's no real acknowledgement of honest disagreement, just suspicion of ulterior motives or sin.

We are in ironic agreement with Lamoureux on this point: "The spiritual state of an individual also shapes his or her view of design" (pp. 74-75). Yes, it does. When your spiritual state makes such condescending piety ooze out of your pen, your view of things becomes very clouded indeed.

Shoveling After the Parade

What motivates historical apologetics, attempts to vindicate the gospels as reliable eyewitness accounts of Jesus, his miracles, and his resurrection? They come into play only once the fundamentalist's own faith has been shaken. Faith is no longer placid and implicit, so one looks for proof and supposes he has found it in the arguments of apologists like Josh McDowell, N.T. Wright, and William Lane Craig. One is too easily satisfied with their reasonings, accepting them like headache pills from the nurse at the infirmary. Most apologetics are really aimed at doubters within the fold. Sure, their arguments are sometimes used as an evangelistic tool to convince non-Christian skeptics, but that is not the primary purpose.

Francis Collins was converted in this way by reading the apologetical writings of C.S. Lewis, many of whose arguments he repeats in the course of his own book, *The Language of God*. In both cases, the house is founded upon sand. Faith based on indoctrination, family loyalty, or an emotional crisis experience does not arise from intellectual factors and thus cannot serve as a guarantee in intellectual matters. Faith based on dubious arguments fares no better. As soon as you encounter good counter-arguments, you either retreat into closed-mindedness, with the attendant pressure build-up of a guilty conscience, or your placidity is disturbed by knowing there is and always will be another side to the question. The assurance of faith is gone once you admit that such propositions can be no more than *probably* true, at best. Maybe it looks true to you now, but haven't you ever had to admit you were wrong before? If you admit you might be wrong this time, too, you've seen the last of "the peace that passes understanding."

We have already seen how evangelical evolutionists reassure their readers that, even though they must give up a historical Adam, there is no danger they will ever have to part with their precious historical Jesus because the factors are not the same. Francis Collins thinks he is safe once he gets to the gospels: no more myths, just eyewitness reports of a man walking on water and rising from the dead. Cunningham observes that

> science-savvy Christians, who hope to find scientific support for the Bible, are faced with the task of continually reinterpreting texts to make them consistent with the latest scientific discoveries. Collins is correct in pointing out the dangers of this quest for scientific support but defends his faith by claiming that the supernatural is beyond the scope of science. [2010, 152-53]

Collins just separates scientific knowledge and religious faith without demonstrating that they are compatible (Cunningham 2010, 153). He "sees that once you start down this slippery slope, you can 'eviscerate the real truths of faith' and will need to decide 'where to place a sensible stopping point.'" Cunningham sees that arbitrary stopping point for what it is: "the use of flawed and unreliable written reconstructions of alleged eyewitnesses to the life of Jesus" (p. 153).

What makes Collins's arguments shoddy and unable to support the weight he places on them? Briefly, the notion of eyewitness authorship of the gospels arises not from the texts themselves, but from second-century statements written by church officials who came along too late in the game to know who said what. The gospels certainly do not bear the formal marks of eyewitness recollections, e.g., table talk. The texts contradict each other at many important points, so much that we could not begin to do justice to the topic here.[83]

Let's just consider one glaring example, Jesus' resurrection appearances. The resurrection accounts of Matthew, Luke, and John are predicated on a rejection of Mark's ending and their substitution of new ones. Mark (the original ending, before scribal redaction centuries later) concludes with *no* resurrection appearances, and the "eyewitness" women at the tomb fail to tell the disciples that it was empty. Matthew, Luke, and John all feature appearances and have the women communicate their news. One can hardly imagine a more severe contradiction. The three later writers do not so much supplement Mark's ending as rewrite his story to fit the way they, and Christians today, wish it had ended. This isn't the way you write history.

The dating of the gospels within the first century is gratuitous and based on the apologetical need to have them be as close to the events as possible, not on any evidence that they actually *are* so close. And even the traditional dating means that the events were written down decades after the fact. Would eyewitnesses of Jesus' life and ministry not have debunked any embellishments of the story? This assumes they would have had the chance to see them in the first place, or that any protests they lodged about inaccuracy would have been taken seriously by Christian propagandists. Were the apostles in charge of clamping the lid on inauthentic sayings and stories of Jesus? Were they set up like a Snopes.com debunking bureau? The

---

83. See Randel Helms's slim but illuminating book *Gospel Fictions* (Amherst: Prometheus, 1988), or the various writings of your co-author Robert M. Price on the subject, e.g., *The Incredible Shrinking Son of Man* (Prometheus, 2003). For a quick layman's observation of various New Testament issues, see co-author Ed's own *An Examination of the Pearl* (2012), Section 7, available online at examinationofthepearl.org/html/section-0021.html.

notion is anachronistic and simply fantasizes that early Christians were trying to make things easier for apologists two millennia later.

Lamoureux again sinks to the level of dismissing as unscholarly whatever historical results he doesn't like, anything that would get in the way of his having a little talk with Jesus every morning. "Most believers are aware that suspicious theologies and false doctrines have come out of universities and seminaries," he says, and cites the "faith disemboweling beliefs of the so-called 'Jesus Seminar'" as an example. He complains about theology that "dismisses the Lord's divinity and rejects his bodily resurrection from the grave," but is "confident that men and women in a personal relationship with Jesus can discern whether or not the voice of God exists in the writings of any theologian" (Lamoureux 2008, 375).

He doesn't realize he sounds just as ignorantly strident (or is that "stridently ignorant"?) on the historical Jesus issue as he did on the origins issue. The Jesus Seminar is largely made up of individuals who followed the same costly and painful path of inquiry into the facts, considered with rigorous methodology. They started from a naïve faith, as Lamoureux did with creationism. He is quick to practice the very same *ad hominem* mudslinging he so resents when others treat him the same way. Here he takes offense at the supposed "scientism" of evil, secular historians:

> Often called "the scientific study of history," this perspective asserts that human history is driven by "nothing but" social, political, and economic forces. The notion that God is behind history or involved in it is not even a consideration. It is clear that these historians retroject a *personal commitment* to religious unbelief into their writing of history. [p. 244]

But all they are doing is applying to the study of history the very same methodology that Lamoureux tries to sell to his pious readers. "Today, the scientific method and its instruments limit the investigation of the cosmos and living organisms to only natural causes and processes" (p. 265). Why can't he see that the same is true for historical method as well? We know why. He is willing to trade Adam for Jesus.

Insofar as Lamoureux, Ramm, Francis Collins, and others appeal to subjective feelings and lame historical apologetics to secure some objectivity, they are building on sand, on a rotten floor, a crumbling foundation. And even if these apologetics were not shoddy and bankrupt, surely one cannot presume to settle matters of science by invoking an alien discipline.

Groping for Gaps

> *When God is supposed to fill in the gaps in man's knowledge, then his place becomes more negligible with each discovery that has managed to explain another bit of what was formerly inexplicable. These "survival niches" of the Creator became smaller and smaller, and the more successes the natural sciences had, the more certain of victory many in the "scientific community" became, claiming that the "God hypothesis" will someday be quite unnecessary.*
>
> —Cardinal Christoph Schönborn, in *Creation and Evolution*

In the beginning was the Word, and the Word was with God. When that opening line of John was penned, there wasn't much else to go on *but* the Word, and the explanation for pretty much everything in the natural world was "God did it." How much things have changed since then! One by one, scientists have been fitting the pieces into the naturalistic puzzle, leaving ever fewer gaps for the pious to fill with God. It was bad enough for the church to lose its picture of divine wrath and power as science explained the true nature of earth's most fearsome spectacles—comets, earthquakes, lightning. Now, with evolution filling in the last gaps of our knowledge about the very nature of life, there appears to be little left for theology to worship and speculate about except itself.

Gasp! Gaps!

Of the many thought-provoking notes in Dietrich Bonhoeffer's collection *Letters and Papers from Prison*, his brief discussion of the "God of the gaps" has been perhaps the most influential.

> Weizsäcker's book on the world view of physics is still keeping me busy. It has brought home to me how wrong it is to use God as a stopgap for the incompleteness of our knowledge. For the frontiers of knowledge are inevitably being pushed back further and further, which means that you only think of God as a stop-gap. He also is being pushed back further and further, and is in more or less continuous retreat. We should find God in what we do know, not in what we don't; not in outstanding problems but in those we have already solved. [Bonhoeffer 1962, 190-91]

Bernard Ramm was unashamed to appeal to these gaps in much the same way Benjamin W. Warfield appealed to the hypothetical inerrant texts of the lost biblical autograph manuscripts. "The gaps in the geological record are gaps because vertical progress takes place only by creation" (Ramm 1954, 191). Of course what he refers to is the alleged lack of transitional forms between species. That is probably what pushed him toward progressive creationism and away from theistic evolution, which otherwise he considered a viable option for evangelicals.

C. John Collins, a more recent writer, seems as heedless as Ramm of Bonhoeffer's warning. He cites John Bloom's "On Human Origins: A Survey" (*Christian Scholars Review* 27/2, 1997, pp. 199-200) as positing

> two important gaps in the available data. The first occurs with the appearance of anatomically modern humans around 130,000 BC. The second gap occurs when culture appears, around 40,000 BC. At this point we find art, and "the complexity and variety of artifacts greatly increases." As Bloom observes, "At present, either of these transitions seems sharp enough that we can propose that the special creation of man occurred in one of these gaps and that it was not bridged by purely natural means." [Collins 2011, 117-18]

Other recent writers on the subject appear at first to have learned the lesson and to want to eschew the God of the gaps. Francis Collins echoes Bonhoeffer, "Advances in science ultimately fill in those gaps, to the dismay of those who attached their faith to them. Ultimately a 'God of the gaps' religion runs a huge risk of simply discrediting faith" (Collins 2006, 193). Then, amazingly, he falls right into the gap like it's an unseen sink hole. He lists six premises of a "typical version" of theistic evolution, all of which are in accord with scientific consensus except perhaps "fine-tuning" (we'll discuss that in a minute) and this last one:

> But humans are unique in ways that defy evolutionary explanation and point to our spiritual nature. This includes the existence of the Moral Law (the knowledge of right and wrong) and the search for God that characterizes all human cultures throughout history. [Collins 2006, 200]

Likewise, Kenneth Miller first disdains the God of the gaps with eloquent language:

> As humans began to find material explanations for ordinary events, the gods broke into retreat. As they lost one battle after another, a pattern was set up. The gods fell backwards into ever more distant phenomena until finally, when all of nature seemed to yield, conventional wisdom might have said that the gods were finished. All of them. [Miller 1999, 193]

"Could there be anything left for God to do?", he asks (p. 195), and speaks of the resultant picture of "this sad specter of God, weakened and marginalized" (p. 288).

"The trend of science is to discover and to explain," Miller notes. He warns believers not to get run over as that progress continues, saying "it would be foolish to pretend that religious faith must be predicated on the inability of science to cross such a line" (p. 276). He is criticizing young earth creationists and Intelligent Design charlatans, but then, incredibly, he joins them by drawing an uncrossable line of his own, "a boundary around our ability to grasp reality," whose presence we can't explain. "But that does not make the boundary any less real," Miller claims, "or any less consistent with the idea that it was the necessary handiwork of a Creator who fashioned it to allow us the freedom and independence necessary to make our acceptance or rejection of His love a genuinely free choice" (p. 213).

This is the very hypostatization of the Gap; it *becomes* the God! Miller sounds like he is teaching the Gnostic doctrine of the Limit of the Pleroma beyond which Sophia dared not seek. And he confuses a leap in the dark with free choice, a blind date with our freedom to consider a marriage proposal. Do we have freedom to become intellectually dishonest and pretend we know something we don't and can't know?

Korsmeyer joins the chorus, finding God's presence "absolutely essential" for evolution. The reason? "Order requires an Orderer. Chance alone is incapable of producing a creative advance to higher levels of being" (1998, 105). That's just assigning teological (and t*h*eological) labels ("creative," "advance," "higher") to a mindless process and then calling it God. It doesn't work. Emergence of complexity is a real and entirely naturalistic phenomenon, as we saw in "Let There be Replicators."

To these verbal tricks, Haught adds his own "retreat to the gap" God. He laments that "materialist evolutionism leaves out any satisfactory account of how or why subjective experience and eventually consciousness entered into the cosmic picture and became so dominant" (Haught 2000, 166). The poor unenlightened materialists ignore "the possibility that the cosmos includes a pervasive dimension of subjectivity" and thus their "conventional scientific 'explanation' of evolution pushes out of view the very feature of nature that a theology of evolution might point to as the primary zone of divine influence" (p. 167). He much prefers "a Whiteheadian idiom," where "nature's receptivity to the novel informational possibilities proposed to it by God could consist—at least in some extended or analogous sense—of a *subjective* capacity to experience and *respond* to the pull of these possibilities" (p. 167).

What does Haught mean by "nature's receptivity," which enables it to respond to God's allurements? Whitehead wrote of the "prehension" even of rocks and pine cones. He posited a rudimentary form of pan-psychism, or what Jacques Monod bluntly labels "animism." Look at it this way: Even a lousy rock *experiences* being shattered or dropped on the ground, or eroded in a stream. One can picture these changes and impacts from the standpoint of the rock insofar as it is a subject of change. Sure, why not? But does that really entail any subjectivity worthy of the name?

At any rate, Haught significantly admits that science does not and cannot tell us this, which might make us question why we should believe it other than the fact that the notion may tickle our fancy. And that, plainly, is Haught's epistemology, if not Whitehead's. He finds God hiding in the gap.

Lamoureux criticizes progressive creationism (Ramm's position) on account of the gap approach. Its greatest problem

> is that it is a God-of-the gaps model of origins. This position envisions divine creative acts in "gaps" or "discontinuities" throughout the history of life. Old earth creationists claim that natural processes are insufficient and cannot produce living organisms. [Lamoureux 2008, 27]

Whenever "physical processes are discovered to fill a gap once claimed to be where God acted," God's "purported intervention vanishes in the advancing light of science" (p. 28). And that raises "serious pastoral concerns":

> As these gaps fill and close through increasing scientific knowledge, many have assumed that there is less reason to believe in God. He then appears to be forced further and further back into the dark recesses of human ignorance. And yes, the frightening thought arises that the Creator might only be a resident of uninformed minds. [p. 28]

The scientist in Lamoureux is uncomfortable with gap apologetics, too, because that approach

> disrupts and even destroys science. Imagine the implications for medical research. If one asserts that direct divine intervention causes AIDS, Asian bird flu, or mad cow disease, then there is no reason in trying to understand the natural processes through which these diseases arose. [p. 61]

A First Cause to Believe?

Despite his warnings, Lamoureux does not hesitate to leap into the gap when it suits him to do so. In one of them he finds plenty of theistic evolutionists to keep him company: the "ontological design parameter" that "recognizes the engineered features in the universe" (p. 73). They all understand the flaws with Paley's old Natural Theology argument for the design of organisms. Once the Big Bang produced its stew of superheated quarks from (apparent) nothingness, spring-loaded with just the right mix of forces to produce light elements and stars and planets and finally sophisticated theologians, everything else happened with no sign or apparent need of intervention by a creator. But Lamoureux comes across a newer version of Paley's watch, nature itself, which "is constructed like an intricate and finely tuned instrument, displaying complexity and functionality" (p. 73).

This is the ultimate Intelligent Design argument: The universe and its constants are finely tuned in such a way that any small variance would have made the cosmos incapable of spawning or maintaining life. To Miller, "It almost seems, not to put too fine an edge on it, that the details of the physical universe have been chosen in such a way as to make life possible" (1999, 228). He gives a critical but fair hearing to non-theistic explanations and remains comfortable with the traditional alternative, God (pp. 228-32). Combining the fine-tuning issue with that of the first cause ("Why is there something rather than nothing?"), he says:

> Even as we use experimental science and mathematical logic to reveal the laws and structure of the physical universe, a series of important questions will always remain, including the sources of those laws and the reason for there being a universe in the first place. . . . [I]f we once thought we had been dealt nothing more than a typical cosmic hand, a selection of cards with arbitrary values, determined at random in the dust and chaos of the big bang, then we have some serious explaining to do. [p. 232]

Fair enough. But relying on the improbability of our particular universe to demonstrate the existence of God is fatally circular. It assumes that things might have been different, that there is no particular reason for natural laws and physical constants being as they are. That in turn presupposes a God who created by will alone, who could have done anything but created things this particular way. In other words, science can't (yet) say why things are given as they are, so it tries to fill the gap with myth. Or in Haught's case, with opera:

> Far from being merely stage preparation and rehearsal, the epochs of physical processes that occurred in the vast span of cosmic history before the actual appearance of life almost four billion years ago may be thought of as an overture to the opera. From the very start, the universe's physical constants and initial cosmic conditions had to be

exactly suitable for life to exist and be the dramatic story that it is. [Haught 2010b, 56]

For Lamoureux, an "internal mechanical rationality" is actually becoming "more and more obvious" as science has advanced. "In particular, the mathematical essence of the cosmos is most evident, and points ultimately toward an amazingly logical Mind" (2008, 73). He goes on for several pages about such "anthropic evidence" (pp. 82-90), extending it from the cosmological constants that are widely cited (the strong nuclear, weak nuclear, electromagnetic, and gravitational forces) to various aspects of the evolution of life. Such an extension of the anthropic principle is, he admits, something only a few biologists have ventured to do (p. 87). Shouldn't that tell him something?

Once the cosmic fine-tuner "had fixed the physical nature of our universe, once He had ensured that the constants of nature would create a chemistry and physics that allowed for life," Miller breezily speculates that God "would then have gone about the process of producing the creatures that would share this new world with Him" (1999, 251-52). He is content enough to imagine God operating in a gap into which he, Miller, cannot see. It is when the question turns to how God should bring those creatures into existence that he shows little patience for the "storytellers of intelligent design" who "assert that He could have done it just one way—by assembling every element, every interlocking bit and piece of tissue, cell, organ, and genetic code in single flashes of creative intensity" (p. 252).

Is there really such a difference, though, between Intelligent Design as an explanation for biological wonders and for *cosmological* ones? Miller rightly disdains the former as pseudo-science; he is a molecular biologist, and understands just how real evolution is. He doesn't need ID to explain that. But when our common scientific knowledge finally disappears, as it eventually must, into the inscrutable mystery of the universe's very formation, he and our other scientific theists seem a little too ready to throw up their hands and invoke a Creator to account for things. And that still sounds like ID to us: locating God in the gaps. It is as if Thales had said: "Gee, I can't quite figure out how the water cycle works, so I guess Zeus really *does* just turn on the faucet when it rains." Or maybe Yahweh tips over the water jars of the heavens (Job 38:37).

At this point, we all know better than the writer of Job about the water cycle. Scientists and well-informed laypeople know better about evolution by mutation and natural selection, too. But none of us really knows about the very beginnings of cosmology, though some in that field are proposing interesting ideas (see, e.g., Stenger 2009, 250-62; Krauss 2012). That's what makes it so tempting for gap seekers.

But fine-tuning is just a vision set before thirsty believers in a desert of intellectual justification: enticing yet ultimately illusory, for a couple of reasons. First, the appeal to a supernatural fine-tuner subverts and stultifies the very idea of natural constants upon which it rests. George C. Cunningham, a critic of the religious evolution apologists, gets this point just right. Suppose Francis Collins is correct about God wanting to create intelligent life for fellowship. Then, Cunningham asks, "Could God have used different physical constants and still have ended up with a universe containing humans?" If not, if none could have been changed "without making evolution of human life impossible," then God *had* to use this particular set of constants. Doesn't that make God's power "limited to obeying natural law, which he does not control?" (Cunningham 2010, 115). If, on the other hand, God is not so constrained, then he

> is all-powerful and there is no reason he could not produce humans with any constants he chooses. But that would mean that these constants are not really unique and very fine-tuned. These constants are only one of several different sets of constants that could produce life, so there is no need to explain why only one improbable set of constants is absolutely necessary and eliminates the argument for a fine-tuner. [p. 115]

Our universe's exact, critical combination of these "anthropic coincidences" is certainly improbable (p. 107), but so is your own existence as the product of a single sperm out of hundred of millions produced in a single sexual encounter, during a single ovulation cycle, between one of trillions of plausible combinations of men and women. This makes an absurdity of the common Christian belief that one's conception was divinely ordained ahead of time (see, e.g., Ps. 139:16 and Acts 17:26). God becomes a puppet-master who stage-directs the universe down to its tiniest details. We are all just marionettes in this inane drama whose uncountable trillions of acts and final ending were all pre-ordained, our every movement reflecting a tug of the Deity's invisible strings.[84]

It makes the Eden story look sophisticated and plausible by comparison. Yet our own "fine-tuning" remains as intuitively compelling to us as that of the universe itself.

> Among all the events possible in the universe the *a priori* probability of any particular one of them occurring is next to zero. Yet the universe exists, particular events must occur in it, the probability of

---

84. Adapted from Suominen 2012, §4.7.6. The Bible's support for such a micromanaging God ("the very hairs of your head are all numbered," Matt. 10:30) does nothing to make the idea more tenable.

which (before the event) was infinitesimal. At the present time we have no justification for either asserting or denying that life made one single appearance on earth, and that, as a consequence, before it appeared its chances of occurring were almost nil ... The universe was not pregnant with life nor the biosphere with man. Our number came up in the Monte Carlo game. Is it surprising that, like the person who has just made a million at the casino, we should feel strange and a little unreal? [Monod 1974, 136-37]

There is also the problem of how the fine tuner came to be. Theists happily punt to faith when positing the existence of this God with his amazingly deft touch on the cosmological-constant knobs. God just *is*, and that's that. Then why be so skeptical that the universe we find ourselves in could have "just existed" with those constants? The same issue comes up with regards to the "first cause" argument; "what is the difference between arguing in favor of an eternally existing creator versus an eternally existing universe without one?" (Krauss 2012, loc. 130).

So, too, with the origin of life. In her comprehensive overview *The Emergence of Life on Earth*, Iris Fry criticizes those who find the origin of life so implausible that it *must* have involved some sort of divine agent. The double standard is at work there as well: "Assuming that this agent is at least as complex as its alleged products, then according to the creationists' own logic we are back to the old impasse whereby the emergence of 'irreducibly complex systems' defies any natural explanation" (Fry 2000, 206). It is to this issue that we now turn.

Who's on First?

Evolution had to begin with a mutating replicator of some kind, and there is no getting around the complexity that such an entity must possess. It must be able to obtain energy from somewhere and convert inert matter into a copy of itself, complete with the information for perpetuating that copying process on and on. It is a bit startling to stop and realize that the DNA package coiled up inside your cells contains the instructions for creating every last type of cell and structure in your body, including the DNA molecule itself. It is "a curious example of self-reference which is one of the intriguing features of life" (McFadden 2002, 61). And the almost error-free copying of that DNA molecule is itself a highly evolved process with "multiple echelons of damage-control systems" (Koonin 2011, loc. 4616).[85]

---

[85]. Indeed, it seems that *evolvability* itself has evolved (Koonin 2011, loc. 4696-4772). Extensive studies "leave no doubt that the evolutionary potential of organisms is itself subject to selection and evolves. Evolution of evolvability is directly observable in laboratory experiments with evolving bacterial populations"

Koonin rightly dismisses ID as "malicious nonsense" (2011, loc. 8250), but he acknowledges that the problem of evolving the cell with all its innovations

> still uncomfortably reminds one of "irreducible complexity." Specific explanations are needed, and these are not easy to come up with. For instance, the elaborate nuclear pore complex cannot function and, accordingly, cannot be selected for in the absence of the nuclear envelope, but the latter cannot communicate with the cytosol without nuclear pore complexes. [loc. 3453]

It is hard to explain how evolution might have gotten started. The "notion that self-replication could emerge by chance is rarely, if ever, entertained by origin-of-life researchers" (Fry 2000, 102). One of those researchers, professor of molecular genetics Johnjoe McFadden, writes, "The simplest living cell could not have arisen by chance. Just like the eye, the proto-cell must have evolved from simpler ancestral cells, presumably by a process of natural selection" (2002, 84-85).

> The difficulty of the problem cannot be overestimated. Indeed, all known cells are complex and elaborately organized. The simplest known cellular life forms, the bacterial (and the only known archaeal) parasites and symbionts . . . , clearly evolved by degradation of more complex organisms; however, even these possess several hundred genes that encode the components of a fully fledged membrane; the replication, transcription, and translation machineries; a complex cell-division apparatus; and at least some central metabolic pathways. [Koonin 2011, loc. 5357]

The problem is compounded by the fact that life "on Earth emerged extremely fast: at most during several hundred million years, or probably during a mere few million years—a blink of an eye in geological terms" (Fry 2000, 125). This means we cannot "rely on extremely unlikely scenarios to get us out of our difficulties," says McFadden. "Life's rapid emergence implies that once conditions were suitable, life was probable" (2002, 84).

But even if origin-of-life scientists agree with the ID "claim that complex biological systems—in fact even functional biological polymers of specific

---

(loc. 4761). This leads Koonin to a bold statement about "evolutionary foresight" that he admits makes evolutionary biologists uneasy: "Evolution has the ability to extrapolate from repeated events of the past and to effectively predict generic aspects of the future" (loc. 4764). These are not the musings of some navel-gazing spiritualist. Koonin is a senior investigator (ncbi.nlm.nih.gov/CBBresearch/Koonin) at the National Center for Biotechnology Information, and his conclusions are derived from direct observation of the way that microbes evolve.

sequence—could not have materialized by a 'happy accident' at one stroke," their idea of research is not throwing up their hands and saying God—er, an Intelligent Designer—did it. Rather, it "consists in looking for a naturalistic alternative to the idea of the creation of life by a designer" (Fry 2000, 184).

The search has led to some interesting ideas thus far. Graham Cairns-Smith proposed that the first replicators were made of minerals (pp. 125-30). If that's true, then Job wasn't too far off the mark in saying "I also am formed out of the clay" (Job 33:6):

> Organisms based on organic chemistry developed at a later evolutionary stage on the preexisting clay scaffolding. After the "genetic takeover" by the organic genes, the scaffolding made by mineral genes disappeared without leaving a trace in extant organisms. [Fry 2000, 185-86]

There has also been a revival of the "RNA world" hypothesis, where something simpler than an impossibly complex RNA sequence, but still organic, started things off (pp. 130-45, 197-200). So, too, with the cell and its complexity: There must have been a simpler prototype, and one proposal is that it was a lipid vesicle (pp. 172-78). Lipids are "biological compounds that are waxy or oily and that dissolve in organic solvents" (Levine and Miller 1991, 388). "Polar lipids can associate to form structures like [the] liposome, a sphere bounded by a lipid bilayer" (p. 391). We've all seen a few big bubbles split into numerous smaller ones. The idea is that lipid vesicles did much the same thing, forming the first replicating entities with membranes.

All of this assumes that natural selection was part of the process from the very beginning, but Stuart Kauffman doesn't find that necessary. "He perceives spontaneous self-organization, an innate property of certain complex systems, as a complementary factor in evolution, providing different patterns of order on which natural selection can act" (Fry 2000, 157-58).

> Kauffman's theory of the origin of life postulates as a first step of organization the emergence of a self-reproducing metabolic system consisting of interacting catalytic polymers. The key concept ... is autocatalysis. Whereas a self-replicating molecule, such as an RNA polymer, is a single autocatalytic unit, doubling itself in each replication cycle, Kauffman describes an autocatalytic set of catalytic polymers, in which no single molecule reproduces itself but the system as a whole does. [p. 158]

This "metabolic set" would eventually "produce mutant polymers and new autocatalytic sets on which natural selection could act" (p. 160). As you might imagine, there is considerable skepticism about the idea. But it is more than just a lazy retreat to mystery, the "solution" of theologians that never produced a single contribution to our knowledge of the universe.

Another idea that has encountered understandable skepticism is McFadden's that the first self-replicating molecule was formed as the decoherence of a quantum superposition, which gave rise to life. The simplest organic replicator we know of is a small peptide that David Lee and colleagues designed at the Scripps Research Institute, 32 amino acids long (McFadden 2002, 97). That may not sound very big, but there are still trillions upon trillions of possible combinations of those amino acids. It is vanishingly unlikely that such a thing formed by chance (and Lee's needed some artificial assistance), even with the entire pre-biotic earth as a laboratory operating over millions of years.

After devoting more than half of his book *Quantum Evolution* to an explanation of the relevant biology and physics, McFadden proposes that a single molecule might have reacted with all of the amino acids as a quantum superposition (p. 223). It is, he says, a process that "could have continued to add more and more amino acids until a peptide which was a superposition of all 10^{41} possible thirty-two-amino-acid peptides ... was eventually generated, *so long as the system remained at the quantum level*" (p. 224, his emphasis).

Quantum decoherence occurs on a very small scale of time, temperature, and distance. So that "so long as ..." is no small requirement. But McFadden thinks there were

> many opportunities for primordial chemistry to become trapped inside tiny structures: perhaps inside the pores of a rock or within a chemically generated oil or protein droplet. These nanoscale structures would have served as a kind of proto-cell (or *very* small warm pond) which might have protected the coherence of the quantum states inside. [pp. 225-26]

That is, until "the growing peptide chain hit upon a proto-enzyme that managed to replicate itself." Then, when the self-replicator "amplifie[d] itself to a classical entity," quantum "superposition would have been shattered and the self-replicator would have crashed out of the quantum superposition of billions of possible peptides" (p. 228). Speculative as this is, it still doesn't solve the problem, and McFadden goes on to talk about quantum measurements by progressively better proto-enzymes. He concludes that "the evolution of the highly improbable self-replicator was guided through a prebiotic chemical multi-verse by quantum measurement's ability to capture the quantum states which led to the self replicator" (p. 235). It may seem like we are getting into Deepak Chopra territory here, and

McFadden's idea has been amply criticized.[86] But again, it *is* an idea, and those, like amino acids, eventually evolve into something useful.

It is undeniable that there are few actual results to all this speculation. Nothing but organic gunk is being produced in those Miller-Urey vats of electricity-zapped chemicals, and the simplest laboratory replicators are still way too complex to have been anything like products of chance. The origin of life remains a real problem. But "God did it" is a false answer. It's actually *no* answer, like saying demons are the cause of disease.

This isn't just about atheism vs. theism: Even if you believe in God, isn't it obvious that he works through secondary causes? He works through *means*; there's a set way in which things happen. So what is the way? What's the advantage to saying that there's no naturalistic explanation? You know the answer: Apologists need a gap for God to hide in.

Francis Collins acknowledges that it could be appealing for the believer to hypothesize that God stepped in to initiate the process of chemical self-assembly. But he warns that it is only today's science that remains puzzled about the lack of a naturalistic explanation for the origin of life (2006, 92-93).

> A word of caution is needed when inserting specific divine action by God in this or any other area where scientific understanding is currently lacking. From solar eclipses in olden times to the movement of the planets in the Middle Ages, to the origins of life today, this "God of the gaps" approach has all too often done a disservice to religion (and by implication, to God, if that's possible). Faith that places God in the gaps of current understanding about the natural world may be headed for crisis if advances in science subsequently fill those gaps. [p. 93]

Bravo! As a scientist, Collins knows better than to try special pleading for the unknown.

---

86. In a review of *Quantum Evolution*, Matthew J. Donald of The Cavendish Laboratory, Cambridge, argues "that McFadden's use of quantum theory is deeply flawed," arXiv:quant-ph/0101019. McFadden's view of measurements driving biological goals "has much in common with the idea of a quantum computer." But, Donald says, "the primitiveness of our artificial quantum computers and the difficulty that we have in building them, seems to me to be fairly good evidence that the sort of multi-atom wavefunction control which would be required to make McFadden's ideas work is unlikely to have occurred by chance."

God of the Quantum Gaps

In "Quantum Apologetics," we saw how quantum mechanics has been invoked to support the possibility of God steering the universe along in an entirely undetectable way. Why is he so concerned about being hidden? To prevent even the scrutiny of particle physicists from identifying proof of his existence, so that atheists will keep enough rope to hang themselves? The real reason, of course, is that God has been so diminished by science as to remain plausible only where science leaves a gap. For Kenneth Miller, John Polkinghorne, et al., the quantum gap is appealing because science promises to never fill it in.

> [T]he uncertainties inherent to quantum theory do not arise because of gaps in our knowledge or understanding. It's not as though someday either the precision of our measurements or the level of our understanding will increase to the point where we can predict which photon will go this way, or which one will move another way. [Miller 1999, 201]

You already know where Miller is going with this: to the same safe, attractive God-gap as Giberson and Collins write about. Since "the laws of physics permit quantum events to go in different directions," they find "no reason why God could not work within these processes, shaping evolutionary history." Sure the events look completely random, but they "might actually be the subtle influence of God working within the system of natural law" (Giberson and Collins 2011, 200). How convenient quantum indeterminacies are for supernaturalism! One can never peer behind the curtain to see if there is someone behind it or not.

So, according to these theologians, it isn't real indeterminacy after all, but a window through which God sneaks in. This is a flavor of Process Theology, which we will discuss below in "Passive-Aggressive Creationism." It's really just a way of having your cake and eating it, too: God directs but he doesn't direct. It sounds more like Taoism than the Bible.

The closest of our evolutionary theologians to Process Theology is John Haught. Perhaps surprisingly, then, he also criticizes the effort to insert "divine action into a series of natural causes" (2010b, loc. 2102). It "not only sounds silly to scientifically educated people," but

> also in effect reduces God to being part of nature rather than nature's abyss and ground. Thinking of God, for example, as acting in the observationally hidden domain of quantum events or in random genetic mutations ends up shrinking God's creativity down to the size of a natural cause. [loc. 2103]

Haught's solution, yet again, is depth and drama. "The fact that nature has an inexhaustible depth allows both science and theology to comment on the drama of life without coming into conflict with each other" (loc. 2130). But this is no solution, just "words strung together to explain the silence and absence of God."[87]

Michael Ruse reminds us that this quantum stuff "is very much a 'God of the gaps' kind of argument; if you cannot think of an explanation of how things work, then let us see if you can fit God into the spaces where your understanding fails." Sympathetic as he is to the theistic impulse, Ruse "will dismiss at once the suggestion appealing to modern physics which suggests that God might slip in a directed quantum event as it suits His purpose" (2000, 91). Like Intelligent Design, it is just an invocation of God resulting from the imagination being too small (Ruse 2010, 221).

The Invisible Gardener of Eden

Jacques Monod long ago exposed the fundamental fallacy of "teleological evolution" as espoused by Henri Bergson, Pierre Teilhard de Chardin, and Alfred North Whitehead. Sorry, gents, but the envisioned process of natural selection is *inherently* blind and aimless, random, hit-and-miss. If that is not so, and if there is some Entity guiding the process, why is it not evident? Why is his action so concealed as to be just like the hypothetical caretaker in the parable proposed by John Wisdom and developed by Anthony Flew—an Invisible, Intangible, and Indiscernible Gardener? Here is Flew's famous passage from his 1950 work, "Theology and Falsification":

> Once upon a time two explorers came upon a clearing in the jungle. In the clearing were growing many flowers and many weeds. One explorer says, "Some gardener must tend this plot." The other disagrees, "There is no gardener." So they pitch their tents and set a watch. No gardener is ever seen. "But perhaps he is an invisible gardener." So they set up a barbed-wire fence. They electrify it. They patrol with bloodhounds. (For they remember how H. G. Well's *The Invisible Man* could be both smelt and touched though he could not be seen.) But no shrieks ever suggest that some intruder has received a shock. No movements of the wire ever betray an invisible climber. The bloodhounds never give cry. Yet still the Believer is not convinced. "But there is a gardener, invisible, intangible, insensible to electric shocks, a gardener who has no scent and makes no sound, a gardener who comes secretly to look after the garden which he loves."

---

87. This wonderful phrase is how ex-pastor Greg Horton aptly summarizes "theology, sermons, and faith talk." Silence and Absence. *The Parish*, theparish.typepad.com/parish/2008/04/silence-and-abs.html (accessed January 2013).

At last the Sceptic despairs, "But what remains of your original assertion? Just how does what you call an invisible, intangible, eternally elusive gardener differ from an imaginary gardener or even from no gardener at all?" [Flew 1950][88]

True, we detect no sign at all that the gardener is visiting the flower patch to tend it, but he *might* be! His hand is just so subtle that we cannot see it. Then why posit it in the first place? Lamoureux does because he wants to remain a member of the Born Again Christian club.

If God were exerting or inserting occasional influence on the process anyway, why didn't he just create it in six days (or one nanosecond) and be done with it? To argue that he took the time and trouble to create the world so subtly and slowly, never showing his hand, is like arguing that all the animal species *could too* have fit on Noah's ark if, uh, God had used Brainiac's shrinking ray to reduce them in size during the voyage. Or that in Joshua's battle at Aijalon the sun didn't really stop moving around the earth, but God "just" changed the refractive index.

Do you really think such a vast and superfluous process is in view in the Bible stories, or even underlies them? It is all pathetic rationalization. Why would the Omnipotent One take such an elusive, needlessly complicated approach to creating the world? Obviously, to make it possible for Lamoureux to believe in evolution and Christianity at the same time! But if Jehovah had really been so concerned to save Lamoureux a headache, he should have just created the whole thing in six days and allowed geology to reflect that. Why didn't he?

None of this should surprise us, for it is only an extension of the classical anthropocentrism of biblical creationism. Listen to Bernard Ramm with his claim to understand "the entire plan of creation." It was all formed, you see, "with man as the climax. Over the millions of years of geological history the earth is prepared for man's dwelling" (Ramm 1954, 155). And what feats God had to work just so we could run our power plants and cars! "The vast forests grew and decayed for his coal, that coal might appear a natural product and not an artificial insertion in Nature. The millions of sea life were born and perished for his oil. The surface of the earth was weathered

---

88. Reprinted in *The Unofficial Stephen Jay Gould Archive*, stephenjaygould.org/ctrl/flew_falsification.html. Another reprint of the essay, with some important commentary about Flew's late-in-life "conversion," is at rjosephhoffmann.wordpress.com/2011/03/28/theology-and-falsification-the-hijacking-of-antony-flew/. Mark Oppenheimer's article "The Turning of an Atheist" in the November 4, 2007 issue of the *New York Times Magazine* tells the story of an aging man in mental decline being exploited for the theological agenda of others: nytimes.com/2007/11/04/magazine/04Flew-t.html.

for his forests and valleys." Finally, when this outrageously slow, costly, and circuitous process of stage preparation was done, "when every river had cut its intended course, when every mountain was in its purposed place, when every animal was on the earth according to blueprint, then he whom all creation anticipated is made, MAN, in whom alone is the breath of God" (p. 155).

One might as well deduce that fingers were designed to grasp hammers. Kenneth Miller, not quite as anthropocentrically inclined as Ramm who has his universe snugly tucked in around him, raises the obvious and damning question of which Ramm is seemingly oblivious. Addressing the latter-day creationists of Intelligent Design who "believe that the sole purpose of the Creator was the production of the human species," Miller asks them about this "magician" of theirs "who periodically creates and creates and then creates again throughout the geologic ages." Why, "in order to produce the contemporary world," did he "find it necessary to create and destroy creatures, habitats, and ecosystems millions of times over?" (Miller 1999, 128).

But what is Miller's own alternative to this magician? A spectator to the show, it seems, rather than a performer in it.

> Is there some reason to expect that the God we know from Western theology had to preordain a timetable for our appearance? After 4.5 billion years can we be sure He wouldn't have been happy to wait a few million longer? And, to ask the big question, do we have to assume that from the beginning He planned intelligence and consciousness to develop in a bunch of nearly hairless, bipedal, African primates? If another group of animals had evolved to self-awareness, if another creature had shown itself worthy of a soul, can we really say for certain that God would have been less than pleased with His new Eve and Adam? I don't think so. [p. 274]

This is the rebuke aimed at human arrogance in *Star Trek IV: The Voyage Home*. The crew of the Enterprise returns to earth after a harrowing set of adventures, only to discover that global weather patterns are being whipped into a frenzy by a space probe that cares nothing for humans but demands to talk to its favorite sentient species, humpback whales. The problem coming into focus here, between the lines, is that Miller's way of picturing God's "plan" is increasingly vague. He seems to be eroding the notion of God's "purposefulness." The divine "plan" was apparently not to have one, to just let things take their own course. There is a confusion here that is ubiquitous in our Process-influenced pro-evolutionists.

Francis Collins takes the same approach, explaining that God's purpose in creation or evolution comes out looking suspiciously like randomness. Yes, evolution could look like something "driven by chance," he admits, "but from God's perspective the outcome would be entirely specified." It's a matter of our lowly perspective, fruit flies contemplating the lifespan of a sequoia: "God could be completely and intimately involved in the creation of all species, while from our perspective, limited as it is by the tyranny of linear time, this would appear a random and undirected process" (Collins 2006, 205).

Ian Barbour notes the suggestion of some authors "that divine indeterminacies are the domain in which God *providentially controls the world.*" He highlights the proposal of William Pollard, a physicist and priest, in the 1950s, in which

> such divine action would violate no natural laws and would not be scientifically detectable. God, he suggests, determines which actual value is realized within the range of a probability distribution. The scientist finds no natural cause for the selection among quantum alternatives; chance, after all, is not a *cause*. The believer, on the other hand, may view the selection as God's doing. God would influence the events without acting as a physical force. Since an electron in a superposition of states does not have a definite position, no force would be required for God to actualize one among the set of alternative potentialities. By coordinated guidance of many atoms, God could providentially govern all events. [Barbour 2000, 86-87]

Barbour himself seems to embrace this view: "I find exciting new possibilities in the use of specific ideas in recent science to conceive of God as designer and sustainer of a self-organizing process" (p. 179). But take a step back and look at this! Isn't it a contradiction in terms? Doesn't all this amount to saying, "God's plan is to have no plan"? Doubletalk, no?

Polkinghorne seems to see the underlying problem, which we discussed above in "Peekaboo Deity." Not that he takes it very seriously when it comes to his own proposals. "Without recourse to the particular, there is a danger that the God who does everything will be perceived as the God who does nothing" (Polkinghorne 2005, 22). One must be able to distinguish certain events, "acts of God," from the others that God causes "naturally" to avoid making his presence known. Otherwise he is just the invisible gardener, leaving *no trace at all* of his influence in the riot of vegetation that appears to have grown without any assistance whatsoever. As Mr. Spock once remarked, "A difference which makes no difference is no difference" (Blish 1970, 4).

Giving God a Place to Hide

Is there a hideout, a cleft in the rock, where God can't be found to get served his pink slip? Richard Dawkins, in his blurb for Victor Stenger's book *The Fallacy of Fine-Tuning: Why the Universe Is not Designed for Us*, pulls no punches: "Darwin chased God out of his old haunts in biology, and he scurried for safety down the rabbit hole of physics .... Victor Stenger drives a pack of energetic ferrets down the last major bolt hole and God is running out of refuges in which to hide" (from Stenger 2011).

The desperation of our biologians is evident in their appeal to superfluous, fifth-wheel causal theories. Kenneth Miller complains that E.O. "Wilson never asks if there might be another way to view the religious impulse, that even if it is more the product of genes than culture, it still is fair to ask whether or not those genes might be the way a Deity ensured His message found receptive ground" (Miller 1999, 183). Sure, anything's possible, but, as Freud remarked with regard to belief in a post-mortem heaven, why think so, except that you want to?

Miller says, "Believers who proclaim their acceptance of evolution are presumed to have found a way to compromise or reconfigure the key elements of their faith in order to sidestep the problems presented by scientific materialism" (1999, 254). He seems to think he has managed to avoid this. But has he? "The common view that religion must tiptoe around the findings of evolutionary biology is simply and plainly wrong" (p. 289). Isn't this precisely what Miller's whole effort is aimed at? What else is he doing, other than educating us about the wonders of evolutionary science? That is really what most of his book is about, written in a way to make the facts palatable to religious readers.

Peaceful Coexistence?

You know the battle is over when one side begins to sue for peace. And that is what evolution theologians are doing: offering terms of their own surrender to their foes. Ian Barbour puts it more politely, noting that today's design arguments are not like those offered by William Paley, who sought to prove there must be a Creator, that a theistic origin was more likely than one without God. No, biologians seek mere co-existence. It's not victory, not by a long shot. But it is certainly better than being destroyed by one's foe, just as it's better to get a hung jury than a guilty verdict against yourself:

> Human beings are clever enough to devise systems in which complexity self-organizes, and the scientific evidence suggests that God has chosen to use self-organized complexity for at least *some* of

cosmological and biological history. But does this prove that God used natural mechanisms to create *all* forms of biological complexity? Does this mean that God *never* miraculously superseded evolution during biological history? Of course not. The jury is still out on that question. [Haarsma and Grey 2003, 308]

Barbour acknowledges that "the Anthropic Principle does not provide a conclusive argument for the existence of God of the sort once sought in natural theology." But, he adds hopefully, "the principle is quite consistent with a theology of nature in which belief in God rests primarily on other grounds" (2000, 59). The authors he reviews "use science not as a proof of God's existence but as a source of new analogies for talking about God" (p. 61). And the latest "God of the quantum gaps" proposal

> is not intended as an argument for the existence of God in the tradition of natural theology. Since uncertainties might still be attributable to human ignorance or to chance, the proposal is offered rather as a theology of nature—that is, a way in which the God in whom we believe *on other grounds* might be conceived in ways consistent with scientific theories. [p. 88, our emphasis]

That is a crucial insight, as one expects from Barbour. Evolution theologians are more modest than the belligerent creationists. Rather than boldly declaring war, today's biologists are just pleading with the bouncer, "Can I come in as long as I promise to stand over there in the corner? I won't bother anybody, honest!"

Back to Bultmann and Bonhoeffer

Shortly before the Nazis hanged him, German pastor Dietrich Bonhoeffer recorded an astute perception of both a past and future of theology:

> Religious people speak of God when human perception (often just from laziness) is at an end, or human resources fail; it is really always the *Deus ex machina* they call to their aid, either for the so-called solving of insoluble problems or as support in human failure—always, that is to say, helping out human weakness or on the borders of human existence. Of necessity, that can only go on until men can, by their own strength, push those borders a little further, so that God becomes superfluous as a *Deus ex machina* ... It always seems to me that in talking thus we are only seeking frantically to make room for God. I should like to speak of God not on the borders of life but at its centre ... God is the 'beyond' in the midst of our life. [Bonhoeffer 1962, 165-66]

And the implied "actions" of God in creation according to theistic evolutionism are exactly like the providence of God according to Bultmann. They are not interruptions *between* worldly events but rather some hidden dimension *within* them. (Whatever that means!) Where does the believer seek to discern God's actions?

> It is not in the infringement of these laws that he expects to see the miracles which God effects. What is truly miraculous for him is the fact that all events which for others are part and parcel of the normal course of things, betray for him the secret working of God. [Bultmann 1960, 165]

Why? Because otherwise one retreats to pre-scientific mythology inappropriate to minds conditioned by modern science. One need not embrace superstition as the precondition for having religious faith.

> In mythological thinking the action of God, whether in nature, history, human fortune, or the inner life of the soul, is understood as an action which intervenes between the natural, or historical, or psychological course of events; it breaks and links them at the same time. The divine causality is inserted as a link in the chain of the events which follow one another according to the causal nexus. [But] in fact the action of God is not [to be] thought of as an action which happens between the worldly actions or events, but as happening within them. [Bultmann 1958, 61]

Think of Luke 17:20-21, where Jesus seems to sweep away the mythic view of God as one more link in the causal nexus: "The kingdom of God is not coming with signs to be observed; nor will they say, 'Lo, here it is!' or 'There!', for behold, the kingdom of God is within you."[89]

Bultmann, like Tillich, anticipated the "non-overlapping magisteria" schema decades before Stephen Jay Gould. "In faith," he realized "that the scientific world-view does not comprehend the whole reality of the world and of human life." But faith, he said,

---

89. This is from the RSV's marginal reading; the main RSV text provides the alternative translation "the kingdom of God is in the midst of you." See Raymond B. Marcin, "The Kingdom of God is Within (Among) (in the Midst of) You," *American Journal of Biblical Theology* 9, No. 32 (2008): "Theological preference aside, analyses of the simple meanings of the preposition ἐντός, its adverbial counterparts, its antonym, and its Septuagint usages in the Old Testament leave little doubt but that the Kingdom of God is "within" us, and provide scant support for the notion that the Kingdom is to be understood as being 'among' us" (pp. 8-9).

> does not offer another general world-view which corrects science in its statements on its own level. Rather, faith acknowledges that the world-view given by science is a necessary means for doing our work within the world. Indeed, I need to see the worldly events as linked by cause and effect not only as a scientific observer, but also in my daily living. In doing so there remains no room for God's working. This is the paradox of faith, that faith "nevertheless" understands as God's action here and now an event which is completely intelligible in the natural or historical connection of events. [p. 65]

Unlike today's Christian evolutionists, Bultmann had at least sorted out the difference between subjectivity and objectivity, gladly embracing the former. In fact, he famously rejected all apologetics as tantamount to seeking salvation by virtue of one's good deeds, both being evasions of pure faith.

> God as acting does not refer to an event which can be perceived by me without myself being drawn into the event as into God's action, without myself taking part in it as being acted upon. In other words, to speak of God as acting involves the events of personal existence . . . When we speak of God as acting, we mean that we are confronted with God, addressed, asked, judged, or blessed by God. [p. 68]

Bultmann was, through and through, an Existentialist. So is Lamoureux, though he cannot seem to see it. The only legitimate statements one can make about God, Bultmann wrote, are those that

> express the existential relation between God and man. Statements which speak of God's actions as cosmic events are illegitimate. The affirmation that God is creator cannot be a theoretical statement about God as *creator mundi* in a general sense. The affirmation can only be a personal confession that I understand myself as a creature which owes its existence to God. It cannot be made as a neutral statement but only as thanksgiving and surrender. [p. 69]

We cannot help thinking this is the insight underlying the positions of Francis Collins and the others but which they do not quite understand. They start out talking like Bultmann, about an added level of meaning. But they wind up digging spider holes for God to hide in, *on the same level as physical phenomenon, a retreat to the God of the gaps*. It becomes very clear that all the indignant blustering about "scientism" ruling out the equally valid and necessary spiritual dimension, religion's stock in trade, is just that: blustering.

The hope of the religious evolutionists is merely to find some way to imagine that religion could still be true even though they themselves admit it adds nothing to our understanding of the cosmos or evolution. Can they at least find a rat hole in the desert for God to hide out, like Saddam Hussein, from the hostile troops of facts gunning for him?

Passive-Aggressive Creationism

That was the ultimate subtlety: consciously to induce unconsciousness, and then, once again, to become unconscious of the act of hypnosis you had just performed.

—George Orwell, *1984*

Perhaps Almighty God could manage to synthesize oil and water, but human beings have never found a way to do it. And that includes our theistic evolutionists. But that is what they try to do when they try to make God take an active hand in guiding evolution while preserving the axiom of evolutionary theory, which says things unfolded with no preferred direction, by fits and starts, by random events and tosses of the dice. How can you make God both the conductor of the orchestra and at the same time a man listening to the barking and caterwauling outside his window as he tries to sleep? There is a name for this passive-aggressive way of dealing with one's impossible doctrines. It is called Process Theology, and among religious evolutionists it is all the rage.

Creationism's Last Stand

Though other bandwagons have come and gone since its heyday in the 1970s, Process Theology still finds itself welcome among Catholic and some evangelical evolution defenders. We need to summarize some of its distinctives before we attempt to trace its contribution to today's Christian evolutionism.

French religious philosopher Blaise Pascal once famously observed that the God of the philosophers is by no means the same as the God of Abraham, Isaac, and Jacob. Whereas the Jehovah of the Bible is a living God, a person who commands, creates, loves, hates, etc., the philosophical God-concept, embraced by theologians as early as Saint Augustine, is a static abstraction paralyzed by his (really "its") perfection. Such a God could only be said to "love" or to "hate" or even to *act* by analogy with human traits, and even then it was a pretty remote analogy.

Anselm, Aquinas, and other classical theologians took this approach, and their God was Being Itself, That than which no greater can be conceived of as existing, the One whose Essence is to Exist, the Ground of Being, the First Cause. Process theologians including Charles Hartshorne, Shubert M. Ogden, David Tracy, John B. Cobb, Norman Pittenger, and John A.T. Robinson embraced the Process metaphysics of philosopher Alfred North

Whitehead. They did so in hopes of moving toward a philosophical God concept that would be less static and impassive, more like the living, acting, caring, and participating God of biblical theism—what one might less sympathetically call a mythological god who is an actor among actors, a cause among causes.

Process theologians also saw in Whitehead a new opportunity to solve the perennial problem of evil: How could it exist, where could it have come from, in a world created by a good God?

Basically, Whitehead posited that ancient thinkers like Heraclitus ("One cannot step twice into the same river, for new waters have replaced those one stepped in a moment before") and Cratylus ("One cannot step into the same river once!") were right in denying that reality was a matter of "being," of unchanging essences. Process theologians follow Whitehead in his contention that abstract concepts such as "substance" and "essence" are not the proper currency of modern metaphysics. All such terms suffer from "the fallacy of misplaced concreteness," and give us a distorted picture of reality. It would be closer to the mark to think of reality as *becoming*. There are no static and self-identical things, but rather channels of unceasing change. A river is not a fixed tank of unmoving water. It is a channel through which new water molecules are constantly moving. So it is with all "things," including God. Each is a succession, a continuity of "drops of experience," and each one forms an ingredient (or "is ingredient" or "is incarnate") in the next. Things are "routings of experience."

God in Process thought is described with some arcane technical terms. *Dipolar theism* denotes that God need not be perfect in every respect. If he were, he could not act, since, a la Aristotle, when one performs any act, one moves from having the *potentiality* to act to the *actuality* of having acted. One does not start out perfect. Perfection traditionally implies perfect completeness, perfect actuality, with all potential realized. God, being perfect, could not act. For Process theologians, only God's ultimate character (his "actuality") needs to be unchanging, i.e., reliable, trustworthy. This is God's *primordial nature*, the first aspect of his dipolar nature. The second aspect is God's *consequent* nature, by which the Process theologian asserts that God's existence *is* changing: God himself is changing, being affected by his creatures and growing by their contributions to him as their actions affect him.

Surrelativism means that God's existence is supremely relative, relative to all his creatures. He experiences everything they experience. Their decisions affect him. This is possible because of *panentheism*, the notion that God is in the world and the world is in God. God is the soul of the world, while the world is the body of God. This is a lot like pantheism ("all is

God-ism"). It also makes God the essence of all things, which are God's manifestations, but there is no personal deity left over in pantheism. In pan*en*theism ("all is *in* God-ism"), however, God transcends the world (including us) as the self transcends the body. My body is me, but I am more than it, more than the sum of my organs.

God, somewhat like Aristotle's Prime Mover, acts in the world as an urge toward *creative transformation*, to lure toward upward progress, as in evolution, or progress in science, social morality, and culture. He seeks to lead humanity toward moral and social perfection, and to attract us toward greater love and maturity. Process Theology thus takes the sting out of the problem of evil by limiting or more closely defining God's "power." It is no longer to be thought of as irresistible coercion but rather as resistably persuasive. After all, coercion destroys an enemy's power by subduing it; persuasion redirects the other's power by appealing to and convincing the other. The person who is made to obey against his will is still an enemy inside, but someone who is persuaded has been reconciled and is no longer an enemy.

Which approach is more powerful? Of course, persuasion does not always work, and Process theologians are quick to admit that. This is their "solution" to theodicy, getting God off the hook. He does the best he can, and the failures are really the fault of those who refuse his invitations. Norman Pittenger is forthright in his admission, "We do not believe any longer in divine intrusions or miraculous deliverances" (1979, 15). That would be an improper (and impossible) use of violence, derailing the precious processes of nature.

Process theologians tend to remain agnostic on the question of any possible afterlife. They are sure that who we have been and what we have done are preserved in eternity, because they have become part of God, who experienced what we experienced. It will all exist, so to speak, in his memory banks forever, though some may consider that cold comfort. Many have hailed Process Theology as the redemption of western theism, but it is easy to remain unmoved and unpersuaded by it (despite the supposed divine lure). For one thing, Whitehead's Process metaphysics appears to be one massive instance of the hypostatization fallacy.

We commit this logical slip-up when we give a name to an abstraction and imagine we have isolated its essence. Mom and dad take junior to a psychologist and ask him why junior is busy setting fires and slashing tires. The shrink's answer? "Your boy suffers from Oppositional Defiant

Disorder." That is actually a diagnosis?[90] Isn't it really just a name for the behavior he engages in? Likewise, when Whitehead asks what causes one particular outcome, not another, as when we flip a coin, he says it is "the principle of concretion" whereby a cloud of potentialities distill into one concrete outcome. Yeah, but what *is* that principle? To give the question a name is not to give it an answer.

And how is Whitehead not just appealing to an imaginary force, like phlogiston? Ian Barbour describes Whitehead's thinking:

> Whitehead ascribes the ordering of potentialities to God as *the primordial ground of order* which structures the potential forms of relationship before they are actualized. In this function God seems to be an abstract and impersonal principle. But God is also *the ground of novelty*, presenting new possibilities among which alternatives are left open. [Barbour 2000, 175]

"But possibilities, as Whitehead notes, must reside somewhere. They could not arise out of sheer nothingness without violating every convention of rationality" (Haught 2000, 71). Talk about the fallacy of misplaced concreteness! Haught makes it sound like possibilities are cars parked in a garage. And God is the garage—not even the attendant!

Another difficulty: On the one hand, Process Theology fails to make God a person and leaves "him" an abstraction, for "becoming" is hardly less an abstraction than "being." And yet, on the other hand, Process theologians speak of God in incurably personalistic terms when they speak of God's luring, drawing, urging, inviting things and people to change for the better. Naturally they hide behind the claim that such language is a mere analogy, just a metaphor or model. But analogous to *what?* Such analogies are a bridge to nowhere. Such talk always implicitly admits, "Well, of course you can't really say God is this or that." *Then why say it?* It is a fancy, wordy way of giving something with one hand and snatching it away with the other, a species of misdirection akin to that practiced by the stage magician.

If this luring, urging, inviting God is persuading, there must be someone or something to be persuaded. That's a real problem when you consider the

---

90. Yes it is, according to the DSM IV, §313.81 (behavenet.com/oppositional-defiant-disorder, accessed February 2013). Ramachandran calls this diagnosis a "notable invention" that "is sometimes given to smart, spirited youngsters who dare to question the authority of older establishment figures, such as psychiatrists. . . . The person who concocted this syndrome, whoever he or she is, is brilliant, for any attempt by the patient to challenge or protest the diagnosis can itself be construed as evidence for its validity! Irrefutability is built into its very definition" (2011, loc. 4690).

vast majority of life—to say nothing of the lifeless universe itself—that has never been remotely conscious. What can "the lure of divine love" (Pittenger) have to do with trilobites, fungi, or even chimpanzees? Korsmeyer espouses the same fairy tale theology, saying, "It is *as though* divinity labored to persuade, to lure creatures who sometimes responded to the invitation, and sometimes did not" (1998, 84, our emphasis). In this fantasy world, the "lure consists of a selected spectrum of possible values for each individual to consider incorporating into the next instant of its existence. The divine call is toward love, goodness, truth and beauty. The whole of creation is the loved one; its response is evolution" (p. 98).

Korsmeyer tells us, "God can only suggest the way evolution should go, and then must experience with creatures the results of the decisions they make" (p. 104). Pray tell, how does he think God makes these "suggestions," these trial balloons? Op-Ed pieces? Motivational speeches? And what "decisions" does an Ankylosaurus make? An octopus? A termite? Someone may point to Korsmeyer's butt-covering use of the phrase "as though" in the earlier quote and say we mustn't take this language any more literally than that of Genesis. It is all what Barbour calls a model or analogy for the unthinkable God and his "actions." But a model for *what?* What's on the other side of it? And note that the idea of divine "actions" is also a model, not a description. Is there any solid footing to be found here?

So, does this hands-off approach that Process theism assigns to God really have anything new or satisfying to say about the problem of evil? Nope. The notion that real power lies in persuasion seems to be a theological version of a pacifism that eschews any use of military force: Diplomacy is always the way to go, and force is a last resort that will never actually be deemed necessary. Such idealism all too often runs up against impossibilities, though, as a man named Neville Chamberlain discovered the hard way when trying to negotiate with Adolf Hitler. Chamberlain's heirs, both political and theological, have not learned the wisdom of Will Rogers: "Diplomacy is the art of saying, 'Nice doggie' until you can find a rock."[91]

Process Theology's theodicy is like Gandhi's advice to European Jews to use nonviolence against Hitler. There is less to this solution to the problem of evil than meets the eye. It is a deification of the impotence of naïve pacifism.

---

91. The reference to Chamberlain is certainly not original to us, even in the context of discussing evolution. In *The God Delusion*, Richard Dawkins referred unflatteringly to the "Neville Chamberlain school of evolutionists" when discussing how American scientists threatened by creationists "bend over backwards in [the] direction" of the "mainstream of clergy, theologians and non-fundamentalist believers" and talk about the non-overlapping magisteria of science and religion (Dawkins 2006, 91). He of course is having none of it; the thesis of *The God Delusion* is that the existence of God really is a question science can address, and has falsified.

Evil is no longer a problem if one declines to deal with it and just lets it have its way. Likewise, our sophisticated theologians want to relieve God of responsibility for the way evolution gets its dirty work done, as we saw in "Intelligent DeSade."

We prefer the honest pathos of Habbakuk. His God had eyes "too pure to approve evil" and could "not look on wickedness with favor." So why, he implored, did this God look "with favor on those who deal treacherously" and remain "silent when the wicked swallow up those more righteous than they?" (Hab. 1:13). Such treachery and swallowing-up has been going on in nature for many millions of years; it is how evolution works. There are no wicked or righteous in its impassive calculations, no good or evil. Just the genetic survivors who walk atop layers and layers of previous contestants, winners and losers alike all dead.

Whiteheaded or Wrong-headed?

In their defenses of evolution, the more liberal Christians explicitly embrace Process Theology and give Alfred North Whitehead the credit. (Conservative evangelical evolutionists are less explicit and less consistent about it, but they seem to be just as process-oriented.) John Haught is eager to redefine God as an abstract Prime Mover:

> The term "God" in this revised metaphysics must once again mean for us, as it did for many of our biblical forbears, the transcendent future horizon that draws an entire universe, and not just human history, toward an unfathomable fulfillment yet to be realized. [2000, 84]

"Unfathomable" is the word, all right. And, in case you didn't notice, he is also indulging in flagrant hypostatization. More:

> But what if "God" is not just an originator of order but also the disturbing wellspring of *novelty*? And, moreover, what if the cosmos is not just an "order" (which is what "cosmos" means in Greek) but a still unfinished *process*? ... And suppose that "God" is less concerned with imposing a plan or design on this process than with providing it opportunities to participate in its own creation. [p. 6]

"Supposition" is another ironically good choice of words. Here is a theologian for whom speculation suffices as epistemology. All he needs is a good story. One is reminded of the bumper sticker: "I have given up on the search for reality and am now looking for a good fantasy."

John Polkinghorne has moved close to Whitehead, too. "The act of creation involved a kenotic action by God in which he accepted the vulnerability implicit in any gift of freedom by love to the beloved" (Polkinghorne 2005,

25). And, as is typical for Process theism, Polkinghorne does not seem to mind shifting erratically between personalism—a God who can readily be described in human terms—and a vague and sterile abstraction. But all this talk about love and persuasion is a mirage. If God is not literally personal, then in what do "his" love and persuasion consist? Nonetheless, a teary-eyed John Haught writes:

> Evolution, according to Process Theology, occurs in the first place only because God's power and action in relation to the world take the form of persuasive love rather than coercive force. In keeping with the notion of grace ... divine love does not compel, but invites. [2000, 41]

For Haught, "a reasonable metaphysical explanation of the evolutionary process" is provided by the "notion of an enticing and attracting divine humility" (p. 53). Kenneth Miller, too, describes God's wishes as if to read his mind:

> Having chosen to base the lives of His creatures on the properties of matter, why not draw the origins of his creatures from exactly the same source? God's wish for consistency in His relations with the natural world would have made this a perfect choice. [1999, 252]

Similarly, Arthur Peacocke "holds that God has endowed the stuff of the world with creative potentialities that are successively disclosed ... Events unfold not according to a predetermined plan but with unpredictable novelty. God is experimenting and improvising in an open-ended process of continuing creation" (in Barbour 2000, 114).

Haught's God appears to be a kind of cosmic interior decorator, not much different from the divine aesthete who fussed about colors and materials of temple curtains (Exod. 26). He "may be pictured as sponsoring the maximizing of beauty and aesthetic enjoyment in the created world" (Haught 2010b, 84), and "is concerned with the transformation of the world into an intensity of beauty whose eventual depth and scale are far beyond our reckoning" (p. 86).

> According to process theology, evolution occurs because God is more interested in adventure than in preserving the status quo.... God's will, apparently, is for the maximization of cosmic beauty. And the epic of evolution is the world's response to God's own longing that it strive toward ever richer ways of realizing aesthetic intensity. [Haught 2000, 42-43]

Korsmeyer agrees: "Evidently, God's purpose is to create value, goodness and beauty. Evolution is God enjoying creation" (1998, 106). This, as anyone but a Process theologian can see, is sheer mythology.

And how does it square with what we discussed earlier in "Intelligent DeSade," all of the cruel bloodshed and destruction that are required to stage the spectacle? Sure, we saw how Haught tries to make God "vulnerable," a sad and humble participant in all the suffering, the cross being its emblem (2010b, 53). He of course is referring to the Christian story, patched together from the gospel accounts, of God incarnate subjecting himself to a day's worth of torture and (temporary) execution. The real sacrifice this God makes is not of himself, but the countless human beings (to say nothing of other creatures) he has subjected to countless thousands of years of disease and death in order to "enjoy" the show of evolution. That shows no more sympathy, is no more worthy of love or worship, than Nero whiling away an afternoon at his front-row seat in the blood-soaked gaudy grandeur of the Roman Colosseum.

FIGURE 16: The Emperor enjoys some drama and aesthetic intensity.

Haught Culture

The idea of the universe being a cosmic drama for God's entertainment seems to be John Haught's favorite theological metaphor. He appeals to Darwin's own work for support, saying that *Origin* "tells the story of a long struggle accompanied by risk, adventure, tragedy, and by what Darwin called 'grandeur,'" and claiming that a "Christian theology of evolution locates this drama within the very heart of God" (Haught 2010b, 53). And what a review Haught gives us of the production!

> Within the ultimate environment that Christians call Trinity, beneath the emblem of suffering that we call the cross, and in the radiance of Christ's resurrection—the evolution of life merges with the great epic of grace, promise, and liberation narrated in the Bible and codified by the doctrines of creation, incarnation, and redemption. Understood theologically, what is really going on in evolution is that the whole of creation, as anticipated by the incarnation and resurrection of Christ, is being transformed into the bodily abode of God. It is not in the design, diversity, and descent, but in the transformative drama of life, that theology finally makes its deepest contact with Darwin's science. [p. 53]

He urges us to lose ourselves in the story rather than fretting about the ticket price, the bad choice of seats, the stale smells. Careful readers of *Origin* will be able to look "beneath Darwin's account of life's design, diversity, and descent," and be "drawn into the subtext of a compelling drama of creativity, loss, suffering, and promise that is especially congenial to theological comment" (p. 54). It is this "deeper drama" that "needs to be the main focus of a theology of evolution," and he proposes "that most of the religiously distressing issues associated with natural selection can achieve a theologically satisfying resolution once we recognize them as essential to the dramatic character of life and the cosmos as a whole" (p. 54).

Over and over, Haught repeats this appeal to drama. We've seen some of it already, in "Peekaboo Deity." The "ID devotees" and "Darwinian materialists" fail to appreciate how much more theologically interesting "the drama underlying evolution" is "than the ephemeral designs that appear fleetingly on the surface of life" (Haught 2010b, 58). The "universe's physical constants and initial cosmic conditions had to be exactly suitable for life to exist and be the dramatic story that it is" (p. 56). Here is some more of the sort of thing Jerry Coyne probably had in mind when he referred to Haught's "unintentionally hilarious attempts to show that evolution is really part of God's plan" (Coyne 2009b):

> The good news that accompanies evolutionary science ... is that all beings—including, in a special way, us humans—are gifted with the

opportunity of making unique and unrepeatable contributions to an ever innovative cosmic adventure, one that aims always toward fresh and more profound forms of beauty. [Haught 2000]

The most important issue in the current debate about evolution and faith is not whether design points to deity but whether the drama of life is the carrier of a meaning. According to rigid design standards, evolution appears to have staggered drunkenly down multiple pathways, leading nowhere. But viewed dramatically, the apparent absence of perfect order at any present moment is an opening to the future, a signal that the story of life is not yet over. [Haught 2009]

[I]f life in our universe is a drama, its present deficiencies in design may not be a sign of an overall senselessness but of narrative nuance instead. Indeed, if life were perfectly designed right now, as Dawkins's implicit theology demands, there could be no drama at all. Perfect design would mean that the work of life has been finalized. There would be no story but only stiff and static structures to talk about. [Haught 2010b, 59]

Since scientific method looks only to the "outside" of things, the deeply interior drama in the domain of life goes unnoticed by the evolutionary grammarian. It is the proper role of a theological reading, however, to point readers of nature to the adventurous narrative of love and liberation at work beneath the surface available to science. [p. 78]

Unfortunately ... the emergent phenomenon of thought remains off the map of the world as most evolutionary naturalists picture it. As a result of this omission, evolutionary naturalism still offers only a diminished picture of what is really going on in the drama of life. [p. 145]

The idea of God sitting in a cosmic theater munching popcorn and watching the universe unfold is not entirely new. Genesis 1 ends each day of creation with a bit of divine applause: "And God saw that it was good." He made everything, yes, but when he steps back and looks at it there seems to be some sense of things having progressed on their own. It's especially apparent in his ovation after the last act, where "God saw every thing that he had made, and, behold, it was very good" (Gen. 1:31). God also says, "Let the earth bring forth the living creatures ..." (v. 24), which Haught cites in pointing out that theology "has not usually been opposed to the belief that the creation itself can keep on creating in the mode of secondary causation," where "God is the primary cause, but God works through the lawfulness and spontaneity of nature" (2010b, 42).[92]

Sometime around the turn of the third century CE, Clement of Alexandria applied the view of God as drama director to the Incarnation itself:

> For it was not without divine care that so great a work was accomplished in so brief a space by the Lord, who, though despised as to appearance, was in reality adored, the expiator of sin, the Saviour, the clement, the Divine Word, He that is truly most manifest Deity, He that is made equal to the Lord of the universe; because He was His Son, and the Word was in God, not disbelieved in by all when He was first preached, nor altogether unknown when, assuming the character of man, and fashioning Himself in flesh, *He enacted the drama of human salvation:* for He was a true champion and a fellow-champion with the creature. And being communicated most speedily to men, having dawned from His Father's counsel quicker than the sun, with the most perfect ease He made God shine on us. [*Exhortation to the Heathen*, Ch. 10, our emphasis]

Who's watching all this drama, anyhow? We, the actors, never get to see anything other than what's visible from our tiny corner of the stage during the brief bit parts we play. The conclusion is seen only by God himself, and he's the only one who understands any kind of a plot to the whole sprawling mess.

> All those things that once made evolution seem non-theistic—the random mutations, the extinctions, the pain, the waste—they're all part of the *drama*! Virtue from necessity! God is just a big playwright, directing a big script that none of us will ever be able to see to its end (or even comprehend), but whose working out surely amuses Him. And isn't all that pain, waste, accident, and extinction so much more *interesting* than the conventional view of creation? [Coyne 2009b]

It's all for an audience of one. And, unless you are willing to throw the idea of divine omniscience out the window, he knows the ending, anyhow! What's the point?

---

92. Later on, of course, God changed his assessment about the supposed pinnacle of all this creation: "The LORD was sorry that He had made man on the earth, and He was grieved in His heart" (Gen. 6:6, NASB). And sometimes the director had to drop the curtain and rewrite the script midway through the show, as with the Ninevites: "When God saw their deeds, that they turned from their wicked way, then God relented concerning the calamity which He had declared He would bring upon them" (Jonah 3:10, NASB).

Refined versus Piecemeal Supernaturalism

William James, the famous Pragmatist philosopher and psychologist, drew a distinction between two types of supernaturalism. One is more characteristic of pantheism, where God is all things, and his only actions are a higher dimension of theirs. The other refers to miraculous acts performed by a personal deity reaching down into the world of mortals. He says, "Refined supernaturalism is universalistic supernaturalism" (1958, 392). As Schleiermacher put it, "To me, all is miracle" (1950, 88). For "the 'crasser' variety" of supernaturalism, James says "'piecemeal supernaturalism' would perhaps be the better name":

> It admits miracles and providential leadings, and finds no intellectual difficulty in mixing the ideal and the real worlds together by interpolating influences from the ideal region among the forces that causally determine the real world's details.... [R]efined supernaturalism cannot get down upon the flat level of experience and interpolate itself piecemeal between distinct portions of nature, as those who believe, for example, in divine aid coming in response to prayer are bound to think it must. [James (1902) 1958, 392-93]

Our new pro-evolutionist Christian writers seem to bounce or shift around, as opportune, between Refined Supernaturalism, Deism, and Piecemeal Supernaturalism, depending on what aspect of each would come in handy. Deism allows them to have God actually intervene in the beginning, whereas elsewhere they implicitly or explicitly take a Process Theology (Refined Supernaturalism) position. When they try to squeeze in prayer and miracles, we have Piecemeal Supernaturalism. When you think about it, Deism itself might be considered a form of piecemeal supernaturalism, in that it does posit divine involvement, God getting his hands dirty at the initial creation of the world, then switching over to a policy of non-intervention. Note such an inconsistency in Bernard Ramm's analysis:

> In Gen. 8:33 God promises the regularity of seed time and harvest, cold and heat, summer and winter, day and night, i.e. all such shall be regular and in order. In Gen. 9:1-17 God vows that Nature will be held in constancy and that never again will such a flood take its toll of human life. Jer. 5:24 attributes the constancy of rain and harvest time to God. Jer. 31:35-36 affirms that the sun, moon, and stars fulfill their function because they function according to the *ordinances* which God controls. The regularity of night and day is called a *covenant* of God which cannot be broken (Jer. 33:20). Job 28:26 indicates the *regularity* of rain and lightning by speaking of a *decree* for rain and a *way* for lightning and thunder. [Ramm 1954, 58-59]

Ramm seems to be saying that the Bible already spoke of natural law, since God had decreed reliable regularity in natural processes. But take a second look at all these passages. If the regularity of nature is the result of discrete orders given by God, are we talking about natural laws enabling the world henceforth to run on automatic pilot, or are we just saying God keeps issuing the same commands over and over again in order to give us reliable seasons, day and night, etc? If the latter, then we are still pretty far from natural law put in place by the Creator.

That looks to be the position of Alvin Plantinga: "God is already and always intimately acting in nature, which depends from moment to moment for its existence upon immediate divine activity; there is not and could not be any such thing as his 'intervening in nature'" (from Ruse 2000, 104). But further ironies lie in store when Ramm approvingly quotes his predecessor J.W. Dawson:

> It is a common but groundless and shallow charge against the Bible that it teaches an "arbitrary supernaturalism." What it does teach is that all of nature is regulated by the laws of God, which like Himself, are unchanging, but which are so complex in their relations and adjustments that they allow of infinite variety, and so do not exclude even miraculous intervention, or what appears to our limited intelligence as such ... But if natural laws are the expression of the divine will, if these laws are multiform and complicated in their relations, and regulate vastly varied causes interacting with each other, and if the action and welfare of man come within the scope of these laws, then there is nothing irrational in the supposition that God, without any capricious or miraculous intervention, may have so correlated the myriad adjustments of His creation, as that, while it is His usual rule that rain falls alike on the evil and on the good, He may make its descent at particular times and places to depend on the needs and requests of His own children. [from Ramm 1954, 61-62]

Can Ramm be oblivious of the implications of this? Dawson seems to be advocating a straight Deistic Rationalism whereby what appear to be miraculous interventions are but rarely seen manifestations of some obscure corollaries built into the system, like a Leap Year. God is not momentarily suspending self-sustaining laws he once enacted; no, the whole Newtonian system continues to run like clockwork. It's just that we are occasionally surprised to see the cuckoo bird come out of the clock. Miracles are on their way out here, though Ramm seems not to notice it.

More liberal theistic evolutionists (and remember, Ramm was not even a theistic evolutionist, but rather a progressive creationist) are quite happy to dispense with interruptive miracles, supernatural intrusions into the natural order. Then they would have the gradual process of evolution, as modern science knows it, left untouched. Kenneth Miller reassures readers that "God has fashioned a self-consistent reality in nature, and He allows us to work within it. This may be frustrating to those who look for signs and wonders, but it's clearly the way life is" (1999, 240). So what about miracles? Well, they

> are not routine subversions of the laws of nature. If they were, then the issue of why so many extinct forms of life preceded us would be a conundrum, since each one would have to be the intentional creation of the mind of God. If each were just another chapter in an unfolding plan driven by the laws and principles of nature, the issue is not important. [pp. 240-41]

Uh, this isn't Deism? And in what sense is it a "plan"? Giberson and Collins are open-eyed about it, saying "we acknowledge that the BioLogos perspective can too easily slip into deism—the view that God starts things off and then leaves them to run on their own" (2011, 191). But Polkinghorne feels he must shoehorn in some possibility that God can answer prayers, as inconsistent as this seems with his otherwise preferred divine "hands-off" approach:

> The room for maneuver that exists for the accomplishment of divine and human ends through cosmic process will surely be enhanced by that collaborative alignment of God's will and ours which lies at the heart of petitionary prayer. [2005, 82]

Here the cards are on the table. It is all high-flying rhetoric trying to justify a theology of a personal God intervening after all. Polkinghorne even tries to justify the virginal conception of Jesus in this way. *This* is something that can emerge from a universal regularity of forces established by a Creator? Yeah, right. These people all shift back and forth as it suits them from Deism and "refined supernaturalism" to "piecemeal supernaturalism."

For Kenneth Miller, too, it all winds up defaulting to piecemeal supernaturalism after all. Recall from "Quantum Apologetics" how he resorts to a gross anthropomorphism, a "clever and subtle God." This sneaky deity can make use of "the indeterminate nature of quantum events" to

> influence events in ways that are profound, but scientifically undetectable to us. Those events could include the appearance of mutations, the activations of neurons in the brain, and even the survival of individual cells and organisms affected by the chance processes of radioactive decay. [Miller 1999, 241]

Denis Lamoureux seeks to reassure his born-again readers that his Process-Deism model of a "hands-off" God when it comes to evolution does not threaten their precious piety. They might complain, "Evolutionary creation rejects personal miracles, signs, and, and wonders." His response?

> Not true. Another popular conflation fuels this misconception. It assumes that because God intervenes dramatically in personal lives, then He must also have created the world with similar spectacular activity. Or inverting this argument, since evolutionary creationists do not believe the Creator intervenes miraculously in evolution, then He does not do so with people. Evolutionary creation underlines that a categorical distinction is vital between "God's activity in the origin of the universe and life" and "God's activity in the lives of men and women." [Lamoureux 2008, 50]

But this distinction itself is an act of fiat creation *ex nihilo* by Lamoureux. Evolution is not a problem for traditional piety because he says so, because he wants it to be that way.

At least we know that an "evolutionary creationist" does not rule out divine interventionism. No Christian can, despite all the fancy talk we've seen about God in process (or hiding). But the more the theist remains consistent with theism, the less he can with hard science. One viewpoint says "God did it," perhaps dressed up with niceties like "is doing it," or "becoming it," or "letting it happen," but all really amounting to the same thing. The other says *it just happens*, and there is never an excuse to invoke God as the cause.

Conclusions

Finally, brethren, whatsoever things are true, whatsoever things are honest, whatsoever things are just, whatsoever things are pure, whatsoever things are lovely, whatsoever things are of good report; if there be any virtue, and if there be any praise, think on these things.

—The Epistle to the Philippians

The Memes Shall Inherit the Earth

> *It is not surprising that religion survives. It has been pruned and revised and edited for thousands of years, with millions of variants extinguished in the process, so it has plenty of features that appeal to people, and plenty of features that preserve the identity of its recipes for these very features, features that ward off or confound enemies and competitors, and secure allegiance.*
>
> —Daniel Dennett, Breaking the Spell

We have seen the negative evidence against Christian doctrine in view of evolution. What's the alternative? Why is Christianity, and religion in general, such an appealing answer for so many? Perhaps surprisingly, evolutionary theory also provides a compelling answer to this question, with constructive explanations rather than just a corrosive force against religious ones.

Doctrinal Darwinism

Revival value

One need not be particularly anti-religious to appreciate that religion is a product of cultural evolution. The "live and let live" Christianity of today is not what it was even fifty years ago. And that was far more benign than the medieval Church of five hundred years earlier, which threatened not just the soul but the body as well.

Christianity's doctrines have evolved along with its attitudes about tolerance and violence. After two thousand years of squabbling, it has emerged with a clarified, more detailed, elaborate, and diverse theology. Who could deny this? Sure, there are hard-core groups who claim that *their* version of Christianity is exactly what was being practiced by the apostles, but does anyone (including them) really believe that? And even the exclusivist pounding the pulpit for his own "one true faith" must acknowledge the existence of thousands of other species of Christianity that are different from his. All those other "false" religions, if not also his own, must have evolved from the supposed original by some cultural process. That is, unless he wants to concoct the absurdity of an incredibly prolific and creative fallen angel churning out new deceits by the day while Almighty God struggles to maintain the head count of his little flock.[93]

All this doctrinal evolution is the result of cultural selection, analogous to the natural selection of biology. Ideas mutate through human creativity and miscommunication, and the resulting memetic rivals battle it out with the pen and sometimes also the sword. As we discussed in "The Soggy Foundation of Original Sin," the doctrine of Original Sin first mutated into existence with Paul, then into several competing forms, one of which (Augustine's) was culturally selected to provide a rationale for an atoning Jesus. During the early Christian centuries when all that was happening, there were vigorous debates about several other aspects of Christian doctrine that led to the great Christological councils and creeds. It was a Cambrian Explosion of doctrinal evolution.

One of the new theological questions that arose was whether the Logos that incarnated in Jesus shares the same nature as God the Father or only one similar to it. It was Athanasian Christology that prevailed: The Logos was fully God. The Arian view (the Logos was a created being lower than God) survived for a while among the Goths at the frontiers of the Roman Empire, but when Islam arose, with a similar view of Jesus, Arianism proved no match. It had no survival value ("selective fitness" is the more technically precise term) and its adherents by and large converted to Islam. Athanasian Orthodoxy was the genus of Christianity that replicated into the many forms existing today: *Athanasius catholicus, Athanasius lutherus, A. lutherus missourus*, and so on.

Sometimes the issue is not so much a mutation of ideas, but a new cognitive environment in which only mutated ideas can survive. White (1895) provides a fascinating account of Christianity's centuries-long struggle with science, showing how it has confronted each new development with denial, hostility, and finally embarrassed evasion of its past position. Here are

---

93. Dennett notes that there are thousands of religions in addition to the few dozen major ones. "Two or three religions come into existence every day," he says, "and their typical lifespan is less than a decade" (2006, 101). The Conservative Laestadian sect from which one of your co-authors (Ed) departed is one of those that views itself as the only true faith, to the exclusion of all other forms of Christianity including all other Lutherans. Even other *Laestadian* groups, formed due to periodic schisms erupting since the original founders did their preaching 150 years ago, are unwittingly heading down the road to hell, despite having mostly the same doctrines and social mores. For the most part, the groups engage in mutual finger-pointing when they bother to even recognize each others' existence, attributing the schisms to the deceits of sin and the corrupt mind of man. (See Suominen 2012, §4.1.6 & §4.2.1.)

some of the erstwhile threats along with a few of White's observations about them.

- **A round earth:** "Basil of Caesarea [4th Century CE] declared it 'a matter of no interest to us whether the earth is a sphere or a cylinder or a disk, or concave in the middle like a fan.' Lactantius [3rd Century CE] referred to the ideas of those studying astronomy as 'bad and senseless,' and opposed the doctrine of the earth's sphericity both from Scripture and reason" (Ch. 2, §2).

- **Humans living at the antipodes:** "St. Basil and St. Ambrose were tolerant enough to allow that a man might be saved who thought the earth inhabited on its opposite sides; but the great majority of the fathers doubted the possibility of salvation to such misbelievers" (Ch. 2, §3).

- **Psychiatry:** "To deny Satan was atheism; and perhaps nothing did so much to fasten the epithet 'atheist' upon the medical profession as the suspicion that it did not fully acknowledge diabolical interference in mental disease" (Ch. 15, §1).

- **Gravity:** "It was vigorously urged against [Newton] that by his statement of the law of gravitation he 'took from God that direct action on his works so constantly ascribed to him in Scripture and transferred it to material mechanism,' and that he 'substituted gravitation for Providence'" (Ch. 1, §1).

- **Plant sexuality:** "[T]he modesty of the Church authorities was so shocked by Linnaeus's proofs of a sexual system in plants that for many years his writings were prohibited in the Papal States and in various other parts of Europe where clerical authority was strong enough to resist the new scientific current" (Ch. 1, §3).

- **Lightning as electricity:** "As late as 1770 religious scruples regarding lightning-rods were still felt, the theory being that, as thunder and lightning were tokens of the Divine displeasure, it was impiety to prevent their doing their full work" (Ch. 6, §4).

- **Epidemiology:** "In 1798 an Anti-vaccination Society was formed by physicians and clergymen, who called on the people of Boston to suppress vaccination, as 'bidding defiance to Heaven itself, even to the will of God,' and declared that 'the law of God prohibits the practice'" (Ch. 10).

The process of theological evolution goes back as far as you care to look. Robert Wright's *The Evolution of God* is a fascinating contemplation of religion as the product of cultural evolution. The Hebrew Bible, he says, "tells the story of a god in evolution, a god whose character changes radically from beginning to end" (Wright 2009, loc. 1649). Just compare the Torah's terrifying view of Yahweh—demanding conquest, ritual slaughter, and compliance with endless rules—to the enlightened Judaism of Micah:

> Does the LORD take delight in thousands of rams, in ten thousand rivers of oil? Shall I present my firstborn for my rebellious acts, the fruit of my body for the sin of my soul? He has told you, O man, what is good; and what does the LORD require of you but to do justice, to love kindness, and to walk humbly with your God? [Mic. 6:7, NASB]

Wright notes that "the Bible's depictions of a vivid, dramatically interventionist Yahweh decline in frequency as the biblical narrative unfolds" (2009, loc. 2113).

> The early scriptures offer a hands-on, anthropomorphic god, who walks through a garden, calls out to people, makes them clothes, courteously closes up an ark before unleashing a lethal flood, and drowns Egyptians by blowing on the sea (through his nose). This god smells the "pleasing odor" of burnt sacrifices. In later scriptures we see less of God in the flesh, and even start to see a god with no flesh at all. [loc. 2118]

It's not just Yahweh who has evolved. The New Testament is a seedbed from which many views of Jesus arose. We find adoptionism (Jesus was a righteous man rewarded with divine honors) in Mark's baptism scene, in Romans 1:3-4, and in Acts 2:36; 3:20; 13:33. A product of both divine and human parentage, the Jesus of Matthew and Luke is more of a "divine man" or demigod like Hercules, Perseus, Theseus, or Pythagoras. The Gospel of John understands Jesus as the incarnation of the divine Logos, existing alongside God from the dawn of creation if not before. All these Christological models (and others—cf., 1 John 4:1-3) had their adherents for a long time, as witness the heresy hunts and Christological councils of the early Christian centuries. The Christology that won the crown of "orthodoxy" had managed to out-compete its rival doctrines in the evolutionary marketplace of ideas. The many others were consigned to the category of "heresy." That is why we find only fragments and fossils of them in scripture.

Accordingly, scripture itself must be viewed as the product of a similar cultural evolution as various gospels, epistles, acts, and apocalypses had various degrees of popularity and influence, competing for spots in the official canon. The losers survive today, if at all, in very few ancient manuscripts recovered by archaeologists. Like the genome with its mishmash of active and "fossil" genes, scripture is both a vehicle and product of evolution, conveying units of culture but also shaped by it.

Origins

Where and why did it all start? Answering that question from an evolutionary perspective is a theme of Daniel Dennett's *Breaking the Spell: Religion as a Natural Phenomenon*. He discusses several alternative explanations, not necessarily exclusive of each other. Religion can offer appealing rewards for the ultimate Darwinian goal of its own cultivation and propagation, analogous to the relationship between prolific sugar beets and the humans who devote such tremendous resources to farming them (Dennett 2006, 58-59 and 82-83). It can have symbiotic benefits for its human hosts, like the bacteria in our guts (pp. 83-85).

Formulating and communicating doctrinal nuances may serve as an expression of mental capacity. Sexual selection occurs when such fitness indicators are used, especially by finicky females who invest a lot in each offspring, for evaluation of potential mates. Dennett speculates that

> there might have been straightforward sexual selection by human females for religion-enhancing psychological traits. Perhaps they preferred males who demonstrated a sensitivity to music and ceremony, which could then have snowballed into a proclivity for elaborate rapture. The females who had this preference wouldn't have had to understand why they had it; it could just have been a whim, a blind personal taste that prompted them to choose, but if the mates they chose just happened to be better providers, more faithful family men, these mothers and fathers would tend to raise more children and grandchildren than others, and both the sensitivity to ceremony and the taste for those who loved ceremony would spread. [pp. 87-88]

This can lead to a "runaway" form of sexual selection, which enhances the selected trait's prominence well beyond its original fitness-signaling function. In that case, the females' whim for prehistoric pastors would "have had a selective advantage *only* because more females shared that whim, so that sons who lacked the fashionable sensitivity to ceremony were passed over by the choosy females" (p. 88).

Religion may simply be what Dennett calls a "good trick," evolved multiple independent times (like the eye) because it is so useful in societies. It makes things more secure and harmonious for everyone, offering comfort to the oppressed and a docile population to the oppressor (p. 90). If nothing else, religion might be viewed as the (sometimes) beautiful byproduct of some unknown trigger, like a pearl that

> begins with a meaningless speck of foreign matter (or, more likely, a parasite), and once the oyster has added layer after beautiful layer, it can become something of coincidental value to members of a species who just happen to prize such things, whether or not this coveting is wise from the point of view of biological fitness. [p. 91]

Ehrlich traces religious ideas "to the evolution of brains large enough to make possible the kind of abstract thought necessary to formulate" the concepts of religion and philosophy. He considers "intense consciousness and knowledge of our own mortality" a side effect of these big brains, which creates anxiety that people have long sought to reconcile through religion. "Beliefs in the supernatural clearly have had—and continue to have—enormous influence on human behavior and the evolution of human societies" (Ehrlich 2000, 214).

We already know from our discussion in "Peekaboo Deity" why the supernatural is such an easy sell for our evolved minds: agency detection. After seeing the impressive track record of science, we are more likely to look to it for explanations, but "evolutionary psychology suggests that a much more natural way to explain anything is to attribute it to a humanlike agent" (Wright 2009, loc. 7685). Darwin himself thought it not "difficult to comprehend" why "the belief in unseen or spiritual agencies" arose: "As soon as the important faculties of the imagination, wonder, and curiosity, together with some power of reasoning, had become partially developed, man would naturally crave to understand what was passing around him, and would have vaguely speculated on his own existence" (1888, 143).

Go Ye Therefore, and Replicate in All Nations

Memes

An ambitious and somewhat speculative viewpoint in the study of cultural evolution is to regard its products as replicators in their own right, rather than materializing for some other underlying reason. These proposed evolutionary units are *memes* rather than genes, propagating themselves through the minds of human beings instead of the gonads. According to memetics proponents like Dennett, "they have their own fitness as replicators, independently of any contribution they may or may not make to the genetic fitness of their hosts, the human vectors" (2006, 350).

Richard Dawkins pioneered the idea in his classic text *The Selfish Gene*, originally published in 1976. In the memetic view of religion's origins, "an organized church, with its architecture, rituals, laws, music, art, and written tradition," could be regarded "as a co-adapted stable set of mutually-assisting memes" (Dawkins 2006, 197). Memetics has gained only modest academic acceptance in the thirty years since Dawkins's proposal, but it makes some interesting points about the evolution of religion that we can't afford to ignore.

We are now well acquainted with the foundational idea of biological evolution: The genes in the DNA recipe for the best-adapted organisms are the ones that wind up replicating the most. Today's evolutionary survivors among the genes are being propagated in beetles and basketball players rather than dodo birds and dinosaurs. Memetics posits memes as cultural equivalents to the biological replicators: Those memes that have replicated the best—via books, videos, blog postings, sermons, gossip, etc.—are the ones that now occupy the most cognitive territory in our brains.

Certainly, the success of the ideas we cherish and spread isn't an accident. It is *these* ideas that won the struggle for our attention, having the right attributes to survive in our brains and replicate from one brain to the next. They are cultural equivalents to the genetic winners who are now alive rather than vanished from the earth with only fossils as their legacy.

Kate Distin, an independent scholar of cultural evolution, views attention as the meme's limited resource, analogous to the limited ecological resources for which genes compete (2005, 14). "There is a struggle for existence because a vast array of memes is competing for the limited resource of human attention, and therefore the fitness of any given meme will be influenced chiefly by its ability to gain and retain attention" (p. 57).

We might say it this way: *Memetic Malthusianism is about minds, not munchies.* If you remember this mnemonic travesty *and* you pass it on (perhaps wincing while saying it out loud to someone else), then it has what it takes to be a successful "organism" of cultural evolution. Copies of it will get lodged in brains, perhaps for a lifetime. They will all bear the memetic code not just for this particular and annoying lineup of the letter "m," but also for that of the underlying "species," the concept of population pressure on ideas due to limited cognitive resources. That says nothing about its inherent literary value, which we admit to be highly dubious. But "the only stocktaking that evolution honors" is "differential reproduction" (Dennett 2006, 62). That is true whether the replicator is biological or cultural.

Brains constitute "a world full of hosts for memes," and there are "far more memes than can possibly find homes," says the memetics pioneer Susan

Blackmore (1999, 37).⁹⁴ So they must be selfish and competitive, like genes; "their success depends on the advantages they confer on *themselves*. In the struggle for brains' attention they must in some way be 'better' than their rivals." This doesn't necessarily have anything to do with the effects the memes "have on the genetic success of their possessors" (Distin 2005, 11). Contraception usage is an example of an idea that has flourished despite the direct and drastic effect it has on the genetic propagation of individuals adopting it.

Often, however, there is indeed a symbiosis between a meme and the person bearing it. Luther's revolutionary theological ideas put his life in danger, but they also greatly impressed the Elector Frederick of Saxony. Spared Rome's wrath by the interventions of this powerful friend, *Der Reformator* had a full lifetime to refine and spread these ideas, and also to procreate his genes: He had six children. There are genetic descendents of Luther walking around today, as well as countless Protestant churches with doctrines that incorporate Lutheran memes of *sola scriptura* and justification by faith alone.

Memetics gets most controversial when it speculates about a causal relationship from memes to genes, culture driving biology. In computers, the complexity of new software bloated with features is a significant impetus for the production of ever more powerful hardware. Blackmore makes the same sort of connection between increasingly complex memes and ever bigger brains to propagate them. She proposes that the

> turning point in our evolutionary history was when we began to imitate each other. From this point on a second replicator, the meme, came into play. Memes changed the environment in which genes were selected, and the direction of change was determined by the outcome of memetic selection. [1999, 74]

The evolutionary driving force, she thinks, was that better imitators did better. They were the ones who could "quickly pick up the new technology of stone flaking," for example, thus having better fed children. They "would also be good at copying ways of making wooden scrapers or baskets, or plaiting their hair, or carrying piles of leaves or fruits, or making warm

---

94. Dawkins, Blackmore, and Dennett deserve most of the credit for spreading the meme about memes, but we want to give credit where else it is due, too. Kate Distin's book *The Selfish Meme* is, as its subtitle says, a useful "critical reassessment" of memetics. Tim Tyler maintains an informative blog *On Memetics* (on-memetics.blogspot.com) that advocates for memetics as a topic of scientific study and discusses some of the controversy about memes. Without using Dawkin's clever word "meme," Jacque Monod had already outlined essentially the same notion in his 1976 book *Chance and Necessity* (pp. 154-56).

clothes," and so on (p. 77). The second replicator was born when early hominids began imitating and the best imitators started being selected, Blackmore says, leading her to make this bold claim: "The enormous human brain has been created by the memes" (pp. 80-81). Ramachandran the neuroscientist (and poet, it seems!) agrees, though without any specific reference to memes:

> By hyper-developing the mirror-neuron system, evolution in effect turned culture into the new genome. Armed with culture, humans could adapt to hostile new environments and figure out how to exploit formerly inaccessible or poisonous food sources in just one or two generations—instead of the hundreds or thousands of generations such adaptations would have taken to accomplish through genetic evolution. Thus culture became a significant new source of evolutionary pressure, which helped select for brains that had even better mirror-neuron systems and the imitative learning associated with them. The result was one of the many self-amplifying snowball effects that culminated in *Homo sapiens*, the ape that looked into its own mind and saw the whole cosmos reflected inside. [2011, loc. 677]

Ehrlich finds the literature on memetics "highly speculative," Blackmore's book extremely so. He prefers "less ambitious, quantitative approaches to cultural evolution" (Ehrlich 2000, 352), though he agrees that cultural evolution involves "processes similar to mutation, migration, and drift" (p. 65). And he still assigns "great significance" to "the co-evolution of genes and culture." There are "many ways in which culture can alter selection pressures," he says (p. 64), citing examples where changes in the cultural environment resulted in different selection pressures on human populations (pp. 64-65).

The Meme's Eye View

What should be uncontroversial, Dennett says, is that "cultural transmission can *sometimes* mimic genetic transmission, permitting competing variants to be copied at different rates, resulting in gradual revisions in features of those cultural items, and *these revisions have no deliberate, foresighted authors*" (2006, 78). Thus cultural evolution offers an explanation for religion, just as biological evolution does for the apparent design of organisms. Looking at religions "from a meme's eye view" allows us to understand their success, says Blackmore:

> These religious memes did not set out with an intention to succeed. They were just behaviours, ideas and stories that were copied from one person to another in the long history of human attempts to understand the world. They were successful because they happened

to come together into mutually supportive gangs that included all the right tricks to keep them safely stored in millions of brains, books and buildings, and repeatedly passed on to more. [1999, 192]

The organisms, religions, and political parties produced by genes and memes don't need to be appealing or useful in the grand scheme of things. Neither "gene nor meme theory has anything to say about the intrinsic value (i.e., 'goodness') of the information that its replicators carry" (Distin 2005, 75). They just need to be good at replicating, and they are happy to use you as a host. It's certainly happening in biology, with "trillions of parasites of thousands of different species inhabiting your gut, your blood, your skin, your hair, your mouth, and every other part of your body. They have been rapidly evolving to survive against the onslaught of your defenses since the day you were born" (Dennett 2006, 65). Just like biological viruses, the memes "have their own fitness as replicators, independently of any contribution they may or may not make to the genetic fitness of their hosts, the human vectors" (p. 350).

The story of Jesus' death and resurrection is an amazing memetic success story. It is a living, thriving meme, preached about passionately in Christian churches around the world and bringing comfort to many millions who earnestly believe it. The other dying and rising Gods like Tammuz and Osiris are just silent memetic fossils, observed only in the rock beds of academic writings and occasionally showcased (with varying degrees of accuracy) in atheist polemics.

What made the difference? Perhaps the Jesus story resonated more with people, offering more of a connection with the struggles of their own lives and providing an antidote to their dread of death. But in some sense, it doesn't matter. Once the cultural package of Christianity spread beyond a critical point—bearing a message of hope and, critically, an evangelical imperative to spread the package further—it ensured its own success.

> Over millennia of human history, religious leaders have hit on social/emotional techniques that work to win converts, just as individual believers have hit on spiritual practices they find satisfying and belief systems that fit how we process information. Techniques that don't trigger powerful spiritual experiences simply die out. Those techniques that do trigger powerful spiritual experiences are refined and handed down. [Loftus 2010, loc. 690]

"Nobody had to invent the witnessing practice," says Dennett. It "just arises, and it *works.*" Thus it gets replicated (2006, 365). Mutation happens when the story changes, however unintentionally. "As stories spread orally, from person to person to person, an overarching dishonesty can take shape without a conscious attempt to mislead" (Wright 2009, loc. 4123).

Defending the Faith

Dennett compares religion's defenses to those that a virus evolves against antibodies. "People of all faiths have been taught that any such questioning," e.g., about anthropomorphic features of God, "is somehow insulting or demeaning to their faith, and must be an attempt to ridicule their views. What a fine protective screen this virus provides—permitting it to shed the antibodies of skepticism effortlessly!" (Dennett 2006, 207). The defenses this replicator erects for its myths are "untestability, threats, and promises" (Blackmore 1999, 192). The meme of blind faith "secures its own perpetuation by the simple unconscious expedient of discouraging rational inquiry" (Dawkins 2006, 198). Dawkins cites the example of Doubting Thomas, whose story "is told, not so that we shall admire Thomas, but so that we can admire the other apostles in comparison. Thomas demanded evidence. Nothing is more lethal for certain kinds of meme than a tendency to look for evidence" (p. 198).

It is not just because of his acknowledged hostility to religion that Dawkins has called it a "virus of the mind."[95] It enters a new brain through words spoken into the ear rather than biological particles in the bloodstream. Once lodged there, usually in the trust and innocence of childhood, the more virulent strains protect themselves against removal by limiting critical inquiry or even exposure to inconvenient facts. Some "children are raised in such an ideological prison that they willingly become their own jailers," Dennett observes, "forbidding themselves any contact with the liberating ideas that might well change their minds" (2006, 324). That is certainly true of the fundamentalist group one of your co-authors (Ed) recently left. Few of its members would openly confess to reading this book or his previous one about the group's history and doctrines (Suominen 2012), and some friends have found themselves being cautioned against even social contact with such a notorious apostate.

Anyone familiar with Christian apologetics, whether from books or in live debates, quickly realizes how vacuous the reasoning is, and how little it changes or improves over the years. Indeed, a common approach is to claim that one's religious ideas are somehow *exempt* from analytical scrutiny:

---

95. Blackmore has backtracked from the "mind virus" comparison, saying in a 2010 *Guardian* editorial "it seems I was wrong and the idea of religions as 'viruses of the mind' may have had its day." The reason for the switch is data showing advantages—reproductive, societal, and individual—to religious belief. We admire how she makes herself an example of "how science (unlike religion) works: in the end it's the data that counts." She still thinks religions "provide a superb example of memeplexes at work, with different religions using their horrible threats, promises and tricks to out-compete other religions, and popular versions of religions outperforming the more subtle teachings of the mystical traditions" (Blackmore 2010).

hidden from the wise and prudent and revealed only to babes (Luke 10:21). When apologists deign to engage human reasoning, their standard arguments on behalf of creationism, biblical inerrancy, the accuracy of the gospels, the factual character of Jesus' resurrection, etc., get refuted again and again. (At least we think so, having seen through them after once believing them ourselves.) They seem to survive by virtue not of refuting objections, but by the expedient of mere repetition (in an almost liturgical fashion) in the hearing of those already inclined to accept them.

The apologist is preaching to the choir, and even some in that pious assembly are peeking at their smartphones while he blathers on. By now, we agree, "the bristling defenses of religions against corrosive doubt begin to look vestigial, like fossil traces of an earlier epoch" (Dennett 2006, 283).

Taming the Beast

There are significant parallels between the cultural evolution of Christianity and our biological evolution as a species. In each case, the present-day entity is a more civilized version of what produced it. We would do well to bear in mind the lessons of both.

As discussed earlier in "Apex or Ex-Ape?", we humans are prone to see ourselves as angels rather than apes. But the very bestiality we deny is what has brought us to this point, since evolution has elevated us, by hook and by crook, to a status higher than our simian ancestors. We fear that we can't stay on this exalted plane unless we manage to forget how we got up here. Part of the fear of acknowledging our evolutionary origins is that we might pop the bubble. In the play *Inherit the Wind*, the fictionalized version of William Jennings Bryan warned that "if man believes he is descended from the beast, he must remain a beast!"

It's certainly no good giving people permission to act out their basest instincts. But that is not the only danger. Isn't it equally perilous to deny that we harbor the apish nature along with the angelic? The beast within us is not going to go away. If we deny its presence, we won't be on guard to take it in hand when next it rears its ugly head, the next time we are seized with fear or rage. (There is also an opposite error to guard against, that of denying the *good* in our nature, thinking "that, since natural selection is a cruel, pitiless process of elimination, it can only have produced cruel and pitiless creatures," de Waal 2006, 58.)

Carl Jung warned of the dangers we invite when we insist we are identical to the Persona, the ideal self we strive to be, denying our lurking Shadow self. The pressure will build up and explode in evil acts we can scarcely believe we have performed, as if we were dumbfounded at the news that our meek co-worker had been revealed as a serial killer: "I never saw it coming! He didn't seem that sort of person!" Or else we will project the lusts and hates of the Shadow within us onto our neighbors outside us, judging and falsely accusing them. "Better him be guilty than me!"

A great parable of the danger is Robert Louis Stevenson's *The Strange Case of Dr. Jeckyll and Mr. Hyde*. Scientist Henry Jeckyll seeks to banish his primitive impulses with a purifying chemical formula. He succeeds in isolating his base nature, but it comes to the fore when the noble scientist transforms into the evil Hyde. There are two classic film adaptations (among very many, since no other horror tale has been filmed as many times as this one). In the 1941 version, Spencer Tracy portrays Hyde as a sly and sadistic sybarite. Bad enough! But in the earlier version from 1932, Frederic March plays Hyde as a re-emergent Neanderthal: beetle-browed with jutting teeth. His first utterance after beholding his apish countenance in the mirror is "*Free!*" He makes quite the sight, parading around the shadowed streets of London in top hat, cape, and tails with that cave man face sticking out!

This is what the Christian anti-evolutionist fears: If there is a savage beast within us, in our heredity, it may escape and rampage. So the creationist Jeckyll denies he is at least potentially Hyde, or chalks it up to a sinful (but forgiven!) nature that was imposed by a mythical Fall. But the danger of not knowing what's inside us is that we'll be surprised again and again when Mr. Hyde emerges.

And does such denial, even when it seems to work, get the anti-evolutionist anywhere? Is the Christian's alternative any better? Is the human predicament alleviated if, instead of Marsh's Hyde, we believe we have Tracy's? Does it make much difference whether we call it a *sinful* nature instead of an *animal* nature? Either way, the moral person accepts the responsibility of keeping watch at the door of a dangerous cage.

Now we arrive at a deeper dimension of the evolutionary character of religion that is admittedly more controversial. Rene Girard has argued that society has created religions again and again as a way of keeping violence suppressed by safely and symbolically reenacting the violence from which a livable social order once emerged through the mob's sacrifice of a scapegoat. The victim might be an arbitrarily accused pariah—an actual goat in Leviticus 16—or a despised, marginal group like the Jews of Medieval Europe.[96] "Cast out the [imagined] scorner and contention will go out" (Pr.

---

96. Rene Girard, *Violence and the Sacred*, trans. Patrick Gregory (Baltimore: Johns Hopkins University Press, 1977); Rene Girard, *The Scapegoat*, trans. Yvonne Freccero (Baltimore: Johns Hopkins University Press, 1986). See also Chapter 6, ("Sacred Scapegoat") of Robert M. Price, *Deconstructing Jesus* (Amherst: Prometheus Books, 2000), pp. 169-211.

22:10). Henceforth this victim of murder-sacrifice, e.g., Jesus Christ, will be sacrificed again and again in representational form, providing a bearer for the hatreds and angers of the worshippers.

But the system works less well as believers come to interpret the ritual sacrifices in more sophisticated, "bloodless" ways, e.g., replacing the Penal Substitution theory of Christ's atonement with the more sentimental Moral Influence theory. Liberal Christians now regard Jesus as a wise teacher instead of a sacrificial savior. Once we finally lose sight of the bloody violence upon which our sophisticated religion rests, there will be no real bulwark to keep deep-seated violence in check. The whole thing continues to work only as long as we remain blind to the violence inherent in our theology and ritual. But when we become so sophisticated that we seek to erase all traces of the underlying violence once we recognize them, the system stops working. This is the stage when we witness guitar masses, jazz liturgy, and attempts to remove from the hymnals lyrics like "Onward, Christian soldiers, marching as to war."

For it is entirely possible to become *too* civilized. At some point, civilization becomes decadence, a decadence that is not an attenuation of morality but rather an inflation of it. It is perhaps hyper-morality that accords human rights to animals; that elevates the interests of murderers above those of their victims and their victims' loved ones; that thinks pacifism superior to going to war against violent tyranny and thus facilitates the tyrant's victory. Liberal religion is thus counter-evolutionary. It seriously lacks survival value and can never hope to overtake virulent religious literalism. This is why theistic evolutionism and Process Theology are doomed to be trod underfoot by crude fundamentalism.

Yes, fundamentalism carries what one might deem a fatal wound of being utterly intellectually bankrupt. But that works to its advantage. Like a strain of *Staphylococcus aureus* mutating defenses to the latest antibiotics, fundamentalism always finds a way to adapt and survive in the face of intellectualism. You don't need much in the way of intellect in order to accept faith; in fact it is better if you don't get bogged down with too much thinking. The weary Apostle Paul recalled a vision of Jesus saying his divine strength was made perfect in Paul's mortal weakness (2 Cor. 12:9), and the strength of fundamentalism is made perfect in the willing or inflicted ignorance of its adherents.

We moderns disdain the idea of holy wars. We are too enlightened for that, well aware of the horrors that zealots of all stripes have inflicted. We are thoroughly repulsed at the idea of coerced "belief." But if Charles Martell hadn't wielded the sword at the Pyrenees, we'd all be speaking Arabic and worshiping Allah, or perhaps just pretending to. Indeed, it still may come to that, as sophisticated intellectuals condemn vigilance against Jihadist terrorism as "Islamophobia" and dismiss terrorism as "workplace violence." The West has imbibed the blood of gentle Jesus meek and mild, and it threatens to make us ripe pickings for more brutal, virulent versions of faith. The truth is that the truth may not yield much survival value. We don't approve of the way Genghis Khan managed to spread his genes, nor should we. But it worked, evolutionarily. His soldiers—and his genes—swept the civilized before him like leaves in a hurricane.

Paradise Lost

> *And the LORD God said, Behold, the man is become as one of us, to know good and evil: and now, lest he put forth his hand, and take also of the tree of life, and eat, and live for ever: Therefore the LORD God sent him forth from the garden of Eden, to till the ground from whence he was taken. So he drove out the man; and he placed at the east of the garden of Eden Cherubims, and a flaming sword which turned every way, to keep the way of the tree of life.*
>
> —The Book of Genesis

There is no getting around it; the idyll and innocence of Eden are gone forever from Christian theology. There are fewer and fewer hold-outs against the "corrupting" knowledge of evolution. And why even use the quotation marks? It *is* corruptive of both faith and theology. That corruption is on display not least in the intellectual dishonesty with which our authors appear to be trying to deceive not so much their readers as *themselves*. "Oh, what a tangled web we weave when first we practice to believe!"

No Turning Back

Even the "fightin' fundies" who still take up arms against evolution are by no means invulnerable to what Harris calls "the hammer blows of modernity that have exposed certain tenets of faith to doubt" (2005, 19). They have hidden bruises. Paul Tillich discerns the great change that has come upon them without their knowledge. Before science raised its troubling questions, he says, literalist believers existed in an Edenic state of "dreaming innocence" as "naive literalists." They were mistaken about the factual truths of the Genesis myths, but that was beside the point. They were interested only in certain religious teachings, much as Lamoureux contends. The creation accounts were important for their implications of world-affirmation.

It was all "very good," an assessment conducive to what William James called the religion of healthy-mindedness. Had Genesis told them the universe was created in ten seconds fifty quadrillion years ago, it wouldn't have made any difference, as long as cherished doctrines like Original Sin were left unscathed. But once the scientific threat arose, they perceived an assault on the holy text and the beliefs it had taught them, so they became "reactive literalists." The medium itself became the message. It suddenly mattered to them that the world was made in six days. Oddly, despite their stubborn clinging to literalistic readings, the fundamentalists, too, had lost the innocence of their faith (Tillich 1958, 52-53).

Things get more difficult for the fundamentalist if he peers outside the safety of church society and "healthy" reading materials to glean some awareness of the many other theological problems lurking in the tall grass of science. He may recognize himself (and Jesus!) as an evolved primate, and Original Sin as an absurd doctrine built on unscientific sand. The very rationale of the atonement collapses, along with all those "sins" his pastor carries on about, which come to look like natural, even healthy traits that allowed his ancestors to replicate and eventually produce him. The God of all Creation he once praised while musing over every tree and sunset goes quiet and cold, fading into an impersonal set of laws and forces that forms life out of randomness shaped by countless acts of suffering and death.

It should be no surprise to see so many Eden dwellers turn away from all this and scurry back to retrenchment and denial, the burden of intellectual dishonesty and cognitive dissonance still lighter than the terrifying alternative. The only other options are to water down one's faith with accommodationism, which brings its own dishonesty and dissonance, or abandon it altogether. But science has set forth the flaming sword, and the Garden cannot remain occupied for long.

Given the embarrassments of its history, the Roman Catholic Church should appreciate better than most "the folly of arraying ecclesiastical power against scientific discovery" (White 1895, Ch. 4, §1). At the end of a conference with Pope Benedict XVI on creation and evolution, Cardinal Christoph Schönborn expressed what seems like a forlorn hope: "We need a genuine creation theology that can see eye-to-eye with the natural sciences on the intellectual level" (Horn and Wiedenhofer 2008, loc. 1717). Somehow, it seems doubtful that he's found what he's looking for in the lofty deist obscurantism we've seen from his fellow Catholics.

We agree with one of them, Kenneth Miller, that it is time to grow up and leave the Garden, though we decline to follow him and his theistic evolutionist colleagues into an equally implausible faith-fairyland where "there might even be a smile on the Creator's face—that at long last His creatures have learned enough to understand His world as it truly is" (1999, 56). No, there is no divine face on which to place either a beatific smile or a judgmental scowl. Nor is there any sense in which this evolved world driven by randomness and natural processes can be said to be "His," any more than you can claim the bacteria churning in your gut as "yours." Nor is there any mechanism—other than the kind of grossly anti-scientific supernaturalism that our sophisticated theologians decry—by which a deity could observe and have some emotional reaction to your finally getting the point of it all.

These can seem like harsh realities, to be sure. Miller quotes Dawkins's famous observation that the universe has "precisely the properties we should expect if there is, at bottom, no design, no purpose, no evil and no good, nothing but blind, pitiless indifference," and then says there is little wonder that people conclude evolution to be the enemy (Miller 1999, 171). Dawkins doesn't have much of a bedside manner for those whose religious convictions are on life support, but that doesn't detract from the truth of his statement.

Philip Kitcher says much the same thing, but as part of a book that shows a great deal of compassion for the disillusioned believer:

> When we understand the messiness of the processes through which life unfolds, any design must be judged as largely unintelligent, any Creator as, at best, whimsical and capricious. Providential religion can only be sustained by supposing that God's design is an unfathomable mystery. [Kitcher 2007, 149]

Providentialism has tried to retreat, to make the sort of "vague gesture toward unknowable purposes" (p. 149) we've seen so much from our sophisticated theologians. But the retreat "can only be sustained if there's some ground for supposing that appearances are deceptive, that, behind the muddle of life, there is a Creator with deeper intentions" (pp. 149-50).

Creationists want to make the scientists out to be the ones who are deceptive, or at least deceived. The theistic evolutionists at least recognize the evidence for what it is. To varying degrees, they also acknowledge that God is not quite as forthcoming as he might be about his role in nudging evolution along. But aside from some cautious hope about fine-tuning and the origins of life, they just can't argue with Dawkins's basic point: No matter how closely we look, we never find a teleological stamp saying "Made in Heaven" on any of it.

After a heartfelt discussion with a creationist he had debated the day before, Miller found himself surprised at the depth of his opponent's convictions. "Scripture tells us what the right conclusion is," the "shaken" but unswayed creationist told Miller. "And if science, momentarily, doesn't agree with it, then we have to keep working until we get to the right answer" (Miller 1999, 172-73). This taught Miller that "the appeal of creationism is emotional, not scientific" (p. 173). But that is also true of the theistic evolutionism advocated by Miller and everyone else we've covered here. Like this, for example:

> When creatures evolved who were ready to know Him, beings worthy of souls, His work would have reached a pinnacle of power and subtlety. That is the point, the religious person might say, where God

decided to reveal Himself to us. And that revelation, in all traditions, is understood as persuasive without being forceful, compelling and not coercive. The Western God stands back from His creation, not to absent Himself, not to abandon His creatures, but to allow His people true freedom. [Miller 1999, 252-53]

One does not find speculations like these coming from objective scientific institutions, only from those who are compelled (or paid) to work industriously at getting "the right answer" with Scripture. There is no scientific problem with evolution that cries out for a theological solution. It is, and always has been, entirely the other way around. In his *Lecture on Ghosts*, Ingersoll asked, "Is science indebted to the Church for a single fact? Let us know what it is." Over a hundred years later, after an explosion of scientific knowledge, it is evident that there will never be a reply to that challenge. Miller himself said it best, even if his intent was a sarcastic summary of materialism: "Science calls the tune, and religion dances to its music" (p. 288).

With our knowledge that the sunset's blazing colors are caused by photons of varying wavelengths scattering differently off of air molecules and dust particles, the heavens don't seem to declare the glory of God so loudly, nor the firmament to show his handiwork (Ps. 19). Our scientific knowledge has made us see God's absence rather than his presence. The theologians stare and stare, either looking for ever smaller God-gaps or concealing their disappointment with appeals to the "beauty" of evolution, its unseen "drama."

Annie Dillard's meditations at Tinker Creek put her up close with a brutal, silent beauty of nature that inspired her to quote Pascal about his *Deus Absconditus*: "Every religion that does not affirm that God is hidden is not true" (1974, 146). Nonetheless, she offered her own cautious paean to this hidden God: "It could be that God has not absconded but spread, as our vision and understanding of the universe have spread, to a fabric of spirit and sense so grand and subtle, so powerful in a new way, that we can only feel blindly of its hem" (p. 144). This divinity of hers "is not playful. The universe was not made in jest but in solemn incomprehensible earnest. By a power that is unfathomably secret, and holy, and fleet. There is nothing to be done about it, but ignore it, or see" (p. 275).

Such prose is like a newborn with a suspicious smell arising from its diaper: so beautiful and pure that you are inclined to smile and well up with tears, even as you contemplate the underlying messiness and the unpleasantness further investigation will bring.

Magnanimous Materialism

We have come full circle. Religion originally provided the explanation, but now *it* is what requires so much explaining from its frustrated adherents.

The occasion calls for grace rather than gloating. Kitcher concludes his book with a warning to his fellow secularists. Yes, "overwhelming evidence favors the apparently menacing claims of Darwinism," and it is indeed a genuine threat to our "providentialist and supernaturalist religions" (2007, 151). But, "though they speak with the tongues of men and of angels, the voices of reason" should remember what's at stake for the religious, and

> not expect to carry the day. The conclusion they draw deprives religious people of what they have taken to be their birthright. In its place, they offer a vision of a world without providence or purpose, and, however much they may celebrate the grand human adventure of understanding nature, that can only appear, by comparison, to be a mess of pottage. Often, the voices of reason I hear in contemporary discussions of religion are hectoring, almost exultant that comfort is being stripped away and faith undermined; frequently, they are without charity. And they are always without hope. [pp. 154-55]

The incredible scientific knowledge that has been gained about our evolutionary origins, only a tiny tip of the iceberg having been revealed in this book, has an almost mystical fascination for your co-authors. But, Kitcher reminds us, "Celebrations of the human accomplishment in fathoming nature's secrets are less likely to thrill those who have only a partial understanding of what has been accomplished, and who recognize that they will not contribute, even in the humblest way, to the continued progress of knowledge" (p. 156).

Some people have found solace in what Kitcher calls "spiritual religion." For "spiritual Christians" that means abandoning almost everything about the standard Jesus story—virgin birth, miracles, resurrection—and focusing on "the teachings, the precepts and parables, and the eventual journey to Jerusalem and the culminating moment of the Crucifixion" (p. 152). "Spiritual" believers see the values of "the stories of the scriptures not in their literal truth but in their deliverances for self-understanding, for improving ourselves and for shaping our attitudes and actions toward others" (p. 153). Of course, such a viewpoint requires one to avert one's eyes from all the atrocities, absurdities, and authoritarianism that stain the pages of those scriptures.

We suspect that spiritual religion is all our more sophisticated theologians really have left. In some cases their beliefs are, as Dawkins characterizes the liberal Episcopalian John Shelby Spong, "so advanced as to be almost unrecognizable to the majority of those who call themselves Christians" (Dawkins 2008, 269). They have pushed God out of the line of fire, with heroic intentions but relegating him to having "a merely ceremonial role to play" (Dennett 1995, 310) or, worse yet, reducing him to the nebulous abstraction of Process Theology.

As a secular humanist, Kitcher isn't buying it himself, noting the challenge spiritual religion faces "of providing more content than the exhortations to, and examples of, compassion and social justice that humanists enthusiastically endorse, without simultaneously reverting to supernaturalism" (2007, 154). While he doesn't personally see how that challenge can be met, he doesn't want to shut the door on polite discourse. Neither do we.

> Religion is, and has been, central to the lives of most people who have ever lived. From what we know of the history of the growth and spread of particular creeds, its pervasiveness is understood in terms of the social purposes it serves, and nobody should expect it to disappear without a struggle, under the impact of what proclaims itself—accurately, I believe—as reason. For the benefits religion promises to the faithful are obvious, and obviously important, perhaps most plainly so when people experience deep distress. Darwin doesn't provide much consolation at a funeral. [p. 155]

As former believers ourselves, we get it. But evolution—*our* evolution—is the truth, and there's no point trying to deny it or the devastating consequences it has for a theology born in an age of scientific innocence. The writer of Ephesians put it well, even if his conclusion was quite the opposite of ours: "We are no longer to be children, tossed here and there by waves and carried about by every wind of doctrine, by the trickery of men, by craftiness in deceitful scheming; but speaking the truth in love" (Eph. 4:14-15, NASB). The truth is different from what we thought, but the inherent value of truth over deceit remains the same.

Walking Upright

Our ancestors got up off their forelimbs and started walking upright more than two million years ago (Gibbons 2007). Now it's time for us to walk upright intellectually and accept our origins and place in the universe for what they are. We are fearfully and wonderfully made, not by the micromanaging deity of Psalm 139 but by a fascinating and elegant naturalistic process.

Think of it! Undirected, random variation rises upward from the mindless froth at the floor of an indeterminate universe and percolates through the screen of selection. That filter—natural and sexual selection—is a roulette wheel of replication probability whose numbers are determined by physical constraints and the products of previous evolution. It's all chance and necessity, as far back as we can see. No deity compatible with evolutionary science is triggering the mutations, spinning the wheel, or determining the odds.

Yes, our existence is fleeting, and can seem insignificant. We are, each of us, just a single one of the uncounted trillions of organisms resulting from evolution, and the longest of our lifetimes will span the tiniest fraction of the billions of years that life has existed on this planet. Dillard faced that reality with the same profound elegance as her many other observations at Tinker Creek: "I am a sacrifice bound with cords to the horns of the world's rock altar, waiting for worms." That is our fate, too, and we might as well accept it with the same equanimity: "I take a deep breath, I open my eyes. Looking, I see there are worms in the horns of the altar like live maggots in amber, there are shells of worms in the rock and moths flapping at my eyes. A wind from no place rises. A sense of the real exults me; the cords loose; I walk on my way" (Dillard 1974, 246).

Her courage in facing the void was shared by the Preacher of Ecclesiastes. He acknowledged that the dead "know not any thing, neither have they any more a reward; for the memory of them is forgotten." Their love, and their hatred, and their envy—all are now perished. Eternal reward? God's ultimate plan for our souls? Forget it, says this Bible writer: The dead will not have "any more a portion for ever in any thing that is done under the sun" (9:5-6).

The Preacher's conclusion (Eccl. 9:7-10) is pragmatic, but cheerful. Go your way, eat your bread with joy, and drink your wine with a merry heart, advises the Preacher, for God now accepts your works. Let your clothes be always white, and let your head lack no ointment. Live joyfully with the wife whom you love all the days of your fleeting life, which he has given you under the sun. *That* is your reward in life, for there is "no work, nor device, nor knowledge, nor wisdom, in the grave, whither thou goest."

References

Al-Khalili, Jim. 2004. *Quantum: A guide for the perplexed.* London: Weidenfeld and Nicolson.

Alexander, Denis. 2008. *Creation or evolution: Do we have to choose?* Oxford: Monarch/Grand Rapids: Kregel.

Aus, Mike. 2012. Conversion on mount improbable: How evolution challenges Christian dogma. *The Richard Dawkins Foundation for Reason and Science.* richarddawkins.net/articles/645853-conversion-on-mount-improbable-how-evolution-challenges-christian-dogma (accessed May 2012).

Babinski, Edward T. Interpretations of biblical cosmology. etb-cosmology.blogspot.com/2012/03/interpretations-of-biblical-cosmology.html (accessed June 2012).

———, ed. 2003. *Leaving the fold: Testimonies of former fundamentalists.* Amherst, NY: Prometheus Books.

———. 2010. The cosmology of the Bible. In Loftus 2010, loc. 1298-1848.

———. 2012. Personal communication.

Balantekin, A.B. and N. Takigawa. 1998. Quantum tunneling in nuclear fusion. *Rev. Mod. Phys.* 70, 77-100.

Barbour, Ian G. 2000. *When science meets religion: Enemies, strangers, or partners?* New York: HarperCollins.

Barr, James. 1977. *Fundamentalism.* Philadelphia: Westminster Press.

Beegle, Dewey M. 1963. *The inspiration of scripture.* Philadelphia: Westminster.

Blackmore, Susan. 1999. *The meme machine.* Oxford University Press.

———. 2010. Why I no longer believe religion is a virus of the mind. *The Guardian.* September 16. guardian.co.uk/commentisfree/belief/2010/sep/16/why-no-longer-believe-religion-virus-mind (accessed February 2013).

Blish, James. 1970. *Spock must die!* New York: Bantam Books.

Bishop, Robert. 2009. Chaos. *The Stanford Encyclopedia of Philosophy*. Ed. Edward N. Zalta. plato.stanford.edu/archives/fall2009/entries/chaos (accessed December 2012).

Boardman, William W., William W. Boardman (Jr.), Robert Frank Koontz, and Henry Madison Morris. 1973. *Science and creation*. N.p.: Creation-Science Research Center.

Bonhoeffer, Dietrich. 1962. *Letters and papers from prison*. Trans. Reginald H. Fuller. Ed. Eberhard Bethge. New York: Macmillan. (Orig. German pub. 1951 by Christian Kaiser Verlag.)

Bright, John. 1971. *The authority of the Old Testament*. Nashville: Abingdon Press.

Bultmann, Rudolf. 1958. *Jesus Christ and mythology*. New York: Charles Scribner's Sons.

———. 1960. *This world and the beyond: Marburg sermons*. Trans. Harold Knight. London: Lutterworth Press.

———. 1961. New Testament and mythology. In *Kerygma and Myth: A Theological Debate*. Ed. Hans Werner Bartsch, trans. Reginald H. Fuller, 1-44. New York: Harper & Row Torchbooks.

Burton, Robert A. 2008. *On being certain: Believing you are right even when you're not*. New York: St. Martin's Press.

Carroll, Sean B. 2005. *Endless forms most beautiful: The new science of evo devo*. New York: W.W. Norton & Co.

Collins, C. John. 2011. *Did Adam and Eve really exist?* Wheaton, IL: Crossway.

Collins, Francis S. 2006. *The language of God: A scientist presents evidence for belief*. New York: Simon & Schuster.

Collins, Robin. 2003. Evolution and original sin. In Miller 2003, 469-501.

Coyne, Jerry. 2009. *Why evolution is true*. New York: Penguin Books.

———. 2009. John Haught's "sophisticated" theology: Evolution is God's drama. *Why Evolution is True* website. whyevolutionistrue.wordpress.com/2009/12/03/john-haughts-sophisticated-theology-why-evolution-is-really-part-of-gods-plan (accessed August 2012).

———. 2011. Making religious virtues from scientific necessities. *Why Evolution is True* website. whyevolutionistrue.wordpress.com/2011/07/17/making-religious-virtues-from-scientific-necessities (accessed May 2012).

———. 2011. How big was the human population bottleneck? Another staple of theology refuted. *Why Evolution is True* website. whyevolutionistrue.wordpress.com/2011/09/18/how-big-was-the-human-population-bottleneck-not-anything-close-to-2 (accessed July 2012).

———. 2012. Science, religion, and society: The problem of evolution in America. *Evolution*, doi:10.1111/j.1558-5646.2012.01664.x.

Craig, D.P. and T. Thirunamachandran. 1998. *Molecular quantum electrodynamics*. Mineola, NY: Dover.

Cunningham, George C. 2010. *Decoding the language of God: Can a scientist really be a believer?* Amherst, NY: Prometheus Books.

Cupitt, Don. 1988. *The sea of faith*. Cambridge: Cambridge University Press.

Darwin, Charles. 1859. *On the origin of species by means of natural selection*. London: Murray.

———. 1888. *The descent of man, and selection in relation to sex*. 2nd ed. London: Murray.

Dawkins, Richard. 2006. *The selfish gene: 30th anniversary edition with a new introduction by the author*. New York: Oxford University Press.

———. 2008. *The God delusion*. New York: Houghton Mifflin.

———. 2009. *The greatest show on earth*. New York: Simon and Schuster.

de Waal, Frans. 2002. Evolutionary psychology: The wheat and the chaff. *Current Directions in Psychological Science* 2, no. 6 (December): 187-91.

———. 2006. *Primates and philosophers: How morality evolved*. Princeton, NJ: Princeton University Press.

DeBruine, Lisa M. 2002. Facial resemblance enhances trust. *Proceedings of the Royal Society: Biological Sciences* 269, no. 1498 (July 7): 1307-12.

Deméré, T.A., M.R. McGowen, A. Berta, and J. Gatesy. 2008. Morphological and molecular evidence for a stepwise evolutionary transition from teeth to baleen in mysticete whales. *Systematic Biology* 57:15-37.

Dennett, Daniel C. 1995. *Darwin's dangerous idea: Evolution and the meanings of life*. New York: Simon and Schuster.

———. 2006. *Breaking the spell: Religion as a natural phenomenon*. New York: Penguin Group.

Dillard, Annie. 1974. *Pilgrim at Tinker Creek*. New York: HarperCollins.

Distin, Kate. 2005. *The selfish meme: A critical reassessment*. New York: Cambridge University Press.

Domning, Daryl P. 2001. Evolution, evil and original sin. *America: The National Catholic Weekly*. November 12. americamagazine.org/content/article.cfm?article_id=1205 (accessed June 2012).

Ebeling, Erich. 1915-23. *Keilschrifttexte aus Assurreligiösen Inhalts*. Wissenschaftliche Veröffentlichung der deutschen orientgesellschaft 28. Leipzig: C. Hinrichs.

Ehrlich, Paul R. 2000. *Human natures: Genes, cultures, and the human prospect*. Washington: Island Press.

Eliade, Mircea. 1969. *Yoga: Immortality and freedom*. Trans. Willard R. Trask. Bollingen Series, vol. 61. Princeton: Princeton University Press.

Enns, Peter. 2012. *The evolution of Adam*. Grand Rapids, MI: Brazos Press.

Everett, Daniel L. 2008. *Don't sleep, there are snakes: Life and language in the Amazonian jungle*. New York: Pantheon Books.

Fagan, Brian. 2010. *Cro-Magnon: How the Ice Age gave birth to the first modern humans*. New York: Bloomsbury Press.

Fairbanks, Arthur, ed. and trans. 1898. *The first philosophers of Greece*. London: K. Paul, Trench, Trubner.

Fairbanks, Daniel J. 2007. *Relics of Eden: The powerful evidence of evolution in human DNA*. Amherst, NY: Prometheus Books.

Flew, Anthony. 1950. Theology and falsification. Reprinted on *The Unofficial Stephen Jay Gould Archive*. stephenjaygould.org/ctrl/flew_falsification.html (accessed April 2012).

Foley, Jim. 2008. Comparison of all skulls. *Fossil Hominids: The Evidence for Human Evolution*. talkorigins.org/faqs/homs/compare.html (accessed October 2012).

Ford, Kenneth W. 2005. *The quantum world: Quantum physics for everyone*. Cambridge, MA: Harvard University Press.

Forster, Lucy, Peter Forster, Sabine Lutz-Bonengel, Horst Willkomm and Bernd Brinkmannet. 2002. Natural radioactivity and human mitochondrial DNA Mutations. *Proceedings of the National Academy of Sciences* 99, no. 21 (October 15): 13950-13954.

Freud, Sigmund. 1964. *The future of an illusion*. Trans. W.D. Robson Scott. Garden City: Doubleday Anchor.

Fry, Iris. 2000. The emergence of life on Earth: A historical and scientific overview. New Brunswick, NJ: Rutgers University Press.

Gibbons, Ann. 2007. *The first human: The race to discover our earliest ancestors*. New York: Random House.

Giberson, Karl W., and Francis S. Collins. 2011. *The language of science and faith*. Downers Grove, IL: InterVarsity Press.

Giovannoni, Stephen J., et al. 2005. Genome streamlining in a cosmopolitan oceanic bacterium. *Science* 309 (August 19), 1242-45.

Gish, Duane T. 1973. *Evolution: The fossils say no!* San Diego: Creation-Life Publishers.

Goren-Inbar, Naama, Nira Alperson, Mordechai E. Kislev, Orit Simchoni, Yoel Melamed, Adi Ben-Nun, and Ella Werker. 2004. Evidence of hominin control of fire at Gesher Benot Ya'aqov, Israel. *Science*, New Series 304, no. 5671 (April 30), 725-27.

Gosse, Philip Henry. 1857. *Omphalos: An attempt to untie the geological knot*. London: John van Voorst.

Graves, Leslie, Barbara L. Horan, and Alex Rosenberg. 1999. Is indeterminism the source of the statistical character of evolutionary theory? *Philosophy of Science* 66 no. 1 (March): 140-57.

Graves, Robert. 1960. *The Greek Myths*. Vol. 1. Baltimore: Penguin Books.

Haarsma, Loren. 2003. Does science exclude God? Natural law, chance, miracles, and scientific practice. In Miller 2003, 71-94.

Haarsma, Loren and Terry M. Gray. 2003. Complexity, self-organization, and design. In Miller 2003, 288-310.

Ham, Ken. 2011. The pope on the big bang. *Around the world with Ken Ham*. blogs.answersingenesis.org/blogs/ken-ham/2011/01/07/the-pope-on-the-big-bang (accessed May 2012).

———. 2012. Beware of those who want the church to compromise. *Around the World with Ken Ham*. blogs.answersingenesis.org/blogs/ken-ham/2012/07/28/beware-of-those-who-want-the-church-to-compromise (accessed August 2012).

Harris, Sam. 2005. *The end of faith: Religion, terror, and the future of reason*. New York: W.W. Norton & Co.

Haught, John F. 2000. *God after Darwin: A theology of evolution*. Boulder, CO: Westview Press.

———. 2009. Darwin, God, and the drama of life. *Washington Post*, November 30. newsweek.washingtonpost.com/onfaith/georgetown/2009/11/darwin_god_and_the_drama_of_life.html (accessed August 2012).

———. 2010. In *Atoms and Eden: Conversations on religion and science*, ed. Steve Paulson, 83-98. New York: Oxford University Press.

———. 2010. *Making sense of evolution: Darwin, God, and the drama of life*. Louisville, KY: Westminster John Knox Press.

Hawks, John. 2011. Mailbag: Y chromosome Adam. *John Hawks weblog*. johnhawks.net/weblog/mailbag/y-chromosome-adam-2011.html (accessed July 2012).

Horn, Stephan Otto and Siegfried Wiedenhofer, eds. 2008. *Creation and evolution: A conference with Pope Benedict XVI in Castel Gandolfo*. San Francisco: Ignatius Press.

Hulsbosch, Ansfridus. 1966. *God in creation and evolution*. Lanham, MD: Sheed and Ward.

Hurd, James P. 2003. Hominids in the garden? In Miller 2003, 208-33.

Hyers, M. Conrad. 1984. *The meaning of creation: Genesis and modern science*. Atlanta: John Knox Press.

Ingersoll, Robert G. [1833-1899]. All cited lectures and interviews are from *The Complete Lectures and Interviews of Robert G. Ingersoll*. Kindle edition.

James, William. 1958. *The varieties of religious experience: A study in human nature*. New York: New American Library/Mentor Books. (Orig. pub. 1902.)

Jaspers, Karl. 1953. *The origin and goal of history*. Trans. Michael Bullock. New Haven, CT: Yale University Press.

Kitcher, Philip. 2007. *Living with Darwin: Evolution, design, and the future of faith.* New York: Oxford University Press.

Kobe, Donald H. 2004. *Luther and Science.* Leadership University. leaderu.com/science/kobe.html (accessed July 2012).

Kolts, Russell. 2011. *The compassionate mind approach to managing your anger.* London: Robinson.

Koonin, Eugene V. 2011. *The logic of chance: The nature and origin of biological evolution.* Upper Saddle River, NJ: FT Press Science.

Kouwenhoven, Arlette P. 1997. World's oldest spears. *Archeology Newsbriefs* 50, no. 3 (May/June), archaeology.org/9705/newsbriefs/spears.html (accessed February 2013).

Korsmeyer, Jerry D. 1998. *Evolution and Eden: Balancing original sin and contemporary science.* Mahwah, NJ: Paulist Press.

Krauss, Lawrence M. 2012. *A universe from nothing: Why there is something rather than nothing.* New York: Free Press.

Kuhn, Thomas S. 1996. *The structure of scientific revolutions.* 3rd ed. Chicago: University of Chicago Press.

Lamoureux, Denis O. 2008. *Evolutionary creation: A Christian approach to evolution.* Eugene, OR: Wipf & Stock.

Lanz von Liebenfels, Jörg. 2001. *Theozoologie: Das Urchristentum neu Erschlossen. Was lehrt die Bibel wirklich?* N.p.: Deutschherrenverlag.

Lévi-Strauss, Claude. 1955. The structural study of myth. *Journal of American Folklore* 68, no. 270, 428-44.

Levine, Joseph S. and Kenneth R. Miller. 1991. *Biology: Discovering life.* Lexington, MA: D.C. Heath and Company.

Loftus, John W, ed. 2010. *The Christian delusion: Why faith fails.* Amherst NY: Prometheus Books.

———. 2011. *The end of Christianity.* Amherst NY: Prometheus Books.

Lovecraft, H.P. 1926. The call of Cthulhu. *The H.P. Lovecraft archive.* hplovecraft.com/writings/texts/fiction/cc.asp (accessed April 2012).

Luper, Steven. 2009. *The philosophy of death.* Cambridge: Cambridge University Press.

Luther, Martin. 1535-1536. *Lectures on Genesis*, Vol. 1. George V. Schick, trans., 1958. Saint Louis, MO: Concordia Publishing House.

Mark, Joshua J. 2011. Enuma Elish—The Babylonian epic. *Ancient History Encyclopedia*. ancient.eu.com/article/225 (accessed January 2013).

McFadden, Johnjoe. 2002. *Quantum evolution: How physics' weirdest theory explains life's biggest mystery*. New York: W.W. Norton & Co.

Mendez, Fernando L., et al. 2013. An African American paternal lineage adds an extremely ancient root to the human Y chromosome phylogenetic tree. *The American Journal of Human Genetics* 92, no. 3 (February 28): 454-59.

Meyer, Klaus-Dieter. 2005. Zur Stratigraphie des Saale-Glazials in Niedersachsen und zu Korrelationsversuchen mit Nachbargebieten. *Eiszeitalter und Gegenwart* 55, no. 1, 25-42.

Meyer, Marvin, ed. 2007. *The Nag Hammadi scriptures*. New York: HarperCollins.

Meyers, Stephen. 1989. A Biblical Cosmology. Th.M. Thesis, Westminster Theological Seminary.

Miller, Keith B., ed. 2003. *Perspectives on an evolving creation*. Grand Rapids, MI: W.B. Eerdmans.

Miller, Kenneth R. 1999. *Finding Darwin's God: A scientist's search for common ground between God and evolution*. New York: HarperCollins.

Monod, Jacques. 1974. *Chance and necessity: An essay on the natural philosophy of modern biology*. Trans. Austryn Wainhouse. London: Collins Fontana Books.

Mohler, Albert, Jr. 2011. False start? The controversy over Adam and Eve heats up. albertmohler.com/2011/08/22/false-start-the-controversy-over-adam-and-eve-heats-up (accessed May 2012).

Morris, Henry M. 1969. *Evolution and the modern Christian*. Phillipsburg, NJ: Presbyterian and Reformed Publishing Co.

Murphy, George L. 2003. Christology, Evolution, and the Cross. In Miller 2003, 370-89.

References

Münzel, S.C., F. Seeberger, and W. Hein. 2002. The Geißenklösterle flute: Discovery, experiments, reconstruction. In *Studien zur Musikarchäologie III. Archäologie früher Klangerzeugung und Tonordnung; Musikarchäologie in der Ägäis und Anatolien*, eds. E. Hickmann, A.D. Kilmer, and R. Eichmann. Orient-Archäologie 10, 107-18. Verlag Marie Leidorf GmbH.

Noll, Mark A., and David Livingstone. 2003. Charles Hodge and B.B. Warfield on Science, the Bible, Evolution and Darwinism. In Miller 2003, 61-71.

Nowak, Martin A. 2012. Why we help. *Scientific American*, July, 34-39.

Otto, Rudolf. 1924. *The idea of the holy: An inquiry into the non-rational factor in the idea of the divine and its relation to the rational*. Trans. John W. Harvey. London: Oxford University Press.

Paget, James Carleton. 1994. *The Epistle of Barnabas*. Tübingen.

Pittenger, Norman. 1979. *The lure of divine love*. New York: Pilgrim.

Polkinghorne, John C. 2005. *Science and Providence: God's interaction with the world*. West Conshohocken, PA: Templeton Foundation Press.

Price, Robert M. 1980. The return of the navel: The "Omphalos" argument in contemporary creationism. *Creation Evolution Journal* 1, no. 2, 26-33. As republished at ncse.com/cej/1/2/return-navel (accessed June 2012).

———. 2006. *The reason-driven life*. Amherst, NY: Prometheus Books.

———. 2007. Biblical criticism. In *The new encyclopedia of unbelief*, ed. Tom Flynn, 123-134. Amherst, NY: Prometheus Books.

———. 2009. *Inerrant the wind: The Evangelical crisis of biblical authority*. Amherst, NY: Prometheus Books.

———. 2010. Apex or Ex-Ape? *The Humanist*, Jan./Feb. thehumanist.org/january-february-2010/apex-or-ex-ape (accessed July 2012).

———. 2011. Myth in the New Testament. *Christian New Age Quarterly* 20, no. 1.

Price, Robert M., and Reginald Finley Sr. Heaven and its wonders, and earth: The world the biblical writers thought they lived in. infidelguy.com/heaven_sky.htm (accessed May 2012).

References

Prothero, Donald R. 2007. *Evolution: What the fossils say and why it matters.* New York: Columbia University Press.

Ramachandran, V.S. 2011. *The tell-tale brain: A neuroscientist's quest for what makes us human.* New York: W.W. Norton & Co.

Ramm, Bernard L. 1954. *The Christian view of science and scripture.* Grand Rapids, MI: W.B. Eerdmans.

Reinikainen, Erkki. 1986. *Näin on kirjoitettu* (Thus it is written). Oulu, Finland: Suomen Rauhanyhdistys. (Quotations based an anonymous English translation.) translated by Ed Suomi

Ridley, Matt. 1996. *The origins of virtue: Human instincts and the evolution of cooperation.* New York: Penguin Books.

Roach, Jared. 2010. Analysis of genetic inheritance in a family quartet by whole-genome sequencing. *Science* 328 (April 30), 636-39.

Robinson, George L. 1913. *Leaders of Israel: A brief history of the Hebrews.* New York: Association Press.

Robson-Brown, Kate. 2011. Hominins. In *Evolution: The human story,* ed. Alice Roberts, 56-173. New York: DK Publishing.

Roughgarden, Joan. 2006. *Evolution and Christian faith: Reflections of an evolutionary biologist.* Washington: Island Press.

Ruse, Michael. 2000. *Can a Darwinian be a Christian?* New York: Cambridge University Press.

———. 2010. *Science and spirituality: Making room for faith in the age of science.* New York: Cambridge University Press.

Russell, Robert John. 2003. Special providence and genetic mutation: A new defense of theistic evolution. In Miller 2003, 335-69.

Schaff, Philip, ed. 1886. Recompiled for Kindle as *The collected works of 46 books by St. Augustine.* Amazon Digital Services. Unless otherwise indicated, source for Augustine: *On marriage and concupiscence* (loc. 154472-156915); *A treatise on the merits and forgiveness of sins, and on the baptism of infants* (loc. 167511-170865).

———, ed. 1885. *Ante-Nicene fathers.* Kindle version: Christian Classics Ethereal Library. Unless otherwise indicated, source for Barnabus, Justin Martyr, and Irenaeus (Vol. 1); Tatian, Theophilus, and Clement of Alexandria (Vol. 2); Tertullian (Vol. 3); Minucius Felix (Vol. 4).

Schleiermacher, Friedrich. 1958. *On religion: Speeches to its cultured despisers*. Translated by John Oman. New York: Harper & Row Torchbooks.

Scholem, Gershom G. 1941. Seventh Lecture: Isaac Luria and his school. In *Major trends in Jewish mysticism*. Jerusalem: Schocken Publishing House. Reprinted frequently by Schocken Books, New York.

Schroeder, Gerald L. 1997. *The science of God: The convergence of scientific and biblical wisdom*. New York: Bantam Doubleday.

Shanahan, Timothy. 2003. The evolutionary indeterminism thesis. *BioScience* 53 no. 2 (February): 163-69.

Shingledecker, Charles P. 2013. Personal communication.

Slevin, Peter . 2005. Teachers, scientists vow to fight challenge to evolution. *Washington Post*. May 5. washingtonpost.com/wp-dyn/content/article/2005/05/04/AR2005050402022_2.html (accessed August 2012).

Sparrow, Giles. *Cosmos: A field guide*. London: Quercus.

Stamos, David N. 2001. Quantum indeterminism and evolutionary biology. *Philosophy of Science* 68 no. 2 (June): 164-84.

Stark, Thom. 2011. *The human faces of God*. Eugene, OR: Wipf & Stock.

Stenger, Victor J. 2009. *Quantum gods: Creation, chaos, and the search for cosmic consciousness*. Amherst, NY: Prometheus Books.

———. 2011. *The fallacy of fine-tuning: Why the universe is not designed for us*. Amherst, NY: Prometheus Books.

Stern, David P. 2005. Quantum tunneling. *From stargazers to starships*. phy6.org/stargaze/Q8.htm (accessed December 2012).

Strong, James. 1979. *Strong's exhaustive concordance of the Bible with Greek and Hebrew dictionaries*. N.p.: Royal Publications.

Suomen Rauhanyhdistysten Keskusyhdistys (Finnish Associations of Peace). *Päivämies* weekly newspaper, from translations published in the Laestadian Lutheran Church's *Voice of Zion* newspaper unless indicated otherwise.

Suominen, Edwin A. 2012. *An examination of the pearl*. Published by the author (also available at examinationofthepearl.org).

Taylor, Arch B. Jr. 2003. The Bible, and What It Means to Me. In Babinski 2003, 153-68.

Teilhard de Chardin, Pierre. 1933. Christology and Evolution. In *Christianity and Evolution*, trans. René Hague. 1971. N.p.: Harcourt. (Kindle ed., pub. 2002).

———. 1934. How I believe. In *Christianity and Evolution*, trans. René Hague. 1971. N.p.: Harcourt. (Kindle ed., pub. 2002).

The 1000 Genomes Project Consortium. 2010. A map of human genome variation from population-scale sequencing. *Nature* 467 (Nov. 28), 1061-73, doi:10.1038/nature09534.

Tillich, Paul. 1958. *Dynamics of faith*. NY: Harper & Row Torchbooks.

Velikovsky, Immanuel. 1950. *Worlds in collision*. Garden City, NY: Doubleday.

Wade, Nicholas. 2006. *Before the dawn: Recovering the lost history of our ancestors*. New York: Penguin Group.

Wellhausen, Julius. 1957. *Prolegomena to the history of ancient Israel*. Translated by A. Menzies. Cleveland: World Publishing Company/Meridian Books.

Wells, Spencer. 2006. *Deep ancestry: The landmark DNA quest to decipher our distant past*. Washington D.C.: National Geographic Society.

Wheless, Joseph. 1926. *Is it God's word? An exposition of the fables and mythology of the Bible and of the impostures of theology*. New York: Alfred A. Knopf.

White, Andrew Dickson. 1895. *History of the warfare of science with theology in Christendom*. Ithaca, NY: Cornell University.

Wilcox, David. 2003. Finding Adam: The genetics of human origins. In Miller 2003, 234-52.

Wiley, Tatha. 2002. *Original sin: Origins, developments, contemporary meanings*. Mahwah, NJ: Paulist Press.

Williams, Sam K. 1975. *Jesus' death as saving event: The background and origin of a concept*. Harvard Dissertations in Religion 2. Missoula: Scholars Press.

Winston, Robert and Don E. Wilson, eds. 2006. *Human*. New York: DK Publishing.

World Health Organization. 2012. Lymphatic filariasis: Fact sheet no. 102. who.int/mediacentre/factsheets/fs102/en (accessed August 2012).

Wright, J. Edward. 2000. *The early history of heaven*. New York: Oxford University Press.

Wright, Robert. 2009. *The evolution of God*. New York: Little, Brown and Company.

Young, Stephen. 1989. Wayward genes play the field. *New Scientist*, September 9, 49-53.

Zimmer, Carl. 2001. *Evolution: The triumph of an idea*. New York: HarperCollins.

———. 2011. *A planet of viruses*. Chicago: The University of Chicago Press.

Illustration Credits

Figure 1: Wikimedia commons, en.wikipedia.org/wiki/File:Darwin_tree.png. Also adapted for the tree-branch section headings.

Figure 2: Author (EAS) illustration using the GIMP image editing software.

Figure 3: The Ancient Hebrew Conception of the Universe, from Robinson 1913, 2.

Figure 4: *The Flammarion engraving*. Unknown artist, 1888. Wikimedia commons, en.wikipedia.org/wiki/File:Flammarion.jpg.

Figure 5: Wikimedia commons, en.wikipedia.org/wiki/File:Goddess_nut.jpg.

Figure 6: From Nicolaus Copernicus, *Revolutions of the Heavenly Spheres* (1543). Wikimedia Commons, en.wikipedia.org/wiki/File:CopernicSystem.png.

Figure 7: *The Confusion of Tongues*. Engraving by Gustave Doré, 1865.

Figure 8: From Foley 2008, reproduced with permission.

Figure 9: Adapted from Wikimedia Commons, commons.wikimedia.org/wiki/File:Spreading_homo_sapiens.jpg.

Figure 10: National Library of Medicine, ghr.nlm.nih.gov/chromosome=2.

Figure 11: Adapted from Wikimedia Commons, commons.wikimedia.org/wiki/File:Tree_of_life_int.svg.

Figure 12: Carl Selby Price, reproduced with permission.

Figure 13: Adapted from the graphic produced by the National Center for Biotechnology Information's DNA sequence viewer. National Library of Medicine, ncbi.nlm.nih.gov/sites/entrez?Db=gene&Cmd=ShowDetailView&TermToSearch=2630 (accessed Nov. 2012).

Figure 14: Adapted from Wikimedia Commons, en.wikipedia.org/wiki/File:Quantum_mechanics_standing_wavefunctions.svg.

Figure 15: Adapted from Wikipedia commons, en.wikipedia.org/wiki/File:Filariasis_01.png.

Figure 16: Detail from *Pollice Verso* by Jean-Léon Gérôme (1872). Wikimedia Commons, en.wikipedia.org/wiki/File:Jean-Leon_Gerome_Pollice_Verso.jpg.

Index

accommodationism 5, 10, 92, 116, 126, 243, 273, 311
Adam and Eve 67
    as ensouled hominids 70, 157, 273
    as metaphor 161
    as representatives 69, 158
    as threat to Yahweh 75, 79
    curse of 84, 190
    disobedience of 174, 184
    dogmatic requirement for 153
    genetic 32
    Greek parallels 83
    impossibility of 67, 71, 160
    sinfulness and 190
adoptionism 167, 170n, 174, 298
Africa 29, 136
agency detection 200, 300
agriculture 18, 70
altruism 141, 188
    for genetic replication 142
    maternal vs. paternal 144
    reciprocal 146
Anaximander 109, 113
ancestors
    Bible and 100, 170n, 172
    common 27, 42, 97, 137, 153
    fish 36
    human 67, 136, 153, 169, 232, 315
        see also: hominids
        abstract thought and 8, 32, 157
        beliefs of 52
        female 31
        forest dwellers 40
        myth and 91
        traits 8, 191, 200, 306
        male 29
    souls of 157
        Original Sin and 173
anthropocentrism 8, 133, 148, 155, 271
apologetics 49, 253
atonement 128, 164, 172, 180, 311
    Moral Influence 188, 308
    substitutionary 186, 308

Babinski, Edward 53, 56, 59
backtracking 129, 296
bacteria 36, 143
Barbour, Ian 251, 272
Barr, James 112
Beegle, Dewey M. 121
Bible
    as product of evolution 298
    central message 80, 116, 122, 161, 237, 273
    contradictions 6, 51, 74, 77, 150, 254, 298
    cosmology of: *see* cosmology–biblical
    devotional reading of 248
    eyewitness testimony 253
    hermeneutical ventriloquism 50, 113, 118
    inerrancy 100, 181
        limited 120
    infallibility 122
    inspiration 121
    literalism 47, 49, 97
    morality and 148, 192
    Old vs. New Testament 114, 123, 128, 162
    revelation
        flawed 117

333

334　INDEX

    of God vs. information 248
    progressive 7, 112, 115
    source criticism 73
    special pleading for 106, 117
BioLogos 25, 164, 291
Blackmore, Susan 301
Bonhoeffer, Dietrich 10, 256, 274
brain
    driven by memes 300
    limitations of 14, 133, 192
    evolution of 32, 36, 69, 300
    genes and 170, 228n
    naturalism of 156
Bultmann, Rudolf 10, 107, 124, 127, 275

Cain 70, 76
camels 20
Catholicism 151, 183, 311
    Benedict XVI 151, 201
chromosome 47, 137, 169, 203
    crossover 19
    in software 16
    paternal vs. maternal 31
        in Jesus 47, 167
    Y 29, 192
Church Fathers 173
    Ambrose 168
    Augustine 99, 102, 168
    Clement of Alexandria 175, 288
    Epistle of Barnabas 99
    Irenaeus 87n, 106n, 174, 199, 207
    Justin Martyr 149, 174
    Origen 175
    Tatian 58
    Tertullian 168, 175
    Theophilus of Antioch 58
cognitive dissonance 10, 27, 71, 126, 129, 243, 311
Collins, C. John 91, 125, 188, 257
Collins, Francis 25
    anthropocentrism and 271
    on biblical revelation 117
    on God of the gaps 257
    on Moral Law 140
    piety of 250, 253
    scientific credentials of 25, 164
Collins, Robin 72, 188, 193
combat, creation by: *see* myth–creation by combat
complexity 33
cooperation: *see* altruism
Copernicus 59, 213
cosmology
    ancient 51, 58
        Greek 52
    biblical 48
        heaven 52, 62, 65n
        stars 56
        support structure 61
    Egyptian 55, 61
    fine-tuning 259
    geocentrism 56, 213
Coyne, Jerry 6n, 18, 163, 286
creationism
    day-age 22, 99
        word meaning and 99, 103
    Gap theory 105
    geographical distribution and 209
    intellectual bankruptcy of 17, 39, 67, 95, 213
    Intelligent Design 24, 187, 212, 252, 264
    Last Thursdayism 106
        see also: Gosse, Philip
    motives for 18, 215, 311
    young-earth 6, 97, 101, 244
Cro-Magnon 32, 69
crossover: *see* chromosome–crossover
cultural evolution 296, 300
    driving biological evolution 302

great leap forward and 32, 134
    meme defense mechanisms 305
    religion as product of 173, 178, 295
    vs. biological evolution 303
Cunningham, George C. 165, 262

Darwin, Charles 4, 27, 42, 133, 204, 286
    humanity and 67, 141, 156, 300
    natural selection and 110, 204
    opposition to 16, 33, 37, 43, 138, 152
    earth's age and 96, 102
dating 29, 96
    see also: mutation–molecular clock
Dawkins, Richard 123, 155
    kin selection and 144
    memes and 300
    criticism of religion 4, 273, 305, 312
    selfish gene concept and 33, 36, 142, 155, 191
Deism 13, 114, 289
denial: *see* science–denial of
Dennett, Daniel 33, 200, 248, 299
design argument 9
    double standard and 206
    origins of 199
    Paley and 201
    problems with 201
Dillard, Annie 233, 313, 316
Distin, Kate 301
Divine action 217, 228, 269
DNA 136, 167
    base pairs 36, 205
    copying of 203
    genetic switches in 38, 170n
    kin selection and 144
    sequencing of 202
    vs. gene 170n
docetism 171, 174
doctrinal change 129, 173, 295
drama 208, 286

dualism 133, 238

Ecclesiastes 150, 316
Eden
    Christian misreading of 84
    invisible gardner of 269
    isolation in Bible 91
    Levitical priesthood origins of 85
    Serpent of 81, 86, 175, 186n
    vs. Africa 31, 68, 69n
Edomites 72
Ehrlich, Paul 19n, 145, 170, 303
Enns, Peter 25, 87, 108, 124
    anthroprocentrism and 151
    biblical inspiration and 114, 121, 124
    biblical literalism and 61, 88, 115
    double standard 165
    on Adam 69, 70
        vs. Jesus 181
    on Original Sin 187
evolutionary psychology 145
ex nihilo 32, 47, 69, 106, 110n, 160, 167, 202, 251

Fall 8, 67
    Church Fathers on 174
    death and 71, 88, 190
    Eve and 83, 190n
    historical vs. myth 92
    Jewish indifference to 173, 183
    knowledge and 80, 82, 90
    of humans only 194
    Original Sin and 172
    Prometheus 81
    sinfulness and 102, 170, 174, 182, 189, 190, 307
    Yahweh's deception 86
fine-tuning: *see* cosmology–fine-tuning
first cause 263
flat earth 48, 52, 59

Flew, Anthony 269
fossils 30, 34, 39, 97, 157
Fuller, Daniel P. 120
fundamentalism 129
   dogmatism of 252
   harmonization efforts of 214
   opposition to evolution 15, 99, 160
   strength of 308, 310

Galileo 59, 65, 119
genetic algorithms 15
genetic drift 30, 41, 226
genotype 136, 202
Gnosticism 28, 79, 87, 180, 209, 258
God 8
   see also: polytheism
   as anthropomorphic deity 88
   as deceiver 86, 210
   beyond good & evil 185
   evolution of 11, 297
   hidden 205, 312
   image of 88
   Most High 62
   natural revelation of 206
   of the gaps 140, 256
      quantum mechanics and 268, 274
   of the philosophers 278
   Trinity 168n, 240, 250, 286
   Yahweh 63, 73, 77, 80, 172, 298
      Elohim 75, 83, 90
      vs. El Elyon 79
      vs. Zeus 82, 86
Gosse, Philip 106, 169, 209
gradualism 33, 36, 136
grandeur 29, 42, 314
group selection 147

Ham, Ken 22, 101, 179
Harris, Sam 134, 206, 310

Haught, John 26
   anthroprocentrism and 153
   drama and 208, 234, 260, 286
   grasp of science 26, 158
   on aesthetics 284
   on Darwin 286
   on fine-tuning 260
   on God of the gaps 258
   on naturalism 162
   on souls 158
   Process Theology and 153, 268, 281
   theodicy and 235, 239, 285
hominids 70, 136
   fossils 67, 211
   Homo erectus 68
   Homo heidelbergensis 69, 233
   Homo sapiens 30, 68, 100
   more than apes 69, 157

Immaculate Conception 169
indeterminacy: *see* quantum mechanics–indeterminacy
Ingersoll, Robert 40, 102, 173, 313
intermediate steps 34, 202

James, William 289, 310
Jesus
   as second Adam 160, 253
   as Logos 166, 296
   as meme 304
   as scapegoat 307
   competing early ideas of 20, 171
   demythologizing of 160, 181, 253
   evolutionary origins of 8, 166
   evolving doctrine of 298
   omphalos and 169
   rationale for 172
   required beliefs about 123
   Trinity and 168, 240

kin selection 143
Kitcher, Philip 96, 137, 312, 314
Kitzmiller v. Dover 24, 154
Koonin, Eugene 34n, 35n, 235, 263n, 264
Korsmeyer, Jerry 123, 183, 188, 248, 258

Laestadianism 99, 305
Lamoureux, Denis 24, 124
    anthropocentrism and 135, 151
    biblical inerrancy and 120
    biblical inspiration and 114
    biblical revelation and 116, 118, 123, 161, 164
    double standard 24, 206, 245, 255
    hermeneutical brakes 162
    on Adam 24, 161, 187
    on Divine action 292
    on Genesis 80, 92, 104, 127
    on God of the gaps 259
    on naturalism 292
    on Original Sin 184, 187
    on progressive creationism 259
    on theodicy 237
    piety of 250, 252, 255
Lewis, C.S. 93, 135, 151
literary genre 86, 115, 125, 163
Luther, Martin 21
    anthropocentrism and 149, 151
    biblical literalism and 97, 100
    faith and 75
    memetic and genetic success of 301
    on biblical cosmology 58
    on faith vs. reason 75, 98
    on reproduction 169
    on the Fall 180, 190
    on women 190n

Malthusianism 301
Mary 37, 47, 166, 170n, 174
materialism: *see* naturalism

memes 300
    see also: cultural evolution
Middle Ages 65, 191, 218
Miller, Keith 207
Miller, Kenneth 25
    accommodationism and 273
    anthropocentrism and 135, 271
    double standard 208
    Humani Generis and 153
    on revelation
        biblical 115
        of God 115, 249, 284, 291
    on creationism 212, 311
    on fine-tuning 260
    on God of the gaps 257
    on naturalism 248, 291, 311
    on quantum indeterminacy 226, 268
    on theodicy 237
    scientific advocacy by 23, 152, 212
miracles 41, 57, 162
mitochondria 31, 36
Monod, Jacques 154, 212, 240, 259, 269
monogenism 153
mutation 69, 202, 207, 315
    as evolutionary experiment 34
    base pair deletions 205
    divinely directed 228
    molecular clock 97
    mostly damaging 232
    randomness of 9, 18, 33
    sickle cell 232
    speciation from 111
mythology
    Atrahasis Epic 80, 84
    Babylonian 74
    creation by combat 77, 108
    Egyptian 55
    Enuma Elish 33
    ethnological 72, 90, 195

etiological 80, 90
Gilgamesh Epic 80, 84
legitimization story 91
New Testament and 128
Prometheus 81
scribal legend 105
Sumerian 55
vs. ancient science 107
vs. folklore 79

natural selection 110, 204
as blunt instrument 170, 232
as filter 18, 20, 316
benign neglect of 38, 40n, 202, 204
Darwin and 110
naturalism 11, 42, 123, 155, 207, 258, 291, 312, 315
New Atheism 4, 246
Newton, Isaac 218
NOMA 115, 151, 243, 275

omphalos: *see* Gosse, Philip
Original Sin 8, 172
absurdity of 172, 175, 177
corruption vs. imitation 177
depravity and 190
importance to Christianity 180, 186, 188
Paul and 178
origins of life 263
God of the gaps and 267
quantum mechanics and 266
RNA world 265
spontaneous self-organization 265
Orthodoxy 177, 189

Paley, William 201, 259
parasites 35, 234
Paul 164, 178, 196, 208
Pelagianism 168, 183, 189
Pentateuch 73, 186

piety 50, 58, 91, 185, 207, 252, 292
Pinnock, Clark H. 120, 125
Polkinghorne, John 225, 238, 251, 272, 283, 291
polytheism 64, 75, 79, 89, 149n
predestination 262
Preterism 126
primates
behavorial comparisons 138
chimpanzees 139, 157, 205
family tree 205
genetic similarities 137, 204
humans as 133
probability 97, 152, 211, 260, 316
quantum mechanics and 219, 225, 272
Process Theology 10, 154, 234, 239, 268, 278, 308
Prothero, Donald 40, 98, 205
pseudogenes 202, 204

quantum mechanics 217, 268
and origins of life 266
indeterminacy 218, 228, 224
issues with 225n, 226, 229
tunneling 220
wave function 219, 221, 227

Ramachandran, V.S. 36, 133, 156, 281n, 303
Ramm, Bernard 22, 127, 133, 252
anthropocentrism and 270
on biblical cosmology 48, 50, 56, 62
on biblical revelation 109, 119, 124
on evolution 119
on God of the gaps 256
on naturalism 289
on progressive creationism 102
Rationalism 50, 126
recurrent laryngeal nerve 36, 38
reductionism: *see* naturalism

religion
    anthropocentrism of 149
    as product of evolution 295, 305
    as sole source of controversy 17
    evolutionary explanations for 299
    importance of 12, 315
    providentialism 312
    requiring rather than offering explanation 230, 314
    uniquely human 140
    utility of 299, 307
    vs. materialism 314
replication
    cooperation and 142
    evolutionary driving force 155, 301
    human traits and 191
    of genes 33
        via fertilized egg 142
    of memes 303
    selfish 301, 304, 308
retroviruses 35, 169, 181
revelation 24, 96, 112, 117, 164, 187, 206, 249
Ridley, Matt 147
Roughgarden, Joan 26, 110
Ruse, Michael 200, 246, 269

science 212
    ancient 109, 115
    ascendancy of 129, 247, 273, 312
    denial of 12, 39, 42, 65, 211, 218, 296
    denigration of 40
    vs. revelation 248
Serpent: *see* Eden–Serpent of
sex 8, 19, 74, 80, 84, 90, 175, 191
sexual selection 26, 299
sin
    biblical explanation of 194

    vs. human nature 8, 190, 306
    vs. ritual transgression 185
skyhooks vs. cranes 33
social insects 36, 143
souls 142n, 151, 158
speciation 40, 110
Stenger, Victor 33, 136, 226
stone tools 157
supernaturalism 289, 291, 311
    see also: naturalism
symbiosis 31, 143, 301

Teilhard de Chardin, Pierre 27, 239
teleology 212, 312
telomere 137
theistic evolution 5, 25
    fundamentalist rejection of 15, 179
    honesty of 312
    limited inerrancy and 120
    misgivings about criticism of 12
    progressive revelation and 115
theodicy 231, 279, 282, 285
Tillich, Paul 310
Torah: *see* Pentateuch
transitional forms 34, 39, 67, 211
tree of life 4, 155

vestigial features 36, 202
    in DNA 30, 203
viruses 35
    see also: retroviruses
    memes and 304

wave function: *see* quantum mechanics–wave function
Wellhausen, Julius 108
Whitehead, Alfred North 154, 278, 283
Wiley, Tatha 26, 153, 173, 186

Made in the USA
Lexington, KY
07 January 2014